SCHAUM'S OUTLINE OF

THEORY AND PROBLEMS

of

APPLIED PHYSICS

by

ARTHUR BEISER, Ph.D.

SCHAUM'S OUTLINE SERIES

McGRAW-HILL BOOK COMPANY

New York   St. Louis   San Francisco   Auckland   Bogotá   Düsseldorf   Johannesburg
London   Madrid   Mexico   Montreal   New Delhi   Panama   Paris
São Paulo   Singapore   Sydney   Tokyo   Toronto

0-07-004377-9

5 6 7 8 9 10 11 12 13 14 15 16 17 18 19 20   SH SH 8 7 6 5 4 3 2 1 0

Library of Congress Cataloging in Publication Data

Beiser, Arthur.

   Schaum's outline of theory and problems of
applied physics.
   (Schaum's outline series)
   Includes index.

   1. Physics. I. Title. II. Title: Theory and
problems of applied physics.

QC21.2.B45   530   76-49870

ISBN 0-07-004377-9

Preface

This book is intended to provide students of applied physics with help in mastering those physical principles that underlie modern technology. A wide spectrum of topics is covered, so that the reader may select those which correspond to his particular needs. Both SI (metric) and British units are used.

Each chapter begins with an outline of its subject. The solved problems that follow are of two kinds: those that show how numerical answers are obtained to typical questions, and those that review important facts and ideas. The supplementary problems give the reader both a chance for practice and a means to gauge his progress.

ARTHUR BEISER

CONTENTS

CONTENTS

CONTENTS

CONTENTS

CONTENTS

CONTENTS

CONTENTS

Chapter 1

Useful Math

ALGEBRA

Algebra is the arithmetic of symbols that represent numbers. Instead of being restricted to relationships among specific numbers, algebra can express more general relationships among quantities whose numerical values need not be known.

The operations of addition, subtraction, multiplication, and division have the same meaning in algebra as in arithmetic. Thus the formula

$$\frac{(a+b)c}{d} - e = x$$

means that in order to find the value of x we must first add a and b, next multiply by c, then divide by d, and finally subtract e. The rules for multiplying and dividing positive and negative quantities are as follows:

$$(+a) \times (+b) = (-a) \times (-b) = +ab \qquad (-a) \times (+b) = (+a) \times (-b) = -ab$$

$$\frac{+a}{+b} = \frac{-a}{-b} = +\frac{a}{b} \qquad \frac{-a}{+b} = \frac{+a}{-b} = -\frac{a}{b}$$

EQUATIONS

An equation is a statement of equality: whatever is on the left-hand side of any equation is equal to whatever is on the right-hand side. The symbols in an algebraic equation usually cannot have arbitrary values if the equality is to hold. To *solve* an equation is to find the possible values of these symbols. The solution of the equation $5x - 10 = 20$ is $x = 6$ because only when x is 6 is this equation a true statement.

The algebraic procedures that can be used to solve an equation are all based on the principle that any operation performed on one side of an equation must be performed on the other side as well. Thus an equation remains valid when the same quantity is added to or subtracted from both sides or is used to multiply or divide both sides. Other operations, such as raising to a power or taking a root, also do not change an equality if the same thing is done to both sides.

Two helpful rules follow from the above principle. The first is that any term on one side of an equation may be transposed to the other side by changing its sign. Thus if $a + b = c$, then $a = c - b$, and if $a - d = e$, then $a = e + d$. Second, a quantity that multiplies one side of an equation may be transposed so as to divide the other side, and vice versa. Thus if $ab = c$, then $a = c/b$, and if $a/d = e$, then $a = de$. When each side of an equation consists of a fraction, *cross-multiplication* removes the fractions. For example, if we are to solve the equation

$$\frac{5}{a+2} = \frac{3}{a-2}$$

for the value of a, we first cross-multiply to obtain

$$5(a-2) = 3(a+2)$$

and then proceed as follows:

Multiply out both sides to give $\qquad\qquad 5a - 10 = 3a + 6$

Transpose the -10 and $3a$ to give $\qquad\quad 5a - 3a = 6 + 10$

1

Carry out the indicated subtraction and
   addition to give $2a = 16$
Divide both sides by 2 to give $a = 8$

EXPONENTS

There is a special shorthand way to express a quantity that is to be multiplied by itself one or more times. In this scheme a superscript number called an *exponent* is used to indicate how many times the self-multiplication is to be carried out, as follows:

$$a = a^1 \qquad a \times a = a^2 \qquad a \times a \times a = a^3 \qquad \text{and so on}$$

The quantity a^2 is read as "a squared" because it is equal to the area of a square whose sides are a long, and a^3 is read as "a cubed" because it is equal to the volume of a cube whose edges are a long. Past an exponent of 3 we read a^n as "a to the nth power," so that a^5 is "a to the fifth power." The product of two powers of the same quantity, say a^n and a^m, is that quantity raised to the sum of the two exponents: $a^n \times a^m = a^{n+m}$. Thus $a^2 \times a^5 = a^7$.

Reciprocal quantities are expressed according to the above scheme but with negative exponents:

$$\frac{1}{a} = a^{-1} \qquad \frac{1}{a^2} = a^{-2} \qquad \frac{1}{a^3} = a^{-3} \qquad \text{and so on}$$

In general, $1/a^n = (1/a)^n = a^{-n}$. A quantity raised to the zeroth power, a^0 for instance, is always equal to 1: $a^0 = 1$. To see why, we note that $a/a = 1$ can also be written $a/a = a^1 \times a^{-1} = a^{1-1} = a^0$.

It is not necessary that an exponent be a whole number. A fractional exponent signifies a *root* of a quantity. The "square root of a," customarily written $\sqrt{a}$, is that quantity which, multiplied by itself once, is equal to a: $\sqrt{a} \times \sqrt{a} = a$. Using exponents we would write the square root of a as $\sqrt{a} = a^{1/2}$, because $a^{1/2} \times a^{1/2} = a^{1/2+1/2} = a^1 = a$. In general, the nth root of any quantity is indicated by the exponent $1/n$.

Some rules concerning exponents and roots are the following:

$$(ab)^n = a^n b^n \qquad (a^n)^m = a^{n \times m} \qquad \left(\frac{a}{b}\right)^n = \frac{a^n}{b^n}$$

$$\sqrt[m]{ab} = \sqrt[m]{a} \times \sqrt[m]{b} \qquad \sqrt[m]{\frac{a}{b}} = \frac{\sqrt[m]{a}}{\sqrt[m]{b}} \qquad \sqrt[m]{a^n} = (a^n)^{1/m} = a^{n/m}$$

POWERS OF TEN

Very small and very large numbers are common in science and engineering and are best expressed with the help of powers of 10. A number in decimal form can be written as a number between 1 and 10 multiplied by a power of 10:

$$834 = 8.34 \times 10^2 \qquad 0.00072 = 7.2 \times 10^{-4}$$

The powers of 10 from 10^{-6} to 10^6 are as follows:

$$10^0 \ = 1 \qquad = 1 \text{ with decimal point moved 0 places}$$
$$10^{-1} = 0.1 \qquad = 1 \text{ with decimal point moved 1 place to the left}$$
$$10^{-2} = 0.01 \qquad = 1 \text{ with decimal point moved 2 places to the left}$$
$$10^{-3} = 0.001 \qquad = 1 \text{ with decimal point moved 3 places to the left}$$
$$10^{-4} = 0.0001 \qquad = 1 \text{ with decimal point moved 4 places to the left}$$

$$10^{-5} = 0.00001 \quad = 1 \text{ with decimal point moved 5 places to the left}$$
$$10^{-6} = 0.000001 \quad = 1 \text{ with decimal point moved 6 places to the left}$$

$$10^0 \quad = 1 \qquad\qquad = 1 \text{ with decimal point moved 0 places}$$
$$10^1 \quad = 10 \qquad\quad\; = 1 \text{ with decimal point moved 1 place to the right}$$
$$10^2 \quad = 100 \qquad\quad = 1 \text{ with decimal point moved 2 places to the right}$$
$$10^3 \quad = 1000 \qquad\; = 1 \text{ with decimal point moved 3 places to the right}$$
$$10^4 \quad = 10,000 \qquad = 1 \text{ with decimal point moved 4 places to the right}$$
$$10^5 \quad = 100,000 \quad\; = 1 \text{ with decimal point moved 5 places to the right}$$
$$10^6 \quad = 1,000,000 \; = 1 \text{ with decimal point moved 6 places to the right}$$

When numbers written in powers-of-10 notation are to be added or subtracted, they must all be expressed in terms of the *same* power of 10:

$$3 \times 10^2 + 4 \times 10^3 = 0.3 \times 10^3 + 4 \times 10^3 = 4.3 \times 10^3$$

To multiply two powers of 10, add their exponents; to divide one power of 10 by another, subtract the exponent of the latter from that of the former:

$$10^n \times 10^m = 10^{n+m} \qquad \frac{10^n}{10^m} = 10^{n-m}$$

Reciprocals follow the pattern

$$\frac{1}{10^n} = 10^{-n}$$

The rules for finding powers and roots of powers of 10 are

$$(10^n)^m = 10^{n \times m} \qquad \sqrt[m]{10^n} = (10^n)^{1/m} = 10^{n/m}$$

In taking the *m*th root, the power of 10 should be chosen to be a multiple of *m*. Thus

$$\sqrt{10^{15}} = \sqrt{10} \times \sqrt{10^{14}} = \sqrt{10} \times 10^7 = 3.16 \times 10^7$$

UNITS

Units are algebraic quantities and may be multiplied and divided by one another. To convert a quantity expressed in a certain unit to its equivalent in a different unit of the same kind, we use the fact that multiplying or dividing anything by 1 does not affect its value. For instance, 12 in. = 1 ft, so 12 in./ft = 1, and we can convert a length s expressed in feet to its value in inches by multiplying s by 12 in./ft:

$$4 \text{ ft} = 4 \, \cancel{\text{ft}} \times 12 \, \frac{\text{in.}}{\cancel{\text{ft}}} = 48 \text{ in.}$$

Conversion factors for the most common British and SI (metric) units are given in Appendix B.

Subdivisions and multiples of metric units are designated by prefixes according to the corresponding power of 10.

Prefix	Power	Abbreviation	Example
pico-	10^{-12}	p	1 pf = 1 picofarad = 10^{-12} farad
nano-	10^{-9}	n	1 ns = 1 nanosecond = 10^{-9} second
micro-	10^{-6}	μ	1 μA = 1 microampere = 10^{-6} ampere
milli-	10^{-3}	m	1 mm = 1 millimeter = 10^{-3} meter
centi-	10^{-2}	c	1 cl = 1 centiliter = 10^{-2} liter
kilo-	10^3	k	1 kg = 1 kilogram = 10^3 grams
mega-	10^6	M	1 MW = 1 megawatt = 10^6 watts
giga-	10^9	G	1 GeV = 1 gigaelectron-volt = 10^9 electron-volts

Solved Problems

1.1. The quantity a is given by the formula

$$a = \frac{x(y - 2z)}{x + y} - 8x$$

Find the value of a when $x = -2$, $y = \frac{1}{2}$, and $z = 4$.

The procedure is to substitute the given values of x, y, and z in the formula and then to carry out the indicated operations:

$$a = \frac{-2(\frac{1}{2} - 2 \times 4)}{-2 + \frac{1}{2}} - 8 \times (-2) = \frac{-2(\frac{1}{2} - 8)}{-1\frac{1}{2}} + 16 = \frac{-2(-7\frac{1}{2})}{-1\frac{1}{2}} + 16 = -\frac{15}{1\frac{1}{2}} + 16 = -10 + 16 = 6$$

1.2. Examples of exponents.

$$a^2 a^{-3} = a^{2-3} = a^{-1} \qquad (a^2)^{-4} = a^{2 \times (-4)} = a^{-8} \qquad (a^3)^{-1/3} = a^{-1/3 \times 3} = a^{-1}$$

$$a^{-1} a^{-4} = a^{-1-4} = a^{-5} \qquad (a^{-3})^{-2} = a^{-3 \times (-2)} = a^6 \qquad a^6 a^{1/2} = a^{6+1/2} = a^{13/2}$$

$$(a^{-2})^4 = a^{-2 \times 4} = a^{-8} \qquad (a^{1/2})^6 = a^{6 \times 1/2} = a^3 \qquad (a^6)^{1/2} = a^{1/2 \times 6} = a^3$$

1.3. Examples of powers-of-10 notation.

$$20 = 2 \times 10 = 2 \times 10^1$$
$$3043 = 3.043 \times 1000 = 3.043 \times 10^3$$
$$8,700,000 = 8.7 \times 1,000,000 = 8.7 \times 10^6$$
$$0.22 = 2.2 \times 0.1 = 2.2 \times 10^{-1}$$
$$0.000035 = 3.5 \times 0.00001 = 3.5 \times 10^{-5}$$

1.4. Examples of addition and subtraction.

$$6 \times 10^2 + 5 \times 10^4 = 0.06 \times 10^4 + 5 \times 10^4 = 5.06 \times 10^4$$
$$2 \times 10^{-2} + 3 \times 10^{-3} = 2 \times 10^{-2} + 0.3 \times 10^{-2} = 2.3 \times 10^{-2}$$
$$7 + 2 \times 10^{-2} = 7 + 0.02 = 7.02$$
$$6 \times 10^4 - 4 \times 10^2 = 6 \times 10^4 - 0.04 \times 10^4 = 5.96 \times 10^4$$
$$3 \times 10^{-2} - 5 \times 10^{-3} = 3 \times 10^{-2} - 0.5 \times 10^{-2} = 2.5 \times 10^{-2}$$
$$7 \times 10^{-5} - 2 \times 10^{-4} = 0.7 \times 10^{-4} - 2 \times 10^{-4} = -1.3 \times 10^{-4}$$
$$6.23 \times 10^{-3} - 6.28 \times 10^{-3} = -0.05 \times 10^{-3} = -5 \times 10^{-5}$$

1.5. Examples of multiplication and division.

$$10^5 \times 10^{-2} = 10^{5-2} = 10^3 \qquad \frac{10^4}{10^{-3}} = 10^{4-(-3)} = 10^{4+3} = 10^7$$

$$\frac{10^3}{10^6} = 10^{3-6} = 10^{-3} \qquad \frac{10^5 \times 10^{-7}}{10^2} = 10^{5-7-2} = 10^{-4}$$

1.6. A sample calculation.

$$\frac{460 \times 0.00003 \times 100,000}{9000 \times 0.0062} = \frac{(4.6 \times 10^2) \times (3 \times 10^{-5}) \times (10^5)}{(9 \times 10^3) \times (6.2 \times 10^{-3})} = \frac{4.6 \times 3}{9 \times 6.2} \times \frac{10^2 \times 10^{-5} \times 10^5}{10^3 \times 10^{-3}}$$

$$= 0.25 \times \frac{10^{2-5+5}}{10^{3-3}} = 0.25 \times \frac{10^2}{10^0} = 25$$

1.7. Examples of powers of numbers.

$$(10^2)^4 = 10^{2 \times 4} = 10^8$$

$$(10^{-3})^5 = 10^{-3 \times 5} = 10^{-15}$$

$$(10^{-4})^{-3} = 10^{-4 \times (-3)} = 10^{12}$$

$$(3 \times 10^3)^2 = 3^2 \times (10^3)^2 = 9 \times 10^6$$

$$(4 \times 10^{-5})^3 = 4^3 \times (10^{-5})^3 = 64 \times 10^{-15} = 6.4 \times 10^{-14}$$

$$(2 \times 10^{-2})^{-4} = \frac{1}{2^4} \times (10^{-2})^{-4} = \frac{1}{16} \times 10^8 = 0.0625 \times 10^8 = 6.25 \times 10^6$$

1.8. Examples of square roots.

$$\sqrt{20} = \sqrt{4 \times 5} = \sqrt{4} \times \sqrt{5} = 2\sqrt{5}$$

$$3\sqrt{2} \times 4\sqrt{5} = (3 \times 4) \times (\sqrt{2} \times \sqrt{5}) = 12\sqrt{10}$$

$$\sqrt{5x^4} = \sqrt{5} \times \sqrt{x^4} = \sqrt{5}\, x^2$$

$$\frac{\sqrt{15}}{\sqrt{5}} = \sqrt{\frac{15}{5}} = \sqrt{3}$$

$$\frac{\sqrt{6}}{5\sqrt{3}} = \frac{1}{5}\sqrt{\frac{6}{3}} = \frac{\sqrt{2}}{5}$$

Even powers of 10:

$$\sqrt{10^6} = 10^{6/2} = 10^3$$

$$\sqrt{5 \times 10^4} = \sqrt{5} \times \sqrt{10^4} = 2.24 \times 10^2$$

Odd powers of 10:

$$\sqrt{3 \times 10^5} = \sqrt{30 \times 10^4} = \sqrt{30} \times \sqrt{10^4} = 5.48 \times 10^2$$

$$\sqrt{0.000025} = \sqrt{2.5 \times 10^{-5}} = \sqrt{25 \times 10^{-6}} = \sqrt{25} \times \sqrt{10^{-6}} = 5 \times 10^{-3}$$

1.9. Examples of cube roots.

$$\sqrt[3]{27x^2} = \sqrt[3]{27} \times \sqrt[3]{x^2} = 3x^{2/3} \qquad \sqrt[3]{\frac{64a}{b^3}} = \frac{\sqrt[3]{64}\,\sqrt[3]{a}}{\sqrt[3]{b^3}} = \frac{4\sqrt[3]{a}}{b}$$

$$\sqrt[3]{10^9} = 10^{9/3} = 10^3$$

$$\sqrt[3]{10^8} = \sqrt[3]{10^2 \times 10^6} = \sqrt[3]{100} \times \sqrt[3]{10^6} = 4.64 \times 10^2$$

$$\sqrt[3]{3.8 \times 10^{19}} = \sqrt[3]{38 \times 10^{18}} = \sqrt[3]{38} \times \sqrt[3]{10^{18}} = 3.36 \times 10^6$$

$$\sqrt[3]{2.7 \times 10^{-5}} = \sqrt[3]{27 \times 10^{-6}} = \sqrt[3]{27} \times \sqrt[3]{10^{-6}} = 3 \times 10^{-2}$$

1.10. Rome is 1440 km by road from Paris. How far is this in miles?

Since 1 km = 0.621 mi,

$$1440 \text{ km} = 1440 \text{ km} \times 0.621 \, \frac{\text{mi}}{\text{km}} = 894 \text{ mi}$$

1.11. A man is 6 ft 2 in. tall. How many cm is this?

Since 1 ft = 12 in. and 1 in. = 2.54 cm,

$$6 \text{ ft } 2 \text{ in.} = (6 \times 12 + 2) \text{ in.} = 74 \text{ in.} \times 2.54 \, \frac{\text{cm}}{\text{in.}} = 188 \text{ cm}$$

1.12. How many ft^2 are there in one m^2?

Since 1 m = 3.28 ft, $1 \text{ m}^2 = 1 \text{ m}^2 \times \left(3.28 \frac{\text{ft}}{\text{m}}\right)^2 = 10.76 \text{ ft}^2$.

1.13. Express a velocity of 60 mi/hr in ft/s.

There are 5280 ft in a mile and 3600 s in an hour, and so

$$60 \frac{\text{mi}}{\text{hr}} = 60 \frac{\cancel{\text{mi}}}{\cancel{\text{hr}}} \times 5280 \frac{\text{ft}}{\cancel{\text{mi}}} \times \frac{1}{3600 \text{ s}/\cancel{\text{hr}}} = 88 \frac{\text{ft}}{\text{s}}$$

Supplementary Problems

1.14. Evaluate the following:

(a) $\frac{3(x+y)}{2}$ when $x = 5$ and $y = -2$

(b) $\frac{1}{x-y} - \frac{1}{x+y}$ when $x = 3$ and $y = 2$

(c) $\frac{4xy}{y+3x} + 5$ when $x = 1$ and $y = -2$

(d) $\frac{x^2+y^2}{2z} + \frac{z}{x-y}$ when $x = -2$, $y = 2$, and $z = 4$

(e) $\frac{(x+z)^2}{y} - \frac{xy}{2}$ when $x = 2$, $y = 8$, and $z = 10$

1.15. Solve each of the following equations for x:

(a) $\frac{4x-35}{3} = 9(1-x)$ (i) $\frac{x}{y} = \frac{4}{z}$ (n) $\frac{8}{x} = \frac{1}{4-x}$

(b) $\frac{3x-42}{9} = 2(7-x)$ (j) $\frac{y}{x} = \frac{x}{5}$ (o) $\frac{x}{2x-1} = \frac{5}{7}$

(c) $x^3 + 27 = 0$

(d) $2x^4 - 32 = 0$ (k) $\frac{1}{x+1} = \frac{1}{2x-1}$ (p) $\frac{y}{x^2} = \frac{z}{x}$

(e) $3x^2 = 6x$

(f) $(x+3)(x-y) = z^2 + x^2$ (l) $\frac{3}{x-1} = \frac{5}{x+1}$ (q) $x^2 = \frac{8}{x}$

(g) $y\sqrt{2x} = 12$ (m) $\frac{1}{3x+4} = \frac{2}{x+8}$ (r) $\frac{2}{x} = \frac{18}{x^3}$

(h) $z = \frac{x+y}{x-y}$

1.16. Evaluate the following:

(a) $a^2 a^5$

(b) $a^4 a^{-2}$ (g) $\frac{a^3 a^{-2}}{a^{-5}}$ (o) $\frac{a}{a^{1/2}}$ (v) $(ab)^3 (a^2 b)^{-2}$

(c) $a^{14} a^{-14}$ (h) $a^2 + a^2$ (p) $\frac{a}{a^{-1/2}}$ (w) $7b^2 \left(\frac{a^2}{b}\right)^3$

(d) $\frac{a^6}{a^2}$ (i) $a^2 - a^2$

                  (j) $a^2 + a^5$ (q) $(a^4)^{1/2}$ (x) $(a^2 b^4)^{1/2}\left(\frac{b^2}{a^3}\right)^3$

(e) $\frac{a^2}{a^6}$ (k) $a^{3/2} a^{1/2}$ (r) $(a^4)^{-1/2}$

                  (l) $a^{1/2} a^{1/2} a^{1/2}$ (s) $(a^{-4})^{1/3}$ (y) $2a\left(\frac{3a}{b}\right)^2 \left(\frac{b^2}{a}\right)^3$

(f) $\frac{a^5 a^{-2}}{a^8}$ (m) $a a^{1/3}$ (t) $a^5 + a^{1/2}$

                  (n) $a^6 a^{-1/2}$ (u) $a^4 (ab)^{-2}$ (z) $(ab)^3 \left(\frac{a^2}{b^8}\right)^{1/4}$

1.17. Express the following numbers in powers-of-10 notation:

(a) 720   (d) 0.000062   (g) 49,527   (j) 49,000,000,000

(b) 890,000   (e) 3.6   (h) 0.002943   (k) 0.000000011

(c) 0.02   (f) 0.4   (i) 0.0014   (l) 1.4763

1.18. Express the following numbers in decimal notation:

(a) 3×10^{-4} (f) 3.2×10^{-2}

(b) 7.5×10^3 (g) 4.32145×10^3

(c) 8.126×10^{-5} (h) 6×10^6

(d) 1.01×10^8 (i) 5.7×10^0

(e) 5×10^2 (j) 6.9×10^{-5}

1.19. Perform the following additions and subtractions:

(a) $3 \times 10^2 + 4 \times 10^3$ (f) $6.32 \times 10^2 + 5$ (k) $4.6 \times 10^5 - 3.2 \times 10^7$

(b) $2 \times 10^4 + 5 \times 10^6$ (g) $4 \times 10^3 - 3 \times 10^2$ (l) $3 \times 10^5 - 2.98 \times 10^5$

(c) $7 \times 10^{-2} + 2 \times 10^{-3}$ (h) $5 \times 10^7 - 9 \times 10^4$ (m) $4.76 \times 10^{-3} - 4.81 \times 10^{-3}$

(d) $4 \times 10^{-5} + 5 \times 10^{-3}$ (i) $3.2 \times 10^{-4} - 5 \times 10^{-5}$ (n) $7 \times 10^3 + 5 \times 10^2 - 9 \times 10^2$

(e) $2 \times 10^1 + 2 \times 10^{-1}$ (j) $7 \times 10^4 - 2 \times 10^5$ (o) $3 \times 10^{-4} + 6 \times 10^{-5} - 7 \times 10^{-3}$

1.20. Perform the following multiplications and divisions using powers-of-10 notation:

(a) 5000×0.005

(b) $\dfrac{5000}{0.005}$

(c) $\dfrac{500,000 \times 18,000}{9,000,000}$

(d) $\dfrac{30 \times 80,000,000,000}{0.0004}$

(e) $\dfrac{30,000 \times 0.0000006}{1000 \times 0.02}$

(f) $\dfrac{0.0001}{60,000 \times 200}$

(g) $\dfrac{200 \times 0.00004}{400,000}$

(h) $\dfrac{0.002 \times 0.00000005}{0.000004}$

(i) $\dfrac{400 \times 0.00006}{0.2 \times 20,000}$

(j) $\dfrac{0.06 \times 0.0001}{0.00003 \times 40,000}$

1.21. Evaluate the following and express the results in powers-of-10 notation:

(a) $(4 \times 10^9)^3$ (e) $(3 \times 10^{-8})^2$

(b) $(2 \times 10^7)^2$ (f) $(5 \times 10^{11})^{-2}$

(c) $(2 \times 10^7)^{-2}$ (g) $(3 \times 10^{-4})^{-3}$

(d) $(2 \times 10^{-2})^5$

1.22. Evaluate the following and express the results in powers-of-10 notation. Note that $\sqrt{4} = 2$, $\sqrt{40} = 6.3$, $\sqrt[3]{4} = 1.6$, $\sqrt[3]{40} = 3.4$ and $\sqrt[3]{400} = 7.4$.

(a) $\sqrt{4 \times 10^6}$ (f) $\sqrt[3]{4 \times 10^{12}}$ (j) $\sqrt[3]{4 \times 10^{-6}}$

(b) $\sqrt{4 \times 10^7}$ (g) $\sqrt[3]{4 \times 10^{13}}$ (k) $\sqrt[3]{4 \times 10^{-7}}$

(c) $\sqrt{4 \times 10^8}$ (h) $\sqrt[3]{4 \times 10^{14}}$ (l) $\sqrt[3]{4 \times 10^{-8}}$

(d) $\sqrt{4 \times 10^{-4}}$ (i) $\sqrt[3]{4 \times 10^{15}}$ (m) $\sqrt[3]{4 \times 10^{-9}}$

(e) $\sqrt{4 \times 10^{-5}}$

1.23. The earth is an average of 9.3×10^7 miles from the sun. How far is this in kilometers? In meters?

1.24. How many cubic feet are there in a cubic meter?

1.25. The speed limit in many European towns is 60 km/hr. How many mi/hr is this?

1.26. The speed of light is 3.00×10^8 m/s. What is this speed in ft/s? In mi/s? In mi/hr?

1.27. A nautical mile is 6076 ft, and a knot is a unit of speed equal to 1 nautical mile/hr. How fast is an 8-knot boat going in mi/hr? In ft/s?

Answers to Supplementary Problems

1.14.   (a) 4.5   (b) 0.8   (c) -3 (d) 0   (e) 10

1.15.

(a)	2	(g)	$72/y^2$	(m)	0
(b)	8	(h)	$-y(1+z)/(1-z)$	(n)	32/9
(c)	-3	(i)	$4y/z$	(o)	5/3
(d)	2	(j)	$\sqrt{5y}$	(p)	y/z
(e)	2	(k)	2	(q)	2
(f)	$(z^2+3y)/(3-y)$	(l)	4	(r)	3

1.16.

(a)	a^7	(h)	$2a^2$	(o)	$a^{1/2}$	(v)	b/a
(b)	a^2	(i)	0	(p)	$a^{3/2}$	(w)	$7a^6/b$
(c)	1	(j)	a^2+a^5	(q)	a^2	(x)	$(b/a)^8$
(d)	a^4	(k)	a^2	(r)	a^{-2}	(y)	$18b^4$
(e)	a^{-4}	(l)	$a^{3/2}$	(s)	$a^{-4/3}$	(z)	$a^{7/2}b$
(f)	a^{-5}	(m)	$a^{4/3}$	(t)	$a^5+a^{1/2}$		
(g)	a^6	(n)	$a^{11/2}$	(u)	a^2/b^2		

1.17.

(a)	7.2×10^2	(d)	6.2×10^{-5}	(g)	4.9527×10^4	(j)	4.9×10^{10}
(b)	8.9×10^5	(e)	3.6×10^0	(h)	2.943×10^{-3}	(k)	1.1×10^{-8}
(c)	2×10^{-2}	(f)	4×10^{-1}	(i)	1.4×10^{-3}	(l)	1.4763×10^0

1.18.

(a)	0.0003	(d)	101,000,000	(g)	4321.45	(j)	0.000069
(b)	7500	(e)	500	(h)	6,000,000		
(c)	0.00008126	(f)	0.032	(i)	5.7		

1.19.

(a)	4.3×10^3	(e)	2.02×10^1	(i)	2.7×10^{-4}	(m)	-5×10^{-5}
(b)	5.02×10^6	(f)	6.37×10^2	(j)	-1.3×10^5	(n)	6.6×10^3
(c)	7.2×10^{-2}	(g)	3.7×10^3	(k)	-3.154×10^7	(o)	-6.64×10^{-3}
(d)	5.04×10^{-3}	(h)	4.991×10^7	(l)	2×10^3		

1.20.

(a)	2.5×10^1	(d)	6×10^{15}	(g)	2×10^{-8}	(j)	5×10^{-6}
(b)	10^6	(e)	9×10^{-4}	(h)	2.5×10^{-5}		
(c)	10^3	(f)	8.3×10^{-12}	(i)	6×10^{-6}		

1.21.

(a)	6.4×10^{28}	(c)	2.5×10^{-15}	(e)	9×10^{-16}	(g)	3.7×10^{10}
(b)	4×10^{14}	(d)	3.2×10^{-9}	(f)	4×10^{-24}		

1.22.

(a)	2×10^3	(e)	6.3×10^{-3}	(i)	1.6×10^5	(m)	1.6×10^{-3}
(b)	6.3×10^3	(f)	1.6×10^4	(j)	1.6×10^{-2}		
(c)	2×10^4	(g)	3.4×10^4	(k)	7.4×10^{-3}		
(d)	2×10^{-2}	(h)	7.4×10^4	(l)	3.4×10^{-3}		

1.23. 1.5×10^8 km; 1.5×10^{11} m

1.24.   35.3 ft^3

1.25.   37 mi/hr

1.26. 9.84×10^8 ft/s; 1.86×10^5 mi/s; 6.72×10^8 mi/hr

1.27.   9.2 mi/hr; 13.5 ft/s

Chapter 2

Vectors

SCALAR AND VECTOR QUANTITIES

A *scalar quantity* has only magnitude and is completely specified by a number and a unit. Examples are mass (a stone has a mass of 2 kg), volume (a bottle has a volume of 12 oz), and frequency (house current has a frequency of 60 cycles/s). Symbols of scalar quantities are printed in italic type (m = mass, V = volume). Scalar quantities of the same kind are added using ordinary arithmetic.

A *vector quantity* has both magnitude and direction. Examples are displacement (an airplane has flown 200 mi to the southwest), velocity (a car is moving at 60 mi/hr to the north), and force (a man applies an upward force of 15 lb to a package). Symbols of vector quantities are printed in boldface type (**v** = velocity, **F** = force) and expressed in handwriting by arrows over the letters ($\vec{v}$, $\vec{F}$). The magnitude of a vector quantity is printed in italic type (F is the magnitude of the force **F**). When vector quantities are added, their directions must be taken into account.

VECTOR ADDITION: GRAPHICAL METHOD

A *vector* is an arrowed line whose length is proportional to a certain vector quantity and whose direction indicates the direction of the quantity.

To add the vector **B** to the vector **A**, draw **B** so that its tail is at the head of **A**. The vector sum **A** + **B** is the vector **R** that joins the tail of **A** and the head of **B** (Fig. 2-1). **R** is usually called the *resultant* of **A** and **B**.

The order in which **A** and **B** are added is not significant, so that **A** + **B** = **B** + **A** (Figs. 2-1 and 2-2).

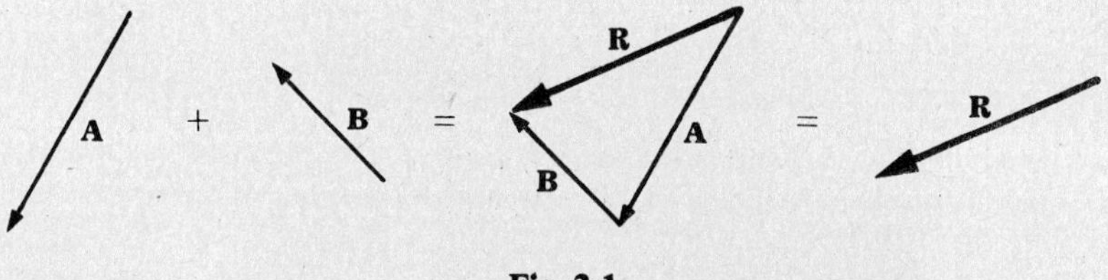

Fig. 2-1

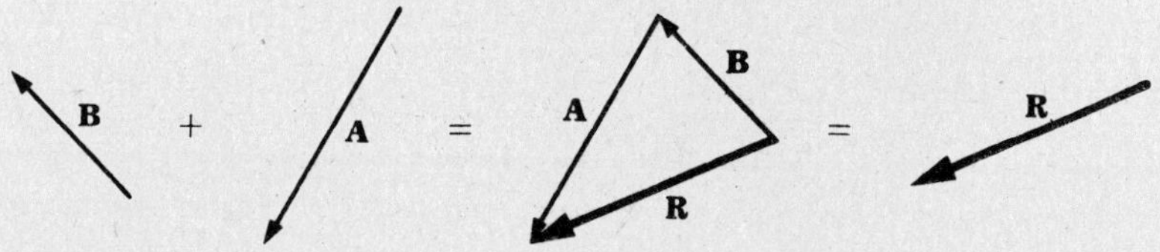

Fig. 2-2

Exactly the same procedure is followed when more than two vectors of the same kind are to be added. The vectors are strung together head to tail (being careful to preserve their correct lengths and directions), and the resultant **R** is the vector drawn from the tail of the first vector to the head of the last. The order in which the vectors are added does not matter (Fig. 2-3).

9

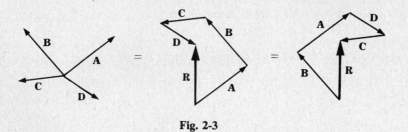

Fig. 2-3

TRIGONOMETRY

Although it is possible to determine the magnitude and direction of the resultant of two or more vectors of the same kind graphically with ruler and protractor, this procedure is not very exact, and for accurate results it is necessary to use trigonometry.

A *right triangle* is a triangle two of whose sides are perpendicular. The *hypotenuse* of a right triangle is the side opposite the right angle, as in Fig. 2-4; the hypotenuse is always the longest side. The three basic trigonometric functions, the sine, cosine, and tangent of an angle, are defined in terms of the right triangle of Fig. 2-4 as follows:

$$\sin \theta = \frac{a}{c} = \frac{\text{opposite side}}{\text{hypotenuse}}$$

$$\cos \theta = \frac{b}{c} = \frac{\text{adjacent side}}{\text{hypotenuse}}$$

$$\tan \theta = \frac{a}{b} = \frac{\text{opposite side}}{\text{adjacent side}} = \frac{\sin \theta}{\cos \theta}$$

$$\sin \theta = \frac{a}{c}$$

$$\cos \theta = \frac{b}{c}$$

$$\tan \theta = \frac{a}{b}$$

$$a^2 + b^2 = c^2$$

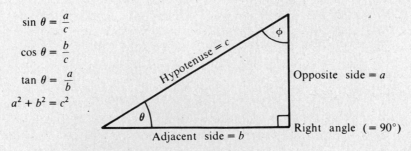

Fig. 2-4

Numerical tables of the sine, cosine, and tangent of angles from 0° to 90° are given in Appendix C. It is possible to extend the definitions of these functions to cover angles from 90° to 360°, and the tables of Appendix C may be used for such angles with the help of the relationships in Table 2-1.

Table 2-1

	θ	$90° + \theta$	$180° + \theta$	$270° + \theta$
sine	$\sin \theta$	$\cos \theta$	$-\sin \theta$	$-\cos \theta$
cosine	$\cos \theta$	$-\sin \theta$	$-\cos \theta$	$\sin \theta$
tangent	$\tan \theta$	$-\dfrac{1}{\tan \theta}$	$\tan \theta$	$-\dfrac{1}{\tan \theta}$

The *Pythagorean theorem* states that the sum of the squares of the short sides of a right triangle is equal to the square of its hypotenuse.   For the triangle of Fig. 2-4,

$$a^2 + b^2 = c^2$$

Hence we can always express the length of any of the sides of a right triangle in terms of the lengths of the other sides:

$$a = \sqrt{c^2 - b^2} \qquad b = \sqrt{c^2 - a^2} \qquad c = \sqrt{a^2 + b^2}$$

Another useful relationship is that the sum of the interior angles of any triangle is 180°.   Since one of the angles in a right triangle is 90°, the sum of the other two must be 90°.   Thus in Fig. 2-4, $\theta + \phi = 90°$.

VECTOR ADDITION: TRIGONOMETRIC METHOD

It is easy to apply trigonometry to find the resultant **R** of two vectors **A** and **B** that are perpendicular to each other.   The magnitude of the resultant is given by the Pythagorean theorem as

$$R = \sqrt{A^2 + B^2}$$

and the angle θ between **R** and **A** (Fig. 2-5) may be found from the relationship

$$\tan \theta = \frac{B}{A}$$

by examining a table of tangents to find the angle whose tangent is closest in value to B/A.

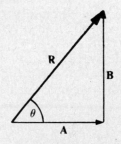

Fig. 2-5

When the vectors to be added are not perpendicular, the method of addition by components described below can be used.   There do exist trigonometric procedures for dealing with oblique triangles (the "law of sines" and the "law of cosines") but these are not necessary since the component method is entirely general in its application.

RESOLVING A VECTOR

Just as two or more vectors can be added together to yield a single resultant vector, so it is possible to break up a single vector into two or more other vectors.   If the vectors **A** and **B** are together equivalent to the vector **C**, then the vector **C** is equivalent to the two vectors **A** and **B** (Fig. 2-6).   When a vector is replaced by two or more others, the process is called *resolving* the vector, and the new vectors are known as the *components* of the initial vector.

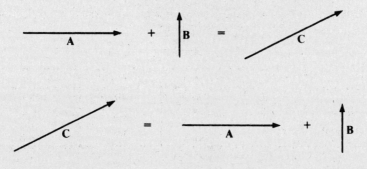

Fig. 2-6

The components into which a vector is resolved are nearly always chosen to be perpendicular to one another. Figure 2-7 shows a wagon being pulled by a man with the force $\mathbf{F}$. Because the wagon moves horizontally, the entire force is not effective in influencing its motion. The force $\mathbf{F}$ may be resolved into two component vectors, $\mathbf{F}_x$ and $\mathbf{F}_y$, where

$$\mathbf{F}_x = \text{horizontal component of } \mathbf{F}$$

$$\mathbf{F}_y = \text{vertical component of } \mathbf{F}$$

The magnitudes of these components are

$$F_x = F \cos \theta \qquad F_y = F \sin \theta$$

Evidently the component $\mathbf{F}_x$ is responsible for the wagon's motion, and if we were interested in working out the details of this motion, we would need to consider only $\mathbf{F}_x$.

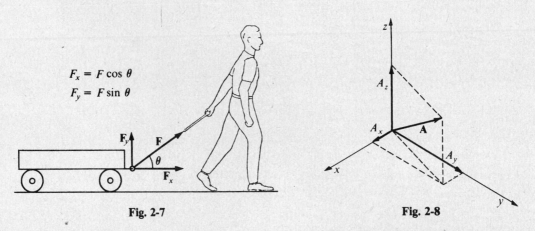

$$F_x = F \cos \theta$$
$$F_y = F \sin \theta$$

Fig. 2-7   **Fig. 2-8**

In Fig. 2-7 the force $\mathbf{F}$ lies in a vertical plane, and the two components $\mathbf{F}_x$ and $\mathbf{F}_y$ are enough to describe it. In general, however, three mutually perpendicular components are required to completely describe the magnitude and direction of a vector quantity. It is customary to call the directions of these components the x, y, and z axes, as in Fig. 2-8. The components of some vector $\mathbf{A}$ in these directions are accordingly denoted $\mathbf{A}_x$, $\mathbf{A}_y$, and $\mathbf{A}_z$. If a component falls on the negative part of an axis, its magnitude is considered negative. Thus if $\mathbf{A}_z$ were downward in Fig. 2-8 instead of upward and its length were equivalent to, say 12 lb, we would write $A_z = -12$ lb.

VECTOR ADDITION: COMPONENT METHOD

To add two or more vectors $\mathbf{A}$, $\mathbf{B}$, $\mathbf{C}$, ... together by the component method, the following procedure is followed:

1. Resolve the initial vectors into components in the x, y, and z directions.

2. Add the components in the x direction together to give $\mathbf{R}_x$, add the components in the y direction together to give $\mathbf{R}_y$, and add the components in the z direction together to give $\mathbf{R}_z$. That is, the magnitudes of $\mathbf{R}_x$, $\mathbf{R}_y$, and $\mathbf{R}_z$ are given by

$$R_x = A_x + B_x + C_x + \cdots$$
$$R_y = A_y + B_y + C_y + \cdots$$
$$R_z = A_z + B_z + C_z + \cdots$$

3. Calculate the magnitude and direction of the resultant $\mathbf{R}$ from its components $\mathbf{R}_x$, $\mathbf{R}_y$, and $\mathbf{R}_z$.

If the vectors being added all lie in the same plane, only two components need be considered.

Solved Problems

2.1.   A man walks eastward for 5 miles and then northward for 10 miles.   How far is he from his starting point?   If he had walked directly to his destination, in what direction would he have headed?

From Fig. 2-9, the length of the resultant vector **R** corresponds to a distance of 11.2 miles and a protractor shows that its direction is 27° east of north.

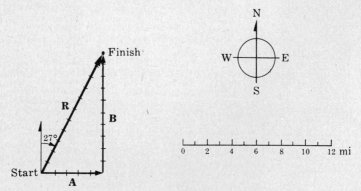

Fig. 2-9

2.2.   Two tugboats are towing a ship.   Each exerts a force of 6 tons, and the angle between the towropes is 60°.   What is the resultant force on the ship?

To add the force vectors **A** and **B**, **B** is shifted parallel to itself so that its tail is at the head of **A**.   The length of the resultant **R** corresponds to a force of 10.4 tons.   (See Fig. 2-10.)

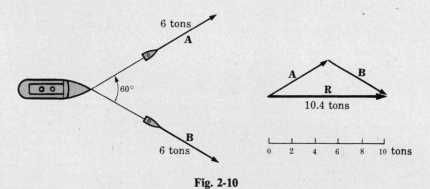

Fig. 2-10

2.3.   A boat moving at 5 mi/hr is to cross a river in which the current is flowing at 3 mi/hr.   In what direction should the boat head in order to reach a point on the other bank of the river directly opposite its starting point?

The procedure here is to first draw the vector that represents v_{river}, the velocity of the current.   Then the vector v_{boat} is drawn from the head of v_{river} so that its head is directly opposite the tail of v_{river} (Fig. 2-11).   A protractor shows the angle between v_{river} and v_{boat} to be 53°.

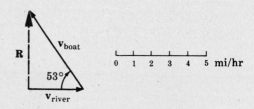

Fig. 2-11

2.4. In going from one city to another, a car whose driver tends to get lost goes 30 miles north, 50 miles west, and 20 miles southeast.   Approximately how far apart are the cities?

The vectors representing the displacements are strung together head-to-tail, and their resultant is found to be 39 miles (Fig. 2-12).

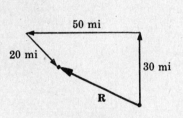

Fig. 2-12

2.5. Use trigonometry to solve Problem 2.1.

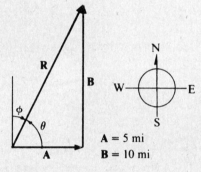

From the vector diagram of Fig. 2-13 we see that **A** and **B** are the sides of a right triangle and **R** is its hypotenuse. According to the Pythagorean theorem, the magnitudes A, B, and R are related by $R^2 = A^2 + B^2$.   Hence the magnitude R is equal to

$$R = \sqrt{A^2 + B^2} = \sqrt{(5 \text{ mi})^2 + (10 \text{ mi})^2} = \sqrt{25 \text{ mi}^2 + 100 \text{ mi}^2}$$

$$= \sqrt{125 \text{ mi}^2} = 11.2 \text{ mi}$$

To find the direction of **R** we note that

$$\tan \theta = \frac{B}{A} = \frac{10 \text{ mi}}{5 \text{ mi}} = 2$$

Fig. 2-13

From a trigonometric table (for instance the one in Appendix C) we find that the angle whose tangent is closest to 2 is $\theta = 63°$.   To express the direction of **R** in terms of north, we see from Fig. 2-13 that the angle ϕ between north and **R**, plus the angle θ between **R** and east, is equal to 90°.   Since $\phi + \theta = 90°$,

$$\phi = 90° - \theta = 90° - 63° = 27°$$

The resultant **R** has a magnitude of 11.2 mi and its direction is 27° east of north.

2.6. The man in Fig. 2-7 exerts a force of 10.0 lb on the wagon at an angle of $\theta = 30°$ above the horizontal.   Find the horizontal and vertical components of this force.

The magnitudes of $\mathbf{F}_x$ and $\mathbf{F}_y$ are respectively

$$F_x = F \cos \theta = 10.0 \text{ lb} \times \cos 30° = 8.66 \text{ lb}$$
$$F_y = F \sin \theta = 10.0 \text{ lb} \times \sin 30° = 5.00 \text{ lb}$$

We note that $F_x + F_y = 13.66$ lb although **F** itself has the magnitude $F = 10.0$ lb.   What is wrong?   The answer is that nothing is wrong; because F_x and F_y are just the *magnitudes* of the vectors $\mathbf{F}_x$ and $\mathbf{F}_y$, it is meaningless to add them together.   However, we can certainly add the *vectors* $\mathbf{F}_x$ and $\mathbf{F}_y$ to find the magnitude of their resultant **F**.   Because $\mathbf{F}_x$ and $\mathbf{F}_y$ are perpendicular,

$$F = \sqrt{F_x^2 + F_y^2} = \sqrt{(8.66 \text{ lb})^2 + (5.00 \text{ lb})^2} = \sqrt{75.00 \text{ lb}^2 + 25.00 \text{ lb}^2} = \sqrt{100.00 \text{ lb}^2} = 10.0 \text{ lb}$$

as we expect.

2.7. A car weighing 3000 lb is on a hill that makes an angle of 20° with the horizontal. Find the components of the car's weight parallel and perpendicular to the road.

The weight of an object is the gravitational force with which the earth attracts it, and this force always acts vertically downward (Fig. 2-14). Because **w** is vertical and F_2 is perpendicular to the road, the angle θ between **w** and $\mathbf{F}_2$ is the same as the angle θ between the road and the horizontal. Hence

$$F_1 = w \sin \theta = 3000 \text{ lb} \times \sin 20° = 3000 \text{ lb} \times 0.342 = 1026 \text{ lb}$$
$$F_2 = w \cos \theta = 3000 \text{ lb} \times \cos 20° = 3000 \text{ lb} \times 0.940 = 2820 \text{ lb}$$

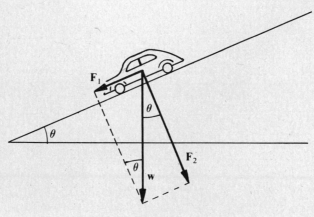

Fig. 2-14

2.8. A boat is headed north at a speed of 8.0 mi/hr. A strong wind is blowing whose pressure on the boat's superstructure causes it to move sideways to the west at a speed of 2.0 mi/hr. There is also a tidal current present that flows in a direction 30° south of east at a speed of 5.0 mi/hr. What is the boat's velocity relative to the earth's surface?

The first step is to establish a suitable set of coordinate axes, such as the one shown in Fig. 2-15(a). Next we draw the three velocity vectors **A**, **B**, and **C**, and calculate the magnitudes of their x and y components. We find these values:

x components	y components
$A_x = 0$	$A_y = 8.0 \text{ mi/hr}$
$B_x = -2.0 \text{ mi/hr}$	$B_y = 0$
$C_x = C \cos 30°$	$C_y = -C \sin 30°$
$\quad = 5.0 \text{ mi/hr} \times 0.866$	$\quad = -5.0 \text{ mi/hr} \times 0.500$
$\quad = 4.3 \text{ mi/hr}$	$\quad = -2.5 \text{ mi/hr}$

These components are shown in Fig. 2-15(b).

Now we add up the values of the x components to get R_x and add up the values of the y components to get R_y:

$$R_x = A_x + B_x + C_x = 0 - 2.0 \text{ mi/hr} + 4.3 \text{ mi/hr} = 2.3 \text{ mi/hr}$$
$$R_y = A_y + B_y + C_y = 8.0 \text{ mi/hr} + 0 - 2.5 \text{ mi/hr} = 5.5 \text{ mi/hr}$$

The magnitude of the resultant **R** is therefore

$$R = \sqrt{R_x^2 + R_y^2} = \sqrt{(2.3 \text{ mi/hr})^2 + (5.5 \text{ mi/hr})^2} = \sqrt{35.54} \text{ mi/hr} = 6.0 \text{ mi/hr}$$

The boat's speed relative to the earth's surface is 6.0 mi/hr.

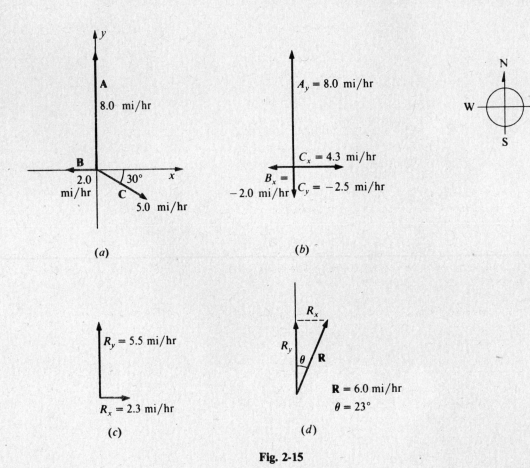

Fig. 2-15

The direction in which the boat is moving relative to the earth's surface can be given in terms of the angle θ between $\mathbf{R}$ and the $+y$ axis, which is north. Since

$$\tan \theta = \frac{R_x}{R_y} = \frac{2.3 \text{ mi/hr}}{5.5 \text{ mi/hr}} = 0.418 \qquad \theta = 23°$$

the boat's direction of motion is actually 23° to the east of north even though it is headed north [Fig. 2-15(d)].

Supplementary Problems

2.9. Two forces, one of 10 lb and the other of 6 lb, act on a body.   The directions of the forces are not known. (*a*) What is the minimum magnitude of the resultant of these forces? (*b*) What is the maximum magnitude?

2.10. A man drives 10 mi to the north and then 20 mi to the east.   What is the magnitude and direction of his displacement from the starting point?

2.11. Find the magnitude and direction of the resultant force produced by a vertically upward force of 40 lb and a horizontal force of 30 lb.

2.12. Find the vertical and horizontal components of a 50-lb force that is directed 50° above the horizontal.

2.13. A man pushes a lawnmower with a force of 20 lb. If the handle of the lawnmower is 40° above the horizontal, how much downward force is being exerted on the ground?

2.14. An airplane is heading northeast at a speed of 550 mi/hr. What is the northward component of its velocity? The eastward component?

2.15. A boat heads northwest at 10 mi/hr in a river that flows east at 3 mi/hr. What is the magnitude and direction of the boat's velocity relative to the earth's surface?

2.16. A boat moving at 12 mi/hr is crossing a river in which the current is flowing at 4 mi/hr. In what direction should the boat head if it is to reach a point on the other side of the river directly opposite its starting point?

2.17. An airplane flies 400 mi west from city A to city B, then 300 mi northeast to city C, and finally 100 mi north to city D. How far is it from city A to city D? In what direction must the airplane head to return directly to city A from city D?

2.18. Find the magnitude and direction of the resultant of a 5-lb force that acts at an angle of 37° clockwise from the +x axis, a 3-lb force that acts at an angle of 180° clockwise from the +x axis, and a 7-lb force that acts at an angle of 225° clockwise from the +x axis.

2.19. Find the magnitude and direction of the resultant of a 60-lb force that acts at an angle of 45° clockwise from the +y axis, a 20-lb force that acts at an angle of 90° clockwise from the +y axis, and a 40-lb force that acts at an angle of 300° clockwise from the +y axis.

Answers to Supplementary Problems

2.9. (*a*) 4 lb   (*b*) 16 lb

2.10. 22 mi at 63° east of north

2.11. 50 lb at 53° above the horizontal

2.12. 38.3 lb; 32.2 lb

2.13. 12.9 lb

2.14. 389 mi/hr; 389 mi/hr

2.15. 8.2 mi/hr at 30° west of north

2.16. 19° upstream of directly across the river

2.17. 364 mi at 31° east of south

2.18. 4.4 lb at 206° clockwise from the +x axis

2.19. 68 lb at 24° clockwise from the +y axis

Chapter 3

Motion in a Straight Line

VELOCITY

The *velocity* of a body is a vector quantity that describes both how fast it is moving and the direction in which it is headed.

In the case of a body traveling in a straight line, its velocity is simply the rate at which it covers distance. If the distance s covered in the time t is proportional to t, the body has the *constant velocity* of

$$v = \frac{s}{t}$$

$$\text{Velocity} = \frac{\text{distance}}{\text{time}}$$

Given the velocity of a body, we can find how far it travels in a time interval t from the formula

$$s = vt$$

provided the velocity does not change in the time interval.

ACCELERATION

A body whose velocity is changing is *accelerated*. A body is accelerated when its velocity is increasing, decreasing, or changing in direction. Accelerations that involve a change in direction are discussed in Chapter 8.

The *acceleration* of a body is the rate at which its velocity is changing. If a body's velocity is v_0 at the start of a certain time interval t and is v at the end, its acceleration is

$$a = \frac{v - v_0}{t}$$

$$\text{Acceleration} = \frac{\text{velocity change}}{\text{time}}$$

A positive acceleration means an increase in velocity; a negative acceleration means a decrease in velocity. Only constant accelerations are considered here.

Velocity has the dimensions of distance/time. Acceleration has the dimensions of velocity/time or distance/time^2. Typical acceleration units are the ft/s^2 and the m/s^2. Sometimes two different time units are used; for instance, the acceleration of a car that can go from rest to 50 mi/hr in 10 s might be expressed as $a = 5$ (mi/hr)/s.

DISTANCE, VELOCITY, AND ACCELERATION

Let us consider a body whose velocity is v_0 when it starts to be accelerated at a constant rate. After the time t the final velocity of the body will be

$$v = v_0 + at$$

$$\text{Final velocity} = \text{initial velocity} + \text{velocity change}$$

How far does the body go during the time interval t? The average velocity $\bar{v}$ of the body is

$$\bar{v} = \frac{v_0 + v}{2}$$

18

and so, since $v = v_0 + at$,

$$\bar{v} = \frac{v_0 + v_0 + at}{2} = v_0 + \tfrac{1}{2}at$$

The distance traveled during t is therefore

$$s = \bar{v}t = v_0 t + \tfrac{1}{2}at^2$$

If the body is accelerated starting from rest, $v_0 = 0$ and

$$s = \tfrac{1}{2}at^2$$

Another useful formula gives the final velocity of a body in terms of its initial velocity, its acceleration, and the distance it has traveled during the acceleration:

$$v^2 = v_0^2 + 2as$$

In the case of a body that starts from rest, $v_0 = 0$ and

$$v^2 = 2as \qquad v = \sqrt{2as}$$

Solved Problems

[*Air resistance is assumed negligible in these problems.*]

3.1. A ship travels 9 mi in 45 min. What is its speed in mi/hr?

Since 45 min = 3/4 hr,

$$v = \frac{s}{t} = \frac{9 \text{ mi}}{3/4 \text{ hr}} = 12 \frac{\text{mi}}{\text{hr}}$$

3.2. The velocity of sound in air at sea level is about 1100 ft/s. If a man hears a clap of thunder 3 s after seeing a lightning flash, how far away was the lightning?

The velocity of light is so great compared with the velocity of sound that the time needed for the light of the flash to reach the man can be neglected. Hence

$$s = vt = 1100 \frac{\text{ft}}{\text{s}} \times 3 \text{ s} = 3300 \text{ ft}$$

3.3. The velocity of light is 3×10^8 m/s. How long does it take light to reach the earth from the sun, which is 1.5×10^{11} m away?

$$t = \frac{s}{v} = \frac{1.5 \times 10^{11} \text{ m}}{3 \times 10^8 \text{ m/s}} = 500 \text{ s} = 8\tfrac{1}{3} \text{ min}$$

3.4. A car travels 270 mi in 4.5 hr. (*a*) What is its average speed? (*b*) How far will it go in 7 hr at this average speed? (*c*) How long will it take to travel 300 mi at this average speed?

(*a*)
$$v = \frac{s}{t} = \frac{270 \text{ mi}}{4.5 \text{ hr}} = 60 \frac{\text{mi}}{\text{hr}}$$

(*b*)
$$s = vt = 60 \frac{\text{mi}}{\text{hr}} \times 7 \text{ hr} = 420 \text{ mi}$$

(*c*)
$$t = \frac{s}{v} = \frac{300 \text{ mi}}{60 \text{ mi/hr}} = 5 \text{ hr}$$

3.5. An airplane whose air speed is 400 mi/hr has a tail wind of 120 mi/hr. How long will it take the airplane to cover 1000 miles relative to the ground?

The ground speed of the airplane is

$$v = 400 \frac{mi}{hr} + 120 \frac{mi}{hr} = 520 \frac{mi}{hr}$$

Hence the time needed to cover 1000 miles over the ground is

$$t = \frac{s}{v} = \frac{1000 \ mi}{520 \ mi/hr} = 1.9 \ hr$$

3.6. A car moves at 40 mi/hr for 2 hr and then at 30 mi/hr for $1\frac{1}{2}$ hr. (a) How far did it go? (b) What was its average speed for the whole trip?

(a) $$s = v_1 t_1 + v_2 t_2 = 40 \frac{mi}{hr} \times 2 \ hr + 30 \frac{mi}{hr} \times 1.5 \ hr = 125 \ mi$$

(b) A total distance of 125 miles was covered in $3\frac{1}{2}$ hours, and so

$$\bar{v} = \frac{s}{t} = \frac{125 \ mi}{3.5 \ hr} = 36 \frac{mi}{hr}$$

3.7. A car starts from rest and reaches a velocity of 40 ft/s in 10 s. (a) What is its acceleration? (b) If its acceleration remains the same, what will its velocity be after 15 s?

Here $v_0 = 0$. Hence

(a) $$a = \frac{v}{t} = \frac{40 \ ft/s}{10 \ s} = 4 \ ft/s^2$$

(b) $$v = at = 4 \frac{ft}{s^2} \times 15 \ s = 60 \frac{ft}{s}$$

3.8. (a) What is the acceleration of a car that goes from 20 mi/hr to 30 mi/hr in 1.5 s? (b) At the same acceleration, how long will it take the car to go from 30 mi/hr to 36 mi/hr?

(a) $$a = \frac{v - v_0}{t} = \frac{30 \ mi/hr - 20 \ mi/hr}{1.5 \ s} = 6.7 \ (mi/hr)/s$$

(b) $$t = \frac{v - v_0}{a} = \frac{36 \ mi/hr - 30 \ mi/hr}{6.7 \ (mi/hr)/s} = 0.9 \ s$$

3.9. A car has an acceleration of 8 m/s². (a) How much time is needed for it to reach a velocity of 24 m/s starting from rest? (b) How far does it go during this period of time?

(a) $$t = \frac{v}{a} = \frac{24 \ m/s}{8 \ m/s^2} = 3 \ s$$

(b) Since the car starts from rest, $v_0 = 0$ and

$$s = \frac{1}{2}at^2 = \frac{1}{2} \times 8 \frac{m}{s^2} \times (3 \ s)^2 = 36 \ m$$

3.10. The brakes of a certain car can produce an acceleration of 6 m/s². (a) How long does it take the car to come to a stop from a velocity of 30 m/s? (b) How far does the car travel during the time the brakes are applied?

(a) $$t = \frac{v}{a} = \frac{30 \ m/s}{6 \ m/s^2} = 5 \ s$$

(b) Here the signs of v_0 and a are important. The initial velocity of the car is $v_0 = +30$ m/s and its acceleration is -6 m/s², so that

$$s = v_0 t + \frac{1}{2}at^2 = 30 \frac{m}{s} \times 5 \ s - \frac{1}{2} \times 6 \frac{m}{s^2} \times (5 \ s)^2 = 75 \ m$$

3.11. A car starts from rest with an acceleration of 5 ft/s². What is its velocity after it has gone 600 ft?

$$v = \sqrt{2as} = \sqrt{2 \times 5 \text{ ft/s}^2 \times 600 \text{ ft}} = 77 \text{ ft/s}$$

3.12. An airplane must have a velocity of 50 m/s in order to take off. What must the airplane's acceleration be if it is to take off from a runway 500 m long?

Since $v^2 = 2as$,

$$a = \frac{v^2}{2s} = \frac{(50 \text{ m/s})^2}{2 \times 500 \text{ m}} = 2.5 \text{ m/s}^2$$

3.13. The brakes of a car whose initial velocity is 30 m/s are applied, and the car receives an acceleration of -2 m/s². (*a*) How far will it have gone when its velocity has decreased to 15 m/s? (*b*) When it has come to a stop?

(*a*) Since $v^2 = v_0^2 + 2as$,

$$s = \frac{v^2 - v_0^2}{2a} = \frac{(15 \text{ m/s})^2 - (30 \text{ m/s})^2}{2 \times (-2 \text{ m/s}^2)} = 169 \text{ m}$$

(*b*) Here $v = 0$, and so

$$s = \frac{0 - (30 \text{ m/s})^2}{2 \times (-2 \text{ m/s}^2)} = 225 \text{ m}$$

Supplementary Problems

3.14. A ship steams at a constant velocity of 15 mi/hr. (*a*) How far does it travel in a day? (*b*) How long does it take to travel 500 miles?

3.15. A car moves at 50 mi/hr for 1/2 hr and then at 60 mi/hr for 2 hr. (*a*) How far did it go? (*b*) What was its average velocity for the entire trip?

3.16. An airplane whose air speed is 500 mi/hr covers a distance of 1000 mi in $2\frac{1}{2}$ hr. How strong was the headwind against it?

3.17. How long does it take an echo to return to a man standing 300 ft from a cliff? The velocity of sound in air is about 1100 ft/s.

3.18. A pitcher takes 0.1 s to throw a baseball, which leaves his hand at a velocity of 90 ft/s. What is the acceleration of the baseball while it is being thrown?

3.19. A car comes to a stop in 6 s from a velocity of 30 m/s. (*a*) What is its acceleration? (*b*) At the same acceleration, how long would it take the car to come to a stop from a velocity of 40 m/s?

3.20. The brakes of a car can slow it down from 60 mi/hr to 40 mi/hr in 2 s. How long will it take to bring the car to a stop from an initial velocity of 25 mi/hr at the same acceleration?

3.21. An object starts from rest with an acceleration of 10 m/s². (*a*) How far does it go in 0.5 s? (*b*) What is its velocity after 0.5 s?

3.22. A car has an initial velocity of 20 ft/s when it begins to be accelerated at 5 ft/s². (*a*) How long does it take to reach a velocity of 50 ft/s? (*b*) How far does it go during this period of time?

3.23. A car whose velocity is 20 mi/hr is given an acceleration of 5 (mi/hr)/s. What is its velocity after it has gone $\frac{1}{4}$ mi?

3.24. A sports car has an acceleration of 10 ft/s². How much distance does it cover while its velocity is increased from 0 to 30 ft/s? From 30 ft/s to 100 ft/s?

3.25. Find the acceleration of a car that comes to a stop from a velocity of 60 ft/s in a distance of 120 ft.

Answers to Supplementary Problems

3.14. (*a*) 360 mi (*b*) 33 hr

3.15. (*a*) 145 mi (*b*) 58 mi/hr

3.16. 100 mi/hr

3.17. 0.55 s

3.18. 900 ft/s²

3.19. (*a*) 5 m/s² (*b*) 8 s

3.20. 2.5 s

3.21. (*a*) 1.25 m (*b*) 5 m/s

3.22. (*a*) 6 s (*b*) 210 ft

3.23. 142 ft/s = 97 mi/hr

3.24. 45 ft; 455 ft

3.25. − 15 ft/s²

Chapter 4

Motion in a Vertical Plane

ACCELERATION OF GRAVITY

All bodies in free fall near the earth's surface have the same downward acceleration of

$$g = 32 \text{ ft/s}^2 = 9.8 \text{ m/s}^2$$

A body falling from rest in a vacuum thus has a velocity of 32 ft/s at the end of the first second, 64 ft/s at the end of the next second, and so forth.   The farther the body falls, the faster it moves.

A body in free fall has the same downward acceleration whether it starts from rest or has an initial velocity in some direction.

The presence of air affects the motion of falling bodies partly through buoyancy and partly through air resistance.   Thus two different objects falling in air from the same height will not, in general, reach the ground at exactly the same time.   Because air resistance increases with velocity, eventually a falling body reaches a *terminal velocity* that depends upon its mass, size, and shape, and it cannot fall any faster than that.

FALLING BODIES

When buoyancy and air resistance can be neglected, a falling body has the constant acceleration g and the formulas for uniformly accelerated motion apply to it.   Thus a body dropped from rest has the velocity

$$v = gt$$

after the time t and has fallen through a vertical distance of

$$h = \tfrac{1}{2} gt^2$$

From the latter formula we see that

$$t = \sqrt{\frac{2h}{g}}$$

and so the velocity of the body is related to the distance it has fallen by $v = gt$ or

$$v = \sqrt{2gh}$$

To reach a certain height h, a body thrown upward must have the same initial velocity as the final velocity of a body falling from that height, namely $v = \sqrt{2gh}$.

PROJECTILE MOTION

In the absence of air resistance, a projectile sent off at an angle of θ above the horizontal with an initial velocity of v_0 has a range of

$$R = \frac{v_0^2}{g} \sin 2\theta$$

The time of flight is

$$T = \frac{2v_0 \sin \theta}{g}$$

Solved Problems

[Air resistance is assumed negligible in these problems.]

4.1.   A stone dropped from a bridge strikes the water 2.5 s later.   (*a*) What is its final velocity in m/s?   (*b*) How high is the bridge?

(*a*) $$v = gt = 9.8 \frac{m}{s^2} \times 2.5 \, s = 24.5 \frac{m}{s}$$

(*b*) $$h = \tfrac{1}{2} gt^2 = \tfrac{1}{2} \times 9.8 \frac{m}{s^2} \times (2.5 \, s)^2 = 30.6 \, m$$

4.2.   A ball is dropped from a window 64 ft above the ground.   (*a*) How long does it take to reach the ground?   (*b*) What is its final velocity?

(*a*) Since $h = \tfrac{1}{2} gt^2$,

$$t = \sqrt{\frac{2h}{g}} = \sqrt{\frac{2 \times 64 \, ft}{32 \, ft/s^2}} = \sqrt{4 \, s^2} = 2 \, s$$

(*b*) $$v = gt = 32 \frac{ft}{s^2} \times 2 \, s = 64 \frac{ft}{s}$$

4.3.   What velocity must a ball have when thrown upward if it is to reach a height of 100 ft?

The upward velocity the ball must have is the same as the downward velocity a ball would have if dropped from that height.   Hence

$$v = \sqrt{2gh} = \sqrt{2 \times 32 \frac{ft}{s^2} \times 100 \, ft} = \sqrt{6400 \frac{ft^2}{s^2}} = 80 \frac{ft}{s}$$

4.4.   A ball is thrown downward with an initial velocity of 20 ft/s.   (*a*) How fast is it moving after 2 s?   (*b*) How far does it fall in 2 s?

(*a*) $$v = v_0 + gt = 20 \frac{ft}{s} + 32 \frac{ft}{s^2} \times 2 \, s = 20 \frac{ft}{s} + 64 \frac{ft}{s} = 84 \frac{ft}{s}$$

(*b*) $$s = v_0 t + \tfrac{1}{2} gt^2 = 20 \frac{ft}{s} \times 2 \, s + \tfrac{1}{2} \times 32 \frac{ft}{s^2} \times (2 \, s)^2$$

$$= 40 \, ft + 64 \, ft = 104 \, ft$$

4.5.   A ball is thrown upward with an initial velocity of 20 ft/s.   (*a*) How fast is it moving after $\tfrac{1}{2}$ s and in what direction?   (*b*) How fast is it moving after 2 s and in what direction?

We consider upward as $+$ and downward as $-$.   Then $v_0 = +20$ ft/s and $g = -32$ ft/s^2, so that

(*a*) $$v = v_0 + gt = 20 \frac{ft}{s} - 32 \frac{ft}{s^2} \times \tfrac{1}{2} s = 20 \frac{ft}{s} - 16 \frac{ft}{s} = 4 \frac{ft}{s}$$

which is positive and hence upward.   After 2 s

(*b*) $$v = v_0 + gt = 20 \frac{ft}{s} - 32 \frac{ft}{s^2} \times 2 \, s = 20 \frac{ft}{s} - 64 \frac{ft}{s} = -44 \frac{ft}{s}$$

which is negative and hence downward.

4.6.   A man in a closed elevator cab with no floor indicator does not know whether the cab is stationary, moving upward at constant velocity, or moving downward at constant velocity.   To try to find out, he drops a coin from a height of 6 ft and times its fall with a stopwatch.   What would he find in each case?

Since the coin has exactly the same velocity as the elevator cab when it is dropped, this experiment would give the same time of fall in each case, namely

$$t = \sqrt{\frac{2h}{g}} = \sqrt{\frac{2 \times 6 \text{ ft}}{32 \text{ ft/s}^2}} = \sqrt{0.375 \text{ s}^2} = 0.61 \text{ s}$$

However, if the elevator cab were *accelerated* upward or downward, the time of fall would be respectively less or more than this.

4.7. An airplane is in level flight at a velocity of 300 mi/hr and an altitude of 5000 ft when it drops a bomb. (*a*) How long does the bomb take to reach the ground? (*b*) How far from the point above which it was dropped will the bomb strike the ground? (*c*) What will the bomb's velocity be on impact?

(*a*) The horizontal velocity of the bomb has no effect on its vertical motion (Fig. 4-1). The bomb therefore reaches the ground at the same time as a bomb dropped from rest at a height of 5000 ft, namely after

$$t = \sqrt{\frac{2h}{g}} = \sqrt{\frac{2 \times 5000 \text{ ft}}{32 \text{ ft/s}^2}} = 17.7 \text{ s}$$

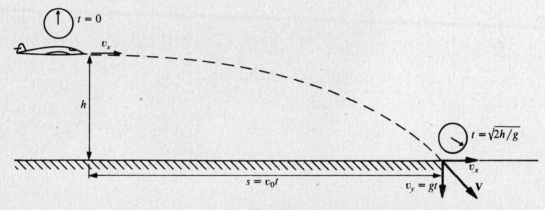

Fig. 4-1

(*b*) A useful conversion relation is 60 mi/hr = 88 ft/s, which may be easier to remember than 1 mi/hr = 1.47 ft/s. The horizontal component of velocity of the bomb is

$$v_x = 300 \text{ mi/hr} \times \frac{88 \text{ ft/s}}{60 \text{ mi/hr}} = 440 \text{ ft/s}$$

In the time $t = 17.7$ s the bomb will travel a horizontal distance of

$$s = v_x t = 440 \text{ ft/s} \times 17.7 \text{ s} = 7790 \text{ ft}$$

(*c*) The final velocity of the bomb has the horizontal component $v_x = 440$ ft/s and the vertical component

$$v_y = gt = 32 \text{ ft/s}^2 \times 17.7 \text{ s} = 566 \text{ ft/s}$$

Hence the magnitude of the final velocity is

$$v = \sqrt{v_x^2 + v_y^2} = \sqrt{(440 \text{ ft/s})^2 + (566 \text{ ft/s})^2} = 717 \text{ ft/s}$$

4.8. A football is thrown with a velocity of 10 m/s at an angle of 30° above the horizontal. (*a*)

How far away should its intended receiver be?   (b) What will the time of flight be?

(a) $$R = \frac{v_0^2}{g} \sin 2\theta = \frac{(10 \text{ m/s})^2}{9.8 \text{ m/s}^2} \times \sin 60° = 8.8 \text{ m}$$

(b) $$T = \frac{2v_0 \sin \theta}{g} = \frac{2 \times 10 \text{ m/s} \times \sin 30°}{9.8 \text{ m/s}^2} = 1.02 \text{ s}$$

4.9. A toy rifle is fired at an angle of 60° above the horizontal.   (a) If the pellet's initial velocity is 40 ft/s, how far does it go?   (b) What is its time of flight?

   (a) Here $2\theta = 120°$ and trigonometric tables only cover angles from 0 to 90°.   However, we recall from Table 2-1 that

$$\sin(90° + \phi) = \cos \phi$$

   and so

$$\sin 120° = \sin(90° + 30°) = \cos 30° = 0.866$$

   Hence

$$R = \frac{v_0^2}{g} \sin 2\theta = \frac{(40 \text{ ft/s})^2}{32 \text{ ft/s}^2} \times 0.866 = 43 \text{ ft}$$

   (b) $$T = \frac{2v_0 \sin \theta}{g} = \frac{2 \times 40 \text{ ft/s} \times \sin 60°}{32 \text{ ft/s}^2} = 2.17 \text{ s}$$

4.10. At what angle should a projectile be fired in order that its range be a maximum?

   The range of a projectile is given by

$$R = \frac{v_0^2}{g} \sin 2\theta$$

   The greatest value the sine function can have is 1.   Since $\sin 90° = 1$, the maximum range occurs when

$$2\theta = 90° \qquad \theta = 45°$$

   Larger and smaller angles than 45° give shorter ranges.   When $\theta = 45°$,

$$R_{max} = \frac{v_0^2}{g}$$

4.11. What minimum initial velocity must a rocket have in order to reach a target 100 mi away?

   The maximum range of a projectile of initial velocity v_0 is $R = v_0^2/g$.   Solving for v_0 yields $v_0 = \sqrt{Rg}$.   Since

$$R = 100 \text{ mi} \times 5280 \text{ ft/mi} = 5.28 \times 10^5 \text{ ft}$$

$$v_0 = \sqrt{Rg} = \sqrt{5.28 \times 10^5 \text{ ft} \times 32 \text{ ft/s}^2} = 4110 \text{ ft/s}$$

   which is

$$v_0 = 4110 \frac{\text{ft}}{\text{s}} \times 0.682 \frac{\text{mi/hr}}{\text{ft/s}} = 2803 \text{ mi/hr}$$

Supplementary Problems

4.12. A stone is dropped from the edge of a cliff.   (a) What is its velocity 3 s later?   (b) How far does it fall in this time?   (c) How far will it fall in the next second?

4.13. A boy throws a ball 60 ft vertically into the air.   (a) How long does he have to wait to catch it on the way down?   (b) What was its initial velocity?   (c) What will be its final velocity?

4.14. The Empire State Building is 1472 ft high. (*a*) How long would it take an object dropped from the top of the building to reach the ground? (*b*) What would the object's final velocity be?

4.15. A body in free fall reaches the ground in 5 s. (*a*) From what height in meters was it dropped? (*b*) What is its final velocity? (*c*) How far did it fall in the last second of its descent?

4.16. A ball is thrown vertically upward with a velocity of 12 m/s. (*a*) At what height is the ball 1 s later? (*b*) 2 s later? (*c*) What is the maximum height the ball reaches?

4.17. A ball is thrown horizontally with a velocity of 12 m/s. (*a*) How far has the ball fallen 1 s later? (*b*) 2 s later?

4.18. A ball is thrown vertically downward with a velocity of 12 m/s. (*a*) How far has the ball fallen 1 s later? (*b*) 2 s later?

4.19. A rifle with a muzzle velocity of 200 m/s is fired with its barrel horizontal at a height of 1.5 m above the ground. (*a*) How long a time is the bullet in the air? (*b*) How far away from the rifle does the bullet strike the ground? (*c*) If the muzzle velocity were 150 m/s, would there be any difference in these answers?

4.20. A ball is rolled off the edge of a table 3 ft high with a horizontal velocity of 4 ft/s. With what velocity does it strike the floor?

4.21. If the initial velocity of a projectile were doubled, how would its maximum range be affected?

4.22. A shell is fired at an angle of 40° above the horizontal at a velocity of 300 m/s. (*a*) What is its range? (*b*) What is its time of flight?

4.23. A golf ball leaves the club at 120 ft/s at an angle of 55° above the horizontal. (*a*) What is its range? (*b*) What is its time of flight?

4.24. A rifle bullet has a muzzle velocity of 1000 ft/s. (*a*) At what angle should the rifle be pointed to achieve the maximum range? (*b*) What is that range? (*c*) What is the time of flight at the maximum range?

Answers to Supplementary Problems

4.12. (*a*) 96 ft/s (*b*) 144 ft (*c*) 112 ft

4.13. (*a*) 3.9 s (*b*) 62 ft/s (*c*) 62 ft/s

4.14. (*a*) 9.6 s (*b*) 307 ft/s

4.15. (*a*) 123 m (*b*) 49 m/s (*c*) 44 m

4.16. (*a*) 7.1 m (*b*) 4.4 m (*c*) 7.3 m

4.17. (*a*) 4.9 m (*b*) 19.6 m

4.18. (*a*) 16.9 m (*b*) 43.6 m

4.19. (*a*) 0.56 s (*b*) 112 m (*c*) the time would be unchanged since it depends only on the height of the rifle, but the distance would be reduced to 84 m.

4.20. 14.4 ft/s

4.21. The maximum range would be four times greater.

4.22. (*a*) 9.044 km (*b*) 39 s

4.23. (*a*) 423 ft (*b*) 614 s

4.24. (*a*) 45° (*b*) 31,250 ft = 5.92 mi (*c*) 44 s

Chapter 5

The Laws of Motion

FIRST LAW OF MOTION

According to Newton's first law of motion, a body at rest will remain at rest and a body in motion will remain in motion at constant velocity in a straight line if no net force acts on it.

This law thus provides a definition of force: a force is any influence that can change the velocity of a body. In order to accelerate something, a net force must be applied to it. Conversely, every acceleration is due to the action of a net force. (Since it is possible for two or more forces acting on the same body to cancel out with a vector sum of zero, a "net force" or "unbalanced force" is required.)

MASS

The property a body has of resisting any change in its state of rest or of uniform motion in a straight line is called *inertia*. The inertia of a body is related to what can be loosely thought of as the "amount of matter" it contains. A quantitative measure of inertia is *mass*: the more mass a body has, the less its acceleration when a net force acts on it.

SECOND LAW OF MOTION

Newton's second law of motion states that the net force acting on a body is proportional to the mass of the body and to its acceleration; the direction of the force is the same as that of the body's acceleration.

In a properly chosen set of units, the proportionality between force and the product of mass and acceleration is an equality, so that

$$F = ma$$
Force = mass × acceleration

The second law of motion is the key to understanding the behavior of moving bodies since it links cause (force) and effect (acceleration) in a definite way.

UNITS OF MASS AND FORCE

In the SI system, the unit of mass is the *kilogram* (kg) and the unit of force is the *newton* (N); a net force of 1 N acting on a mass of 1 kg produces an acceleration of 1 m/s^2.

In the British system, the unit of mass is the *slug* and the unit of force is the *pound* (lb); a net force of 1 lb acting on a mass of 1 slug produces an acceleration of 1 ft/s^2.

The second law of motion in SI and British units is as follows:

SI units: $F(\text{newtons}) = m(\text{kilograms}) \times a(\text{m/s}^2)$

British units: $F(\text{pounds}) = m(\text{slugs}) \times a(\text{ft/s}^2)$

28

WEIGHT AND MASS

The *weight* of a body is the gravitational force with which the earth attracts it. If a person weighs 150 lb, this means that the earth pulls him down with a force of 150 lb. Weight is different from mass, which is a measure of the response of the body to an applied force. The weight of a body varies with its location near the earth (or other astronomical body) whereas its mass is the same everywhere in the universe.

The weight of a body is the force that causes it to be accelerated downward with the acceleration of gravity g. Hence, from the second law of motion with $F = w$ and $a = g$.

$$w = mg$$

Weight = mass × acceleration of gravity

Because g is constant near the earth's surface, the weight of a body there is proportional to its mass—a large mass is heavier than a small one.

System of units	To find mass m given weight w	To find weight w given mass m
SI	$m(\text{kg}) = \dfrac{w(\text{N})}{9.8 \text{ m/s}^2}$	$w(\text{N}) = m(\text{kg}) \times 9.8 \text{ m/s}^2$
British	$m(\text{slugs}) = \dfrac{w(\text{lb})}{32 \text{ ft/s}^2}$	$w(\text{lb}) = m(\text{slugs}) \times 32 \text{ ft/s}^2$

Conversion of units

$1 \text{ kg} = 10^3 \text{ g} = 0.0685 \text{ slug}$ [1 kg corresponds to 2.21 lb in the sense that the *weight* of 1 kg is 2.21 lb]

$1 \text{ slug} = 14.6 \text{ kg}$ [1 slug corresponds to 32 lb in the sense that the *weight* of 1 slug is 32 lb]

1 newton = 0.225 lb

1 lb = 4.45 newtons

THIRD LAW OF MOTION

Newton's third law of motion states that when one body exerts a force on another body, the second exerts an equal force in the opposite direction on the first.

Thus for every action force, there is an equal and opposite reaction force; no force can occur all by itself. Action and reaction forces never balance out because they act on *different* bodies.

Solved Problems

5.1. A book rests on a table. (*a*) Show the forces acting on the table and the corresponding reaction forces. (*b*) Why don't the forces acting on the table cause it to move?

   (*a*) See Fig. 5-1.

   (*b*) The forces that act on the table have a vector sum of zero, so there is no net force acting on it.

5.2. In the process of walking, what force makes a person move forward?

   The person's foot exerts a backward force on the ground; the forward reaction force of the ground on the foot produces the forward motion.

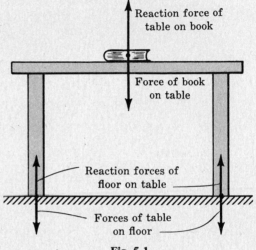

Fig. 5-1

PROBLEMS IN SI UNITS

5.3. (*a*) What is the weight of an object whose mass is 5 kg? (*b*) What is its acceleration when a net force of 100 N acts on it?

   (*a*)
$$w = mg = 5 \text{ kg} \times 9.8 \,\frac{\text{m}}{\text{s}^2} = 49 \text{ N}$$

   (*b*) From the second law of motion, $F = ma$, we have

$$a = \frac{F}{m} = \frac{100 \text{ N}}{5 \text{ kg}} = 20 \,\frac{\text{m}}{\text{s}^2}$$

5.4. A force of 1 N acts on (*a*) a body whose mass is 1 kg, and (*b*) a body whose weight is 1 N. Find their respective accelerations.

   (*a*)
$$a = \frac{F}{m} = \frac{1 \text{ N}}{1 \text{ kg}} = 1 \,\frac{\text{m}}{\text{s}^2}$$

   (*b*) The mass of a body whose weight is w is $m = w/g$. Hence, in general,

$$a = \frac{F}{m} = \frac{F}{w/g} = \frac{F}{w}\,g$$

   Here $F = w = 1$ N, so the acceleration is $a = g = 9.8 \text{ m/s}^2$.

5.5. A 10-kg body is observed to have an acceleration of 5 m/s². What is the net force acting on it?

$$F = ma = 10 \text{ kg} \times 5 \,\frac{\text{m}}{\text{s}^2} = 50 \text{ N}$$

5.6. A force of 80 N gives an object of unknown mass an acceleration of 20 m/s². What is its mass?

$$m = \frac{F}{a} = \frac{80 \text{ N}}{20 \text{ m/s}^2} = 4 \text{ kg}$$

5.7.   A force of 3000 N is applied to a 1500-kg car at rest.   (*a*) What is its acceleration?   (*b*) What will its velocity be 5 s later?

$$(a) \quad a = \frac{F}{m} = \frac{3000 \text{ N}}{1500 \text{ kg}} = 2 \, \frac{\text{m}}{\text{s}^2} \qquad (b) \quad v = at = 2 \, \frac{\text{m}}{\text{s}^2} \times 5 \text{ s} = 10 \, \frac{\text{m}}{\text{s}}$$

5.8.   An empty truck whose mass is 2000 kg has a maximum acceleration of 1 m/s². What will its maximum acceleration be when it carries a load of 1000 kg?

The maximum force available is

$$F = ma = 2000 \text{ kg} \times 1 \, \frac{\text{m}}{\text{s}^2} = 2000 \text{ N}$$

When this force is applied to the total mass of the loaded truck, which is 3000 kg, the resulting acceleration will be

$$a = \frac{F}{m} = \frac{2000 \text{ N}}{3000 \text{ kg}} = \frac{2}{3} \, \frac{\text{m}}{\text{s}^2}$$

5.9.   A 1000-kg car goes from 10 m/s to 20 m/s in 5 s.   What force is acting on it?

$$a = \frac{v - v_0}{t} = \frac{20 \text{ m/s} - 10 \text{ m/s}}{5 \text{ s}} = 2 \, \frac{\text{m}}{\text{s}^2}$$

$$F = ma = 1000 \text{ kg} \times 2 \, \frac{\text{m}}{\text{s}^2} = 2000 \text{ N}$$

5.10.   The brakes of a 1000-kg car exert 3000 N.   (*a*) How long will it take the car to come to a stop from a velocity of 30 m/s?   (*b*) How far will the car travel during this time?

(*a*)   The acceleration the brakes can produce is

$$a = \frac{F}{m} = \frac{3000 \text{ N}}{1000 \text{ kg}} = 3 \, \frac{\text{m}}{\text{s}^2}$$

Here the initial velocity is $v_0 = 30$ m/s, the final velocity is $v = 0$, and the acceleration is -3 m/s², which is negative since the car is slowing down.   Hence

$$v = v_0 + at$$

$$0 = 30 \, \frac{\text{m}}{\text{s}} - 3 \, \frac{\text{m}}{\text{s}^2} \times t$$

$$t = \frac{30 \text{ m/s}}{3 \text{ m/s}^2} = 10 \text{ s}$$

(*b*) $$s = v_0 t + \tfrac{1}{2} a t^2 = 30 \, \frac{\text{m}}{\text{s}} \times 10 \text{ s} - \tfrac{1}{2} \times 3 \, \frac{\text{m}}{\text{s}^2} \times (10 \text{ s})^2 = 300 \text{ m} - 150 \text{ m} = 150 \text{ m}$$

5.11.   A box is sliding down a frictionless plane that is inclined at an angle of 30° with the horizontal. What is the acceleration of the box?

The force **F** that accelerates the box is the component of the box's weight **w** parallel to the plane. From Fig. 5-2 we see that

$$F = w \sin \theta$$

Hence the acceleration of the box is

$$a = \frac{F}{m} = \frac{F}{w/g} = \frac{w \sin \theta}{w/g} = g \sin \theta$$

$$= 9.8 \text{ m/s}^2 \times \sin 30° = 4.9 \text{ m/s}^2$$

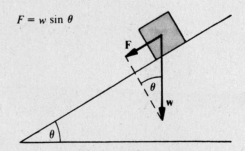

$$F = w \sin \theta$$

Fig. 5-2

PROBLEMS IN BRITISH UNITS

5.12. (a) What is the weight of an object whose mass is 50 slugs?   (b) What is the mass of an object whose weight is 50 lb?

(a)
$$w = mg = 50 \text{ slugs} \times 32\, \frac{\text{ft}}{\text{s}^2} = 1600 \text{ lb}$$

(b)
$$m = \frac{w}{g} = \frac{50 \text{ lb}}{32 \text{ ft/s}^2} = 1.56 \text{ slugs}$$

5.13. (a) What is the mass of a 160-lb man?   (b) With what force is he attracted to the earth?   (c) What is the reaction force to this force?   (d) If he jumps from a diving board, what will his downward acceleration be?

(a)
$$m = \frac{w}{g} = \frac{160 \text{ lb}}{32 \text{ ft/s}^2} = 5 \text{ slugs}$$

(b)   The attractive force exerted on the man by the earth is his weight of 160 lb.

(c)   The reaction force is an upward pull he exerts on the earth of 160 lb.   Since the earth is much more massive than the man, the presence of this reaction force is not obvious, but it nevertheless exists.

(d)   His downward acceleration is the acceleration of gravity, $g = 32 \text{ ft/s}^2$.

5.14. A net force of 75 lb acts on a body of mass 25 slugs which is initially at rest.   (a) Find its acceleration.   (b) How fast will the body be moving 12 s later?

(a)
$$a = \frac{F}{m} = \frac{75 \text{ lb}}{25 \text{ slugs}} = 3\, \frac{\text{ft}}{\text{s}^2}$$

(b)
$$v = at = 3\, \frac{\text{ft}}{\text{s}^2} \times 12 \text{ s} = 36\, \frac{\text{ft}}{\text{s}}$$

5.15. A net force of 150 lb acts on a body whose weight is 96 lb.   What is its acceleration?

The mass of the body is
$$m = \frac{w}{g} = \frac{96 \text{ lb}}{32 \text{ ft/s}^2} = 3 \text{ slugs}$$

Its acceleration is therefore
$$a = \frac{F}{m} = \frac{150 \text{ lb}}{3 \text{ slugs}} = 50\, \frac{\text{ft}}{\text{s}^2}$$

5.16. How much force is needed to bring a 3200-lb car from rest to a velocity of 44 ft/s (30 mi/hr) in 8 s?

The car's mass is
$$m = \frac{w}{g} = \frac{3200 \text{ lb}}{32 \text{ ft/s}^2} = 100 \text{ slugs}$$

and its acceleration is
$$a = \frac{v}{t} = \frac{44 \text{ ft/s}}{8 \text{ s}} = 5.5\, \frac{\text{ft}}{\text{s}^2}$$

Hence
$$F = ma = 100 \text{ slugs} \times 5.5\, \frac{\text{ft}}{\text{s}^2} = 550 \text{ lb}$$

5.17. The brakes of a certain 2400-lb car can exert a maximum force of 750 lb. (a) What is the minimum time needed to slow the car down from 60 ft/s to 20 ft/s? (b) How far does the car travel in this time?

(a) The mass of the car is

$$m = \frac{w}{g} = \frac{2400 \text{ lb}}{32 \text{ ft/s}^2} = 75 \text{ slugs}$$

Its maximum acceleration is therefore

$$a = \frac{F}{m} = \frac{750 \text{ lb}}{75 \text{ slugs}} = 10 \frac{\text{ft}}{\text{s}^2}$$

Here $v_0 = 60$ ft/s, $v = 20$ ft/s, and $a = -10$ ft/s^2. Since $a = (v - v_0)/t$,

$$t = \frac{v - v_0}{a} = \frac{20 \text{ ft/s} - 60 \text{ ft/s}}{-10 \text{ ft/s}^2} = \frac{-40 \text{ ft/s}}{-10 \text{ ft/s}^2} = 4 \text{ s}$$

(b) $$s = v_0 t + \tfrac{1}{2} a t^2 = 60 \frac{\text{ft}}{\text{s}} \times 4 \text{ s} - \tfrac{1}{2} \times 10 \frac{\text{ft}}{\text{s}^2} \times (4 \text{ s})^2 = 240 \text{ ft} - 80 \text{ ft} = 160 \text{ ft}$$

5.18. A 3200-lb elevator cab is supported by a cable in which the maximum safe tension is 4000 lb. (a) What is the greatest upward acceleration the elevator cab can have? (b) The greatest downward acceleration?

(a) The mass of the cab is

$$m = \frac{w}{g} = \frac{3200 \text{ lb}}{32 \text{ ft/s}^2} = 100 \text{ slugs}$$

The maximum net upward force the cable can exert is the difference between the maximum tension of 4000 lb and the 3200-lb weight of the cab, or 800 lb. Hence the greatest upward acceleration is

$$a = \frac{F}{m} = \frac{800 \text{ lb}}{100 \text{ slugs}} = 8 \frac{\text{ft}}{\text{s}^2}$$

(b) The cable is flexible and so cannot push down on the cab. The maximum downward acceleration therefore corresponds to free fall with an acceleration of $g = 32$ ft/s^2.

5.19. A weight of 50 lb and another of 30 lb are suspended by a rope on either side of a frictionless pulley (Fig. 5-3). What is the acceleration of each weight?

The net force acting on the system of two objects is the *difference* between their weights:

$$F = 50 \text{ lb} - 30 \text{ lb} = 20 \text{ lb}$$

The mass to be accelerated is the total mass of the system, namely

$$m = \frac{w_1 + w_2}{g} = \frac{50 \text{ lb} + 30 \text{ lb}}{32 \text{ ft/s}^2} = 2.5 \text{ slugs}$$

Hence the upward acceleration of the system is

$$a = \frac{F}{m} = \frac{20 \text{ lb}}{2.5 \text{ slugs}} = 8 \frac{\text{ft}}{\text{s}^2}$$

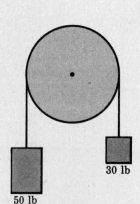

30 lb

50 lb

Fig. 5-3

The 50-lb weight moves downward with this acceleration, and the 30-lb weight moves upward with the same acceleration.

5.20. How much force is required to push a 48-lb wooden block up a frictionless plane that is inclined at an angle of 20° with the horizontal so that it has an acceleration along the plane of 10 ft/s²?

To give the box an upward acceleration of a along the plane, the applied force F must exceed the component F_w of the block's weight w parallel to the plane by the amount ma. From Fig. 5-2, $F_w = w \sin \theta$, and so

$$F = w \sin \theta + ma = w \sin \theta + \frac{w}{g} a$$

$$= 48 \text{ lb} \times \sin 20° + \frac{48 \text{ lb}}{32 \text{ ft/s}^2} \times 10 \text{ ft/s}^2 = 31.4 \text{ lb}$$

Supplementary Problems

5.21. Since action and reaction forces are always equal in magnitude and opposite in direction, how can anything ever be accelerated?

5.22. A horse is pulling a cart. (*a*) What is the force that causes the horse to move forward? (*b*) What is the force that causes the cart to move forward?

5.23. Is it possible for something to have a downward acceleration greater than g? If so, how can this be accomplished?

5.24. (*a*) When a horizontal force equal to its weight is applied to an object on a frictionless surface, what is its acceleration? (*b*) What is its acceleration when the force is applied vertically upward?

5.25. (*a*) What is the weight of 6 kg of potatoes? (*b*) What is the mass of 6 N of potatoes?

5.26. A force of 10 N is applied to (*a*) a body of mass 5 kg, and (*b*) a body of weight 5 N. Find their accelerations.

5.27. (*a*) What is the weight of 2 slugs of salami? (*b*) What is the mass of 2 lb of salami?

5.28. (*a*) How much upward force is needed to support a 20-kg object at rest? (*b*) To give it an upward acceleration of 2 m/s²? (*c*) To give it a downward acceleration of 2 m/s²?

5.29. (*a*) What is the acceleration of a 5-kg object suspended by a string when an upward force of 39 N is applied to the string? (*b*) When an upward force of 49 N is applied to the string? (*c*) When an upward force of 59 N is applied to the string?

5.30. (*a*) How much applied force is needed to give an 8-lb object an upward acceleration of 10 ft/s²? (*b*) A downward acceleration of 10 ft/s²? (*c*) In what direction must the latter force act?

5.31. A net force of 12 N gives an object an acceleration of 4 m/s². (*a*) What net force is needed to give it an acceleration of 1 m/s²? (*b*) An acceleration of 10 m/s²?

5.32. A certain net force gives a 2-kg object an acceleration of 0.5 m/s². What acceleration would the same force give a 10-kg object?

5.33. When an 8-lb rifle is fired, the 0.02-lb bullet receives an acceleration of 10^5 ft/s² while it is in the barrel. (*a*) How much force acts on the bullet? (*b*) Does any force act on the rifle? If so, how much and in what direction? (*c*) The bullet is accelerated for 0.007 s. How fast does it leave the barrel of the rifle?

5.34. How much force is needed to accelerate a train whose mass is 1000 metric tons (1 metric ton = 1000 kg) from rest to a velocity of 6 m/s in 2 min?

5.35. (a) How much force is needed to increase the velocity of a 6400-lb truck from 20 ft/s to 30 ft/s in 5 s? (b) How far does the truck travel in this time?

5.36. (a) How much force is needed to decrease the velocity of a 6400-lb truck from 30 ft/s to 20 ft/s in 5 s? (b) How far does the truck travel in this time?

5.37. A car strikes a stone wall at a velocity of 40 ft/s (27 mi/hr). (a) The car is rigidly built and the 160-lb driver comes to a stop in a time of 0.05 s. How much force acts on him? (b) The car is built so that its front end collapses gradually and the driver comes to a stop in 0.1 s. How much force acts on him in this case?

5.38. A 0.05-kg snail goes from rest to a velocity of 0.01 m/s in 5 s. (a) How much force does it exert? (b) How far does it go during this time?

5.39. A 160-lb man stands on a scale in an elevator cab. What does the scale read when the cab is (a) ascending at a constant velocity of 10 ft/s? (b) Ascending at a constant acceleration of 2 ft/s²? (c) Descending at a constant velocity of 10 ft/s? (d) Descending at a constant acceleration of 2 ft/s²? (e) In free fall because the cable has broken?

5.40. An 80-kg man stands on a scale in an elevator cab. When it starts to move, the scale reads 700 N. (a) Is the cab moving upward or downward? (b) Is its velocity constant? If so, what is it? If not, what is its acceleration?

5.41. Two boxes, one of mass 20 kg and the other of mass 30 kg, are sliding down a frictionless inclined plane that makes an angle of 25° with the horizontal. Find their respective accelerations.

5.42. A force of 50 lb is used to pull a 50-lb crate up a frictionless plane which is inclined at 30° with the horizontal. Find the acceleration of the crate.

Answers to Supplementary Problems

5.21. The action and reaction forces always act on different bodies.

5.22. (a) The reaction force the ground exerts on its feet. (b) The force the horse exerts on it.

5.23. Yes, by an applied downward force in addition to the downward force of gravity.

5.24. (a) g　(b) 0

5.25. (a) 59 N　(b) 0.61 kg

5.26. (a) 2 m/s²　(b) 19.6 m/s²

5.27. (a) 64 lb　(b) 0.0625 slug

5.28. (a) 196 N　(b) 236 N　(c) 156 N

5.29. (a) 2 m/s² downward　(b) 0　(c) 2 m/s² upward

5.30. (a) 10.5 lb　(b) 5.5 lb　(c) upward

5.31. (a) 3 N　(b) 30 N

5.32. 0.1 m/s²

5.33. (a) 62.5 lb　(b) 62.5 lb backward　(c) 700 ft/s

5.34. 5×10^4 N

5.35. (a) 400 lb　(b) 125 ft

5.36. (a) 400 lb　(b) 125 ft

5.37. (a) 4000 lb　(b) 2000 lb

5.38. (a) 10^{-4} N　(b) 0.025 m

5.39. (a) 160 lb　(b) 170 lb　(c) 160 lb (d) 150 lb　(e) 0

5.40. (a) Downward　(b) no; 1.05 m/s²

5.41. 4.14 m/s²; 4.14 m/s²

5.42. 16 ft/s²

Chapter 6

Friction

STATIC AND SLIDING FRICTION

Frictional forces act to oppose relative motion between surfaces that are in contact. Such forces act parallel to the surfaces.

Static friction occurs between surfaces at rest relative to each other. When an increasing force is applied to an object resting on a surface, for example, the force of static friction initially increases as well to prevent motion. Ultimately a certain limiting force is reached which static friction cannot exceed, and the object begins to move. Once the object is moving, the force of *sliding* (or *kinetic*) *friction* remains constant at a value that is usually somewhat smaller than the maximum reached by static friction.

COEFFICIENT OF FRICTION

The frictional force between two surfaces depends upon the normal (perpendicular) force N pressing them together and also upon the natures of the surfaces. The latter factor is expressed quantitatively in the *coefficient of friction* μ (Greek letter "mu") whose value depends upon the materials in contact. The frictional force is experimentally found to be given by the formula

$$F_f = \mu N$$

Frictional force = coefficient of friction $\times$ normal force

In the case of static friction, the coefficient represents the maximum value. The coefficient of static friction μ_s exceeds that of sliding friction μ except for smooth, well-lubricated surfaces where the two are very nearly equal. Thus a typical value of μ_s for wood on wood is 0.5 and that of μ is 0.3, whereas both coefficients might be 0.03 for metal surfaces separated by an oil film.

ROLLING FRICTION

Ideally there should be no rolling friction, because there should be no relative motion between the surfaces in contact. In reality, a wheel or ball is slightly flattened when it rests on a surface, which itself is slightly dented. A resistive force arises when the wheel or ball rolls, partly because it and the surface must be continually deformed and partly because there is some relative motion between them owing to the deformation. Coefficients of rolling friction are nevertheless much smaller than those of sliding friction: for a rubber tire rolling on a concrete road μ is about 0.04, for instance, whereas it is 0.7 for the same tire sliding on the road.

Solved Problems

6.1. How much force is needed to keep a 3000-lb car moving at constant velocity on a level concrete road? Assume that the car is moving too slowly for air resistance to be important, and use $\mu = 0.04$ for the coefficient of rolling friction.

The normal force is the car's weight of 3000 lb.   Hence

$$F = \mu N = \mu w = 0.04 \times 3000 \text{ lb} = 120 \text{ lb}$$

6.2. A force of 200 N is just sufficient to start a 50-kg steel trunk moving across a wooden floor. Find the coefficient of static friction.

The normal force is the trunk's weight of mg.   Hence

$$\mu_s = \frac{F}{N} = \frac{F}{mg} = \frac{200 \text{ N}}{50 \text{ kg} \times 9.8 \text{ m/s}^2} = 0.41$$

6.3. A 100-lb wooden crate is being pushed across a wooden floor with a force of 40 lb.   If $\mu = 0.3$, find the acceleration of the crate.

The applied force of $F_A = 40$ lb is opposed by the frictional force of

$$F_f = \mu N = \mu w = 0.3 \times 100 \text{ lb} = 30 \text{ lb}$$

The net force on the crate is therefore

$$F = F_A - F_f = 40 \text{ lb} - 30 \text{ lb} = 10 \text{ lb}$$

The crate's mass is

$$m = \frac{w}{g} = \frac{100 \text{ lb}}{32 \text{ ft/s}^2} = 3.125 \text{ slugs}$$

and so its acceleration is

$$a = \frac{F}{m} = \frac{10 \text{ lb}}{3.125 \text{ slugs}} = 3.2 \text{ ft/s}^2$$

6.4. A bowling ball with an initial velocity of 3 m/s rolls along a level floor for 50 m before coming to a stop.   What is the coefficient of rolling friction?

We begin by finding the ball's acceleration.   From Chapter 3, $v^2 = v_0^2 + 2as$.   Here $v = 0$, $v_0 = 3$ m/s, and $s = 50$ m, and so

$$a = -\frac{v_0^2}{2s} = -\frac{(3 \text{ m/s})^2}{2 \times 50 \text{ m}} = -0.09 \text{ m/s}^2$$

The minus sign means that the velocity is decreasing, and we can disregard it here.   The force corresponding to this acceleration is ma, which is equal to the frictional force $\mu N = \mu mg$.   Hence

$$\mu mg = ma \qquad \mu = \frac{a}{g} = \frac{0.09 \text{ m/s}^2}{9.8 \text{ m/s}^2} = 0.0092$$

6.5. The coefficient of sliding friction between a rubber tire and a wet concrete road is 0.5.   (a) Find the minimum time in which a car whose initial velocity is 30 mi/hr can come to a stop on such a road.   (b) What distance will the car cover in this time?

(a) The maximum available frictional force is $\mu N = \mu mg$, and so

$$F_f = \mu mg = ma$$

$$a = \mu g = 0.5 \times 32 \text{ ft/s}^2 = 16 \text{ ft/s}^2$$

Since $v = at$ and here

$$v = 30 \text{ mi/hr} \times 1.47 \frac{\text{ft/s}}{\text{mi/hr}} = 44 \text{ ft/s}$$

the time required for the car to come to a stop is

$$t = \frac{v}{a} = \frac{44 \text{ ft/s}}{16 \text{ ft/s}^2} = 2.75 \text{ s}$$

(b) The distance covered by the car in coming to a stop is, since $a = -16$ ft/s^2,

$$s = v_0 t + \tfrac{1}{2} at^2 = 44 \text{ ft/s} \times 2.75 \text{ s} - \tfrac{1}{2} \times 16 \text{ ft/s}^2 \times (2.75 \text{ s})^2 = 60.5 \text{ ft}$$

Another way to obtain this result is to use the formula $v^2 = 2as$:

$$s = \frac{v^2}{2a} = \frac{(44 \text{ ft/s})^2}{2 \times 16 \text{ ft/s}^2} = 60.5 \text{ ft}$$

6.6. A block at rest on an adjustable inclined plane begins to move when the angle between the plane and the horizontal reaches a certain value θ, which is known as the *angle of repose*. How is this angle related to the coefficient of static friction between the block and the plane?

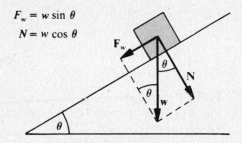

$$F_w = w \sin \theta$$
$$N = w \cos \theta$$

Fig. 6-1

The weight **w** of the block can be resolved into a component $\mathbf{F}_w$ parallel to the plane and another component **N** perpendicular to the plane. From Fig. 6-1 the magnitudes of $\mathbf{F}_w$ and **N** are

$$F_w = w \sin \theta \qquad N = w \cos \theta$$

When the block just begins to move, the downward force along the plane $\mathbf{F}_w$ must be equal to the maximum force $\mu_s N$ of static friction, so that

$$F_w = \mu_s N$$
$$w \sin \theta = \mu_s w \cos \theta$$

$$\mu_s = \frac{\sin \theta}{\cos \theta} = \tan \theta$$

Thus the coefficient of static friction equals the tangent of the angle of repose. The weight of the block has no significance here.

6.7. A lathe mounted on wooden skids is to be slid down a pair of planks placed against the back of a truck. (*a*) If the coefficient of sliding friction is 0.28, what angle should the planks make with the ground in order that the lathe slide down at constant velocity? (*b*) When the planks are at this angle, will the lathe start to slide down of its own accord?

(*a*) When the lathe moves at constant velocity, there is no net force on it, according to Newton's first law of motion. The downward force due to the lathe's weight therefore exactly balances the retarding force of sliding friction, and by the reasoning of Problem 6.6

$$\tan \theta = \mu = 0.28 \qquad \theta = 16°$$

(*b*) Since the coefficient of static friction exceeds that of sliding friction, the lathe will have to be given a push to start it moving.

6.8. A man whose shoes have leather soles and heels is able to stand without slipping on a wooden surface that makes an angle of 25° with the horizontal. What is the minimum coefficient of static friction for leather on wood?

Since the angle of repose is at least 25°,

$$\mu_s \geqslant \tan 25° = 0.47$$

The symbol $\geqslant$ means "equal to or greater than."

6.9. A 5000-ton ship rests on launching ways that slope down to the water at an angle of 10°. If the coefficient of sliding friction is 0.18, how much force is needed to winch the ship down the ways into the water?

The frictional force to be overcome is

$$F_f = \mu N = \mu w \cos \theta = 0.18 \times 5000 \text{ tons} \times \cos 10° = 886 \text{ tons}$$

The component of the ship's weight parallel to the ways is

$$F_w = w \sin \theta = 5000 \text{ tons} \times \sin 10° = 868 \text{ tons}$$

Therefore an additional force of

$$F_f - F_w = 886 \text{ tons} - 868 \text{ tons} = 18 \text{ tons}$$

is required.

Supplementary Problems

6.10. An 80-lb wooden crate rests on a horizontal wooden floor. If the coefficient of static friction is 0.5, how much force is needed to set the crate in motion?

6.11. A force of 300 N is sufficient to keep a 100-kg wooden crate moving at constant velocity across a wooden floor. What is the coefficient of sliding friction?

6.12. A force of 1000 N is applied to a 1200-kg car. If the coefficient of rolling friction is 0.04, what is the car's acceleration?

6.13. The coefficients of static and sliding friction for stone on wood are respectively 0.5 and 0.4. If a 150-lb stone statue is pushed with just enough force to start it moving across a wooden floor and the same force continues to act afterward, find the statue's acceleration.

6.14. A car whose brakes are locked skids to a stop in 200 ft from an initial velocity of 50 mi/hr. Find the coefficient of sliding friction.

6.15. A truck moving at 100 km/hr carries a steel girder which rests on its wooden floor. What is the minimum time in which the truck can come to a stop without the girder moving forward? The coefficient of static friction between steel and wood is 0.5.

6.16. A car with its brakes locked will remain stationary on an inclined plane of dry concrete when the plane is at an angle of less than 45° with the horizontal. What is the coefficient of static friction of rubber tires on dry concrete?

6.17. A steel ramp is to be built for sliding blocks of ice from a refrigeration plant down to ground level. If $\mu = 0.05$, find the angle with the horizontal at which the ice will slide at constant velocity.

6.18. A box slides down a plane 8 m long that is inclined at an angle of 30° with the horizontal. If the box starts from rest and $\mu = 0.25$, find (a) the acceleration of the box, (b) its velocity at the bottom of the plane, and (c) the time required for it to reach the bottom.

6.19. (a) If the box of Problem 6.18 has a mass of 60 kg, how much force is needed to move it up the plane at constant velocity? (b) With an acceleration of 2 m/s²?

6.20. A skier stands on a 5° slope. If the coefficient of static friction is 0.1, does he start to slide down?

Answers to Supplementary Problems

6.10.   40 lb

6.11.   0.306

6.12.   0.44 m/s^2

6.13.   3.2 ft/s^2

6.14.   0.422

6.15.   5.67 s

6.16.   1.0

6.17.   3°

6.18.   (*a*) 2.78 m/s^2 (*b*) 6.67 m/s   (*c*) 2.40 s

6.19.   (*a*) 421 N   (*b*) 541 N

6.20.   No

Chapter 7

Equilibrium

TRANSLATIONAL EQUILIBRIUM

A body is in *translational equilibrium* when no net force acts upon it. Such a body is not accelerated, and either remains at rest or in motion at constant velocity along a straight line, whichever its initial state was.

A body in translational equilibrium may have forces acting upon it, but they must be such that their vector sum is zero. Thus the condition for the translational equilibrium of a body may be written

$$\Sigma\mathbf{F} = 0$$

where the symbol Σ (Greek capital letter *sigma*) means "sum of" and $\mathbf{F}$ refers to the various forces that act on the body.

In analyzing the equilibrium of a particular body, the easiest procedure is usually to establish a set of coordinate axes and to resolve the forces that act on the body into their components along these axes. In this way the vector equation $\Sigma\mathbf{F} = 0$ is replaced by the three scalar equations

$$\Sigma F_x = 0 \qquad \Sigma F_y = 0 \qquad \Sigma F_z = 0$$

A proper choice of directions for the axes often simplifies the calculations. When all the forces lie in a plane, for instance, the coordinate system can be chosen so that the x and y axes lie in the plane and then the two equations $\Sigma F_x = 0$ and $\Sigma F_y = 0$ are enough to express the condition for translational equilibrium.

TORQUE

When the lines of action of the forces that act on a body in translational equilibrium intersect at a common point, they have no tendency to turn the body. Such forces are said to be *concurrent*. When the lines of action do not intersect, the forces are *nonconcurrent* and act to turn the body even though the resultant of the forces is zero (Fig. 7-1).

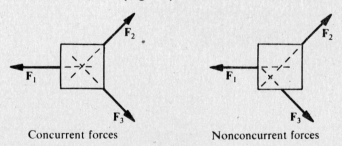

Concurrent forces   Nonconcurrent forces

Fig. 7-1

The *torque* τ (Greek letter "tau") exerted by a force on a body is a measure of its effectiveness in turning the body about a certain pivot point. The *moment arm* of a force $\mathbf{F}$ about a pivot point O is the perpendicular distance L between the line of action of the force and O (Fig. 7-2). The torque τ exerted by the force about O has the magnitude

$$\tau = FL$$
Torque = force × moment arm

41

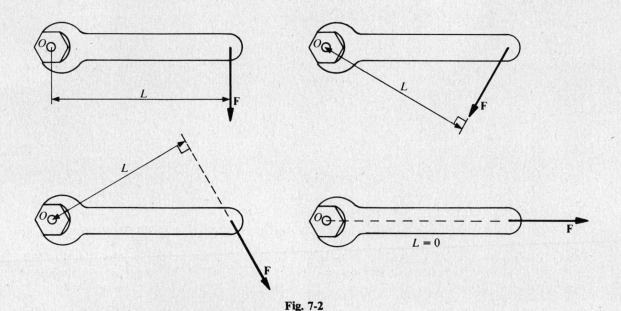

Fig. 7-2

The torque exerted by a force is also known as the *moment* of the force.   A force whose line of action passes through O produces no torque about O because its moment arm is zero.

A torque that tends to cause a counterclockwise rotation when viewed from a given direction is considered positive; a torque that tends to cause a clockwise rotation is considered negative (Fig. 7-3).

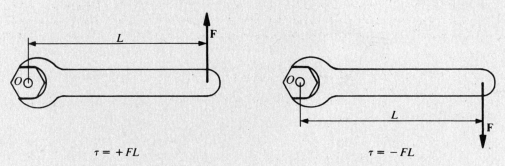

Fig. 7-3

ROTATIONAL EQUILIBRIUM

A body is in *rotational equilibrium* when no net torque acts upon it.   Such a body remains in its initial rotational state, either not spinning at all or spinning at a constant rate.   The condition for the rotational equilibrium of a body may therefore be written

$$\Sigma \tau = 0$$

where $\Sigma \tau$ refers to the sum of the torques acting on the body about any point.

In investigating the rotational equilibrium of a body, any convenient point may be used as the pivot point for calculating torques; if the sum of the torques on a body in translational equilibrium is zero about some point, it is also zero about any other point.

CENTER OF GRAVITY

The *center of gravity* of a body is that point at which the body's entire weight can be regarded as being concentrated. A body can be suspended in any orientation from its center of gravity without tending to rotate. In analyzing the equilibrium of a body, its weight can be considered as a downward force acting from its center of gravity.

Solved Problems

7.1. What are the units of torque in the British and SI (metric) systems?

In the British system, the unit of torque is the lb-ft; in the SI system, it is the newton-meter (N-m).

7.2. A weight is suspended from the middle of a rope whose ends are at the same height. Is it possible for the tension in the rope to be sufficiently great to prevent the rope from sagging at all?

The rope must sag in order for its tension to provide an upward component of force to support the weight. The greater the tension, the less the sag, but it is impossible for the rope to be perfectly horizontal.

7.3. (*a*) Under what circumstances is it necessary to consider torques when analyzing an equilibrium situation? (*b*) About what point should torques be calculated when this is necessary?

(*a*) Torques must be considered when the various forces that act on the body are nonconcurrent; that is, when their lines of action do not intersect at a common point.

(*b*) Torques may be calculated about any point whatever for the purpose of determining the equilibrium of the body. Hence it makes sense to use a point which minimizes the labor involved, which usually is the point through which pass the maximum number of lines of action of the various forces; this is because a force whose line of action passes through a point exerts no torque about that point.

PROBLEMS INVOLVING CONCURRENT FORCES

7.4. A 100-lb box is suspended from two ropes that each make an angle of 40° with the vertical. Find the tension in each rope.

The forces that act on the box are:

T_1 = tension in left-hand rope

T_2 = tension in right-hand rope

w = weight of box, which acts downward from its center of gravity

As shown in Fig. 7-4(*a*) these forces are concurrent, so only translational equilibrium need be considered.

The procedure for working out an equilibrium problem that involves concurrent forces has three steps:

1. Draw a diagram of the forces that act *on* the body; this is called a *free-body* diagram.

2. Select a set of coordinate axes and resolve the various forces into their components along these axes.

3. Set the sum of the force components along each axis equal to zero, so that $\Sigma F_x = 0$, $\Sigma F_y = 0$, $\Sigma F_z = 0$. Then solve the resulting equations for the unknown quantities.

The free-body diagram for this problem is shown in Fig. 7-4(*b*) along with a convenient set of coordinate axes. Since the forces all lie in a plane, we need only *x* and *y* axes. In Fig. 7-4(*c*) the forces are resolved into their *x* and *y* components, whose magnitudes are as follows:

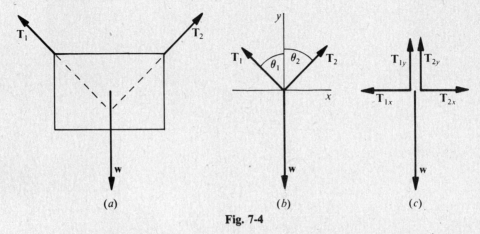

Fig. 7-4

$$T_{1x} = -T_1 \sin \theta_1 = -T_1 \sin 40° = -0.643\,T_1$$
$$T_{1y} = T_1 \cos \theta_1 = T_1 \cos 40° = 0.766\,T_1$$
$$T_{2x} = T_2 \sin \theta_2 = T_2 \sin 40° = 0.643\,T_2$$
$$T_{2y} = T_2 \cos \theta_2 = T_2 \cos 40° = 0.766\,T_2$$
$$w = -100\,\text{lb}$$

Because T_{1x} and **w** are respectively in the $-x$ and $-y$ directions, both have negative magnitudes.

Now we are ready for step 3. First we add up the x components of the forces and set the sum equal to zero. This yields

$$\Sigma F_x = T_{1x} + T_{2x} = 0$$
$$-0.643T_1 + 0.643T_2 = 0$$
$$T_1 = T_2 = T$$

Evidently the tensions in the two ropes are equal. Next we do the same for the y components:

$$\Sigma F_y = T_{1y} + T_{2y} + w = 0$$
$$0.766T_1 + 0.766T_2 - 100\,\text{lb} = 0$$
$$0.766(T_1 + T_2) = 100\,\text{lb}$$
$$T_1 + T_2 = \frac{100\,\text{lb}}{0.766} = 130.5\,\text{lb}$$

Since $T_1 = T_2 = T$,

$$T_1 + T_2 = 2T = 130.5\,\text{lb}$$
$$T = 65\,\text{lb}$$

The tension in each rope is 65 lb.

7.5.   A 500-kg load is suspended from the end of a horizontal boom, as in Fig. 7-5(a). The angle between the boom and the cable supporting its end is 45°. Assuming that the boom's mass can be neglected compared with that of the load, find (a) the tension in the cable, and (b) the inward force the boom exerts on the wall.

(a)   The three forces that act on the end of the boom are the weight **w** of the load, the tension **T** in the cable, and the outward force **F** exerted by the boom. A free-body diagram of these concurrent forces is shown in Fig. 7-5(b). The x and y components of these forces have the magnitudes

$$T_x = -T \cos \theta = -T \cos 45° = -0.707T$$
$$T_y = T \sin \theta = T \sin 45° = 0.707T$$
$$w = -mg = -500\,\text{kg} \times 9.8\,\text{m/s}^2 = -4900\,\text{N}$$
$$F = ?$$

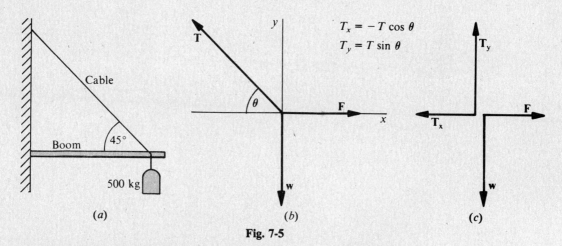

Fig. 7-5

The condition for translational equilibrium in the y (vertical) direction yields

$$\Sigma F_y = T_y + w = 0$$

$$0.707T - 4900 \text{ N} = 0$$

$$T = \frac{4900 \text{ N}}{0.707} = 6930 \text{ N}$$

(b) To find the inward force the boom exerts on the wall, we start with the condition for equilibrium in the x (horizontal) direction:

$$\Sigma F_x = T_x + F = 0$$

$$-0.707T + F = 0$$

$$F = 0.707T = 0.707 \times 6930 \text{ N} = 4900 \text{ N}$$

The inward force on the wall must have the same magnitude as the outward force on the load, hence the inward force is also equal to 4900 N.

7.6. A 50-lb box is suspended by a rope from the ceiling. If a horizontal force of 20 lb is applied to the box, what angle will the rope make with the vertical?

A free-body diagram of the forces acting on the box is shown in Fig. 7-6(a) and the forces are resolved into components in Fig. 7-6(b). The x and y components of the forces are

$$T_x = -T \sin \theta \qquad T_y = T \cos \theta$$

$$F = 20 \text{ lb} \qquad w = -50 \text{ lb}$$

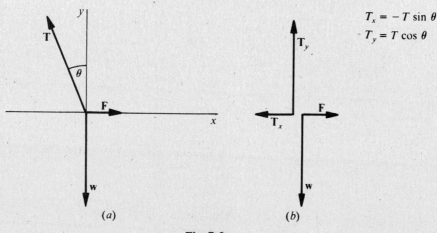

Fig. 7-6

Applying the conditions for equilibrium yields

$$\Sigma F_x = T_x + F = 0 \qquad -T\sin\theta + 20\text{ lb} = 0 \qquad \sin\theta = \frac{20\text{ lb}}{T}$$

$$\Sigma F_y = T_y + w = 0 \qquad T\cos\theta - 50\text{ lb} = 0 \qquad \cos\theta = \frac{50\text{ lb}}{T}$$

If we divide the expression for $\sin\theta$ by that for $\cos\theta$, the T's cancel out to give

$$\frac{\sin\theta}{\cos\theta} = \tan\theta = \frac{20\text{ lb}/T}{50\text{ lb}/T} = \frac{20\text{ lb}}{50\text{ lb}} = 0.40$$

The angle whose tangent is nearest to 0.40 is 22°, and so $\theta = 22°$.

7.7.   In order to move a heavy crate across a floor, one end of a rope is tied to it and the other end is tied to a wall 30 ft away.   When a force of 100 lb is applied to the midpoint of the rope, the rope stretches so that the midpoint moves to the side by 2 ft.   What is the force on the crate?

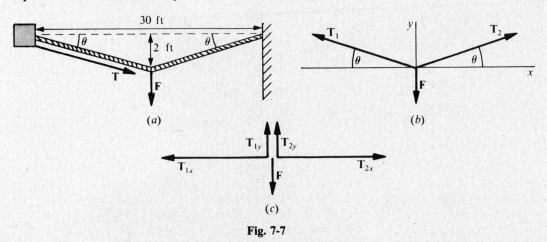

Fig. 7-7

The first step is to find the angle θ between either part of the rope and a straight line between the crate and the point of attachment of the rope to the wall.   With the help of Fig. 7-7(a) we find that

$$\tan\theta = \frac{2\text{ ft}}{15\text{ ft}} = 0.133 \qquad \theta = 7.6°$$

Figure 7-7(b) is a free-body diagram of the forces acting on the midpoint of the rope; T_1 and T_2 are the tensions in the two parts of the rope, and $F = -100$ lb is the applied force.   These forces are resolved in Fig. 7-7(c).   Since $T_1 = T_2 = T$,

$$T_{1y} = T_{2y} = T\sin\theta = T\sin 7.6° = 0.132T$$

At equilibrium

$$\Sigma F_y = T_{1y} + T_{2y} + F = 0 \qquad 2 \times 0.132T - 100\text{ lb} = 0 \qquad T = 379\text{ lb}$$

The tension in the rope provides the force applied to the crate.   We note that the force on the crate exceeds the force applied to the rope, which is why this arrangement is used instead of simply applying the 100-lb force to the crate directly.

7.8.   A boom hinged at the base of a vertical mast is used to lift a weight of 1500 lb, as in Fig. 7-8(a).   Find the tension in the cable from the top of the mast to the top of the boom.

We begin by finding the angles θ_1 and θ_2.   Since the sum of the interior angles of a triangle is always 180°,

$$\theta_1 = 180° - 40° - 65° = 75°$$

Because the mast is vertical, it makes a 90° angle with the ground, and so

$$\theta_2 = 90° - 40° = 50°$$

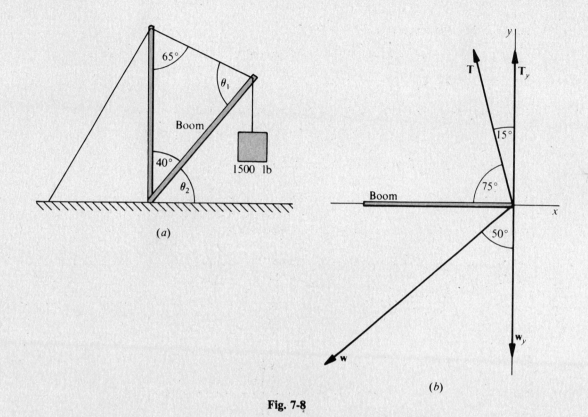

Fig. 7-8

We now let the x axis be in the direction of the boom with the y axis perpendicular to it, as in Fig. 7-8(b). Since the boom is rigid, we need consider only the translational equilibrium of its upper end in the y direction. The angle between $\mathbf{T}$ and the y axis is $90° - 75° = 15°$, and so

$$T_y = T \cos 15° = 0.966T$$

The component of the weight $\mathbf{w}$ in the y direction is

$$w_y = -1500 \text{ lb} \times \cos 50° = -964 \text{ lb}$$

At equilibrium

$$\Sigma F_y = T_y + w_y = 0 \qquad 0.966T - 964 \text{ lb} = 0 \qquad T = 998 \text{ lb}$$

PROBLEMS INVOLVING NONCONCURRENT FORCES

7.9. A beam 10 ft long has a weight of 50 lb at one end and another of 20 lb at the other end. The weight of the beam itself is negligible. Find the balance point of the beam.

When the beam is supported at its balance point, the torques of the two weights cancel each other out, and the beam has no tendency to rotate. The supporting force F exerts no torque since it acts through the balance point. If the balance point is the distance x from the 50-lb weight, as in Fig. 7-9, it is the distance $10 \text{ ft} - x$ from the 20-lb weight. Since the beam is horizontal, the moment arms of the weights are respectively x and $10 \text{ ft} - x$, and the torques the weights exert about the balance point O are

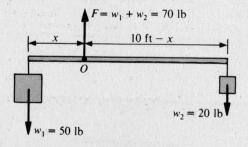

Fig. 7-9

$$\tau_1 = w_1 L_1 = 50x \text{ lb} \qquad \tau_2 = -w_2 L_2 = -20(10 \text{ ft} - x) \text{ lb}$$

The torque τ_1 is positive because it tends to cause a counterclockwise rotation; the torque τ_2 is negative because it tends to cause a clockwise rotation. The condition for rotational equilibrium yields

$$\Sigma\tau = \tau_1 + \tau_2 = 0$$
$$50x \text{ lb} - 20(10 \text{ ft} - x) \text{ lb} = 0$$
$$50x \text{ lb} = 200 \text{ lb-ft} - 20x \text{ lb}$$
$$70x \text{ lb} = 200 \text{ lb-ft}$$
$$x = 2.86 \text{ ft}$$

7.10. A 12-ft wooden platform is suspended from the roof of a house by ropes attached to its ends. A painter weighing 160 lb stands 4 ft from the left-hand end of the platform, whose own weight is 40 lb. Find the tension in each of the ropes.

With the left-hand end of the platform as the pivot point (Fig. 7-10), the condition for rotational equilibrium yields

$$\Sigma\tau = -w_1L_1 - w_2L_2 + T_2L_3 = 0$$
$$-160 \text{ lb} \times 4 \text{ ft} - 40 \text{ lb} \times 6 \text{ ft} + 12T_2 \text{ ft} = 0$$
$$12T_2 \text{ ft} = 640 \text{ lb-ft} + 240 \text{ lb-ft} = 880 \text{ lb-ft}$$
$$T_2 = 73 \text{ lb}$$

To find T_1 we proceed as follows:

$$T_1 + T_2 = w_1 + w_2 \qquad T_1 = 160 \text{ lb} + 40 \text{ lb} - 73 \text{ lb} = 127 \text{ lb}$$

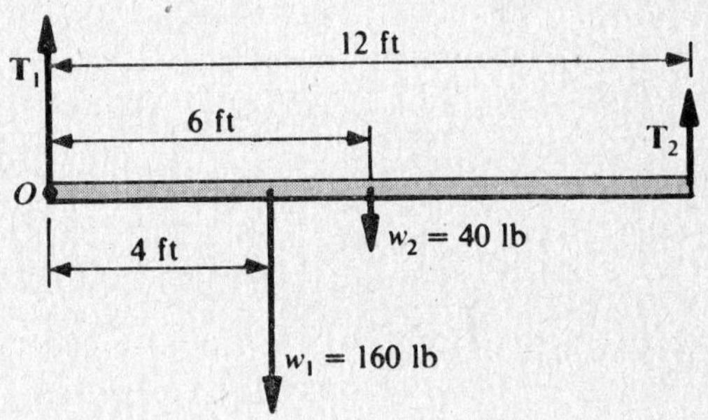

Fig. 7-10 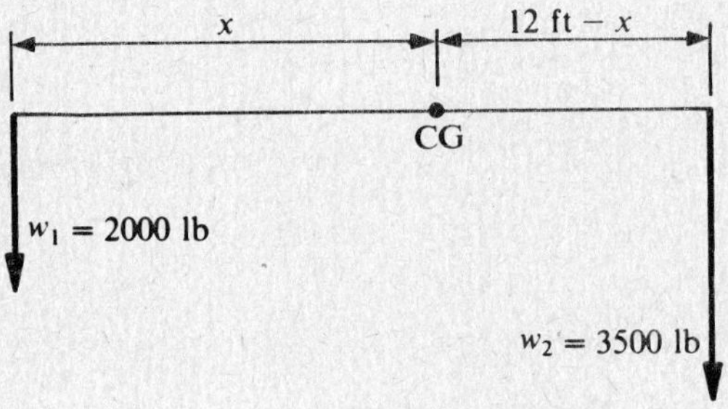   Fig. 7-11

7.11. The front wheels of a truck together support 2000 lb and its rear wheels together support 3500 lb. The axles are 12 ft apart. Where is the center of gravity of the truck located?

With x the distance between the front axle and the center of gravity, as in Fig. 7-11, calculating torques about the center of gravity yields

$$\Sigma\tau = w_1x - w_2(12 \text{ ft} - x) = 0$$
$$2000x \text{ lb} - 3500(12 \text{ ft} - x) \text{ lb} = 0$$
$$5500x \text{ lb} = 42{,}000 \text{ lb-ft}$$
$$x = 7.64 \text{ ft}$$

7.12. A horizontal boom 8 ft long is attached to a wall at its inner end and supported at its outer end by a cable that makes an angle of 30° with the boom. The boom weighs 60 lb and a load of 500 lb is attached to its outer end. Find (a) the tension in the cable and (b) the compression force in the boom.

(a) Four forces act on the boom, as in Fig. 7-12: the load w_1 at its outer end; the boom's own weight w_2 that acts from its center; the tension **T** in the cable; and the force **F** that the wall exerts on the inner end of the boom. We can disregard **F** by calculating torques about the inner end of the boom. It simplifies matters to use the vertical component T_y of the tension in the cable instead of **T**, since the moment arm of T_y is the boom's length L_1. The horizontal component of the tension T_x exerts no

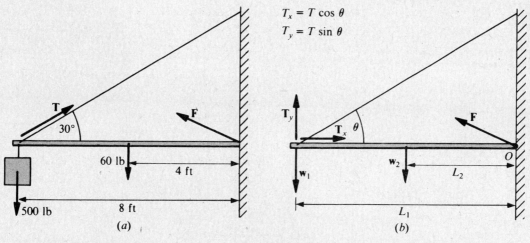

$$T_x = T \cos \theta$$
$$T_y = T \sin \theta$$

Fig. 7-12

torque about O because its line of action passes through O. The torques exerted by the load w_1, the boom's weight w_2, and the vertical component of tension T_y are respectively

$$\tau_1 = w_1 L_1 = 500 \text{ lb} \times 8 \text{ ft} = 4000 \text{ lb-ft}$$
$$\tau_2 = w_2 L_2 = 60 \text{ lb} \times 4 \text{ ft} = 240 \text{ lb-ft}$$
$$\tau_3 = -T_y L_1 = -8T \sin 30° \text{ ft} = -4T \text{ ft}$$

For the boom to be in equilibrium,

$$\Sigma \tau = \tau_1 + \tau_2 + \tau_3 = 0$$
$$4000 \text{ lb-ft} + 240 \text{ lb-ft} - 4T \text{ ft} = 0$$
$$T = \frac{4240 \text{ lb-ft}}{4 \text{ ft}} = 1060 \text{ lb}$$

(b) The compression force $\mathbf{F}_x$ in the boom is equal in magnitude to the horizontal component of the cable tension $\mathbf{T}$. Thus

$$F_x = T \cos \theta = 1060 \text{ lb} \times \cos 30° = 918 \text{ lb}$$

7.13. A gate 6 ft long and 4 ft high has hinges at the top and bottom of one edge. (a) If the entire 70-lb weight of the gate is supported by the lower hinge, find the force the gate exerts on the upper hinge. (b) Find the force the gate exerts on the lower hinge.

(a) The weight of the gate acts from its center of gravity, which we will assume is its geometrical center as in Fig. 7-13(a). The gate exerts a force that is downward and to the right on the lower hinge,

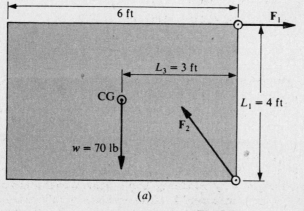
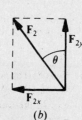

Fig. 7-13

which in response exerts a reaction force $\mathbf{F}_2$ on the gate that is upward and to the left. The gate exerts a force to the left on the upper hinge, which exerts a reaction force $\mathbf{F}_1$ to the right on the gate. To find the magnitude of $\mathbf{F}_1$, it is easiest to calculate torques about the lower hinge. The torques exerted on the gate by the upper hinge and by its own weight are respectively

$$\tau_1 = -F_1 L_1 = -4F_1 \text{ ft}$$
$$\tau_3 = wL_3 = 70 \text{ lb} \times 3 \text{ ft} = 210 \text{ lb-ft}$$

Hence

$$\Sigma\tau = \tau_1 + \tau_3 = 0$$
$$-4F_1 \text{ ft} + 210 \text{ lb-ft} = 0$$
$$F_1 = 52.5 \text{ lb}$$

The force exerted by the gate on the upper hinge has the same magnitude.

(b) In order that the gate be in translational equilibrium, $\mathbf{F}_{2x}$ must be equal and opposite to $\mathbf{F}_1$, and $\mathbf{F}_{2y}$ must be equal and opposite to $\mathbf{w}$. Hence

$$F_{2x} = 52.5 \text{ lb} \qquad F_{2y} = 70 \text{ lb}$$

Since $\mathbf{F}_{2x}$ and $\mathbf{F}_{2y}$ are perpendicular, the magnitude of F is

$$F = \sqrt{F_{2x}^2 + F_{2y}^2} = 87.5 \text{ lb}$$

If θ is the angle between $\mathbf{F}_2$ and the vertical, as in Fig. 7-13(b),

$$\tan\theta = \frac{F_{2x}}{F_{2y}} = 0.75 \qquad \theta = 37°$$

The force the gate exerts on the lower hinge is equal and opposite to $\mathbf{F}_2$, hence it acts at an angle of 37° below the vertical and toward the right.

7.14. A 12-ft ladder that weighs 50 lb rests against a frictionless wall at a point 10 ft above the ground. How much force does the ladder exert (a) on the ground and (b) on the wall?

(a) The forces that act on the ladder are its weight $\mathbf{w}$ acting downward from its center, the horizontal reaction force $\mathbf{F}_1$ of the wall (there is no vertical force component because the wall is frictionless), and the reaction force $\mathbf{F}_2$ of the ground, which has both vertical and horizontal components. Since $\mathbf{F}_{2y}$ and $\mathbf{w}$ are the only vertical forces,

$$F_{2y} = w = 50 \text{ lb}$$

To find F_{2x}, we begin by finding the value of the angle between the ladder and the ground. From Fig. 7-14

$$\sin\theta = \frac{10 \text{ ft}}{12 \text{ ft}} = 0.833 \qquad \theta = 56°$$

Now we calculate the torques produced by $\mathbf{F}_{2y}$, $\mathbf{F}_{2x}$, and $\mathbf{w}$ about the upper end of the ladder:

$$\tau_1 = -F_{2y}L_1 = -50 \text{ lb} \times 12 \text{ ft} \times \cos 56°$$
$$= -336 \text{ lb-ft}$$
$$\tau_2 = F_{2x}L_2 = F_{2x} \times 10 \text{ ft} = 10F_{2x} \text{ ft}$$
$$\tau_3 = wL_3 = 50 \text{ lb} \times 6 \text{ ft} \times \cos 56° = 168 \text{ lb-ft}$$

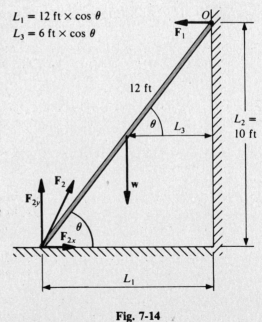

$L_1 = 12 \text{ ft} \times \cos\theta$
$L_3 = 6 \text{ ft} \times \cos\theta$

Fig. 7-14

Applying the condition for rotational equilibrium yields

$$\Sigma\tau = \tau_1 + \tau_2 + \tau_3 = 0$$
$$-336 \text{ lb-ft} + 10F_{2x} \text{ ft} + 168 \text{ lb-ft} = 0$$
$$F_{2x} = 16.8 \text{ lb}$$

The total force the ground exerts on the ladder is

$$F_2 = \sqrt{F_{2x}^2 + F_{2y}^2} = 53 \text{ lb}$$

The force the ladder exerts on the ground has the same magnitude.

(b)　The force the ladder exerts on the wall is equal in magnitude to F_{2x}, namely 16.8 lb.

Supplementary Problems

7.15. A *couple* consists of two equal forces that act along parallel lines of action in opposite directions. If each of the forces in a couple has the magnitude F and their lines of action are d apart, find the torque exerted by the couple.

7.16. The wheels of a certain bus are 2 m apart and the bus falls over when tilted sideways at a 45° angle. How high above the road is the center of gravity of the bus?

7.17. A 40-lb box is suspended from two ropes which each make a 45° angle with the vertical. What is the tension in each rope?

7.18. A 100-kg box is suspended from two ropes, the left-hand one of which makes an angle of 20° with the vertical and the other an angle of 40°. What is the tension in each rope?

7.19. A load of unknown weight is suspended from the end of a horizontal boom whose own weight is negligible. The angle between the boom and the cable supporting its end is 30° and the tension in the cable is 400 lb. Find the weight of the load.

7.20. A 2000-lb steel beam is raised by a crane and a horizontal rope is used to pull it into position in a bridge under construction. What is the tension in the rope when the supporting cable is at an angle of 15° from the vertical?

7.21. A 1-kg pigeon sits on the middle of a clothesline whose supports are 10 m apart. The clothesline sags by 1 m. If the weight of the clothesline is negligible, find the tension in it.

7.22. A 30-lb weight is attached to one end of a 6-ft uniform beam whose own weight is 20 lb. Where is the balance point of the system?

7.23. A pail of water is to be carried by a man and a boy who each hold one end of a 6-ft pole thrust through the pail's handle. Where should the pail be located along the pole so that the man carries twice as much weight as the boy? Neglect the pole's weight.

7.24. A 20-kg child and a 30-kg child sit at opposite ends of a 4-m seesaw that is pivoted at its center. Where should another 20-kg child sit in order to balance the seesaw?

7.25. The axles of a 2400-lb car are 7 ft apart. If the center of gravity of the car is 3 ft behind the front axle, how much weight is supported by each of the car's wheels?

7.26. (*a*) A 50-kg horizontal boom 4 m long is hinged to a vertical mast and its outer end is supported by a cable attached to the mast 3 m above the hinge pin.   Find the tension in the cable.   (*b*) A load of 200 kg is suspended from the outer end of the boom.   Find the new tension in the cable.

7.27. Solve Problem 7.8 by considering the rotational equilibrium of the boom, using the lower end of the boom as the pivot point.

7.28. A door 7 ft high and 3 ft wide has hinges at the top and bottom of one edge.   Half the 40-lb weight of the door is supported by each hinge.   Find the horizontal components of the force the door exerts on each hinge.

7.29. A 16-ft ladder that weighs 60 lb rests against a frictionless wall at an angle of 60° from the ground.   A 150-lb man stands on the ladder 3/4 of the distance from the top.   How much force does the top of the ladder exert on the wall?

Answers to Supplementary Problems

7.15. *Fd*

7.16. 1 m

7.17. 28 lb in each rope

7.18. The tension in the left-hand rope is 727 N and that in the right-hand rope is 387 N.

7.19. 200 lb

7.20. 536 lb

7.21. 26 N

7.22. The balance point is 2.4 ft from the 30-lb weight.

7.23. The pail should be 2 ft from the man.

7.24. The third child should be 1 m from the 20-kg child.

7.25. The front wheels each support 686 lb and the rear wheels each support 514 lb.

7.26. (*a*) 306 N   (*b*) 2756 N

7.27. 998 lb

7.28. 8.6 lb; 8.6 lb

7.29. 110 lb

Chapter 8

Circular Motion and Gravitation

UNIFORM CIRCULAR MOTION

A body that moves in a circular path at a velocity whose magnitude is constant is said to undergo *uniform circular motion.*

CENTRIPETAL ACCELERATION

Although the velocity of a body in uniform circular motion is constant in magnitude, its direction changes continually. The body is therefore accelerated. The direction of this *centripetal acceleration* is toward the center of the circle in which the body moves, and its magnitude is

$$a_c = \frac{v^2}{r}$$

$$\text{Centripetal acceleration} = \frac{(\text{velocity of body})^2}{\text{radius of circular path}}$$

Because the acceleration is perpendicular to the path followed by the body, its velocity changes only in direction and not in magnitude.

CENTRIPETAL FORCE

The inward force that must be applied to keep a body moving in a circle is called *centripetal force.* Without centripetal force, circular motion cannot occur. Since $F = ma$, the magnitude of the centripetal force on a body in uniform motion is

$$\text{Centripetal force} = F_c = \frac{mv^2}{r}$$

GRAVITATION

According to Newton's *law of universal gravitation*, every body in the universe attracts every other body with a force that is directly proportional to each of their masses and inversely proportional to the square of the distance between them. In equation form,

$$\text{Gravitational force} = F = G\,\frac{m_1 m_2}{r^2}$$

where m_1 and m_2 are the masses of any two bodies, r is the distance between them, and G is a constant whose values in SI and British units are respectively

$$\text{SI units:} \qquad G = 6.67 \times 10^{-11}\,\frac{\text{N-m}^2}{\text{kg}^2}$$

$$\text{British units:} \qquad G = 3.44 \times 10^{-8}\,\frac{\text{lb-ft}^2}{\text{slug}^2}$$

Gravitation provides the centripetal forces that keep the planets in their orbits around the sun and the moon in its orbit around the earth. A spherical body behaves gravitationally as though its entire mass were concentrated at its center.

Solved Problems

8.1. A ball is whirled in a horizontal circle 1.5 ft in radius at the rate of one revolution every two seconds. Find the ball's centripetal acceleration.

The distance the ball travels per revolution is

$$s = 2\pi r = 2\pi \times 1.5 \text{ ft} = 9.4 \text{ ft}$$

and so, since each revolution requires 2 s, its velocity is

$$v = \frac{s}{t} = \frac{9.4 \text{ ft}}{2 \text{ s}} = 4.7 \text{ ft/s}$$

The ball's centripetal acceleration is therefore

$$a_c = \frac{v^2}{r} = \frac{(4.7 \text{ ft/s})^2}{1.5 \text{ ft}} = 14.7 \text{ ft/s}^2$$

8.2. How much centripetal force is needed to keep a 0.5-kg stone moving in a horizontal circle of radius 1 m at a velocity of 4 m/s?

$$F_c = \frac{mv^2}{r} = \frac{0.5 \text{ kg} \times (4 \text{ m/s})^2}{1 \text{ m}} = 8 \text{ N}$$

8.3. A centripetal force of 1 N is used to keep a 0.1-kg yoyo moving in a horizontal circle of radius 0.7 m. What is the yoyo's velocity?

Since $F_c = mv^2/r$,

$$v = \sqrt{\frac{F_c r}{m}} = \sqrt{\frac{0.7 \text{ m} \times 1 \text{ N}}{0.1 \text{ kg}}} = \sqrt{7} \text{ m/s} = 2.6 \text{ m/s}$$

8.4. How much centripetal force is needed to keep a 160-lb skater moving in a circle 20 ft in radius at a velocity of 10 ft/s?

The skater's mass is $m = \dfrac{w}{g} = \dfrac{160 \text{ lb}}{32 \text{ ft/s}^2} = 5$ slugs. Hence

$$F_c = \frac{mv^2}{r} = \frac{5 \text{ slugs} \times (10 \text{ ft/s})^2}{20 \text{ ft}} = 25 \text{ lb}$$

8.5. A 1000-kg car rounds a turn of radius 30 m at a velocity of 9 m/s. (*a*) How much centripetal force is required? (*b*) Where does this force come from?

(*a*) $$F_c = \frac{mv^2}{r} = \frac{1000 \text{ kg} \times (9 \text{ m/s})^2}{30 \text{ m}} = 2700 \text{ N}$$

(*b*) The centripetal force on a car making a turn on a level road is provided by the road acting via friction on the car's tires.

8.6. The maximum force a road can exert on the tires of a certain 3200-lb car is 2000 lb. What is the maximum velocity at which the car can round a turn of radius 320 ft?

The car's mass is $m = \dfrac{w}{g} = \dfrac{3200 \text{ lb}}{32 \text{ ft/s}^2} = 100$ slugs. Solving the formula $F_c = mv^2/r$ for v gives

$$v = \sqrt{\frac{F_c r}{m}} = \sqrt{\frac{2000 \text{ lb} \times 320 \text{ ft}}{100 \text{ slugs}}} = \sqrt{6400} \text{ ft/s} = 80 \text{ ft/s}$$

which is

$$80 \text{ ft/s} \times 0.682 \frac{\text{mi/hr}}{\text{ft/s}} = 55 \text{ mi/hr}$$

8.7. A car is traveling at 20 mi/hr on a level road where the coefficient of static friction between tires and road is 0.8. Find the minimum turning radius of the car.

The maximum centripetal force that friction can provide here is

$$F_f = \mu N = \mu w = \mu mg$$

Hence

$$\text{Centripetal force} = \text{frictional force}$$

$$\frac{mv^2}{r} = \mu mg$$

$$r = \frac{v^2}{\mu g}$$

The car's velocity is

$$v = 20 \text{ mi/hr} \times 1.47 \frac{\text{ft/s}}{\text{mi/hr}} = 29.4 \text{ ft/s}$$

and so the minimum turning radius is

$$r = \frac{v^2}{\mu g} = \frac{(29.4 \text{ ft/s})^2}{0.8 \times 32 \text{ ft/s}^2} = 34 \text{ ft}$$

8.8. Highway curves are usually *banked* (tilted inward) at an angle θ such that the horizontal component of the reaction force of the road on a car traveling at the design velocity equals the required centripetal force. The proper banking angle for a car making a turn of radius r at the velocity v is given by $\tan \theta = v^2/gr$. (a) Find the proper banking angle for cars moving at 50 mi/hr to go around a curve 1000 ft in radius. (b) If the curve were not banked, what coefficient of friction would be required between tires and road?

(a) Here $v = 50 \text{ mi/hr} \times 1.47 \text{ (ft/s)/(mi/hr)} = 73.5 \text{ ft/s}$, and so

$$\tan \theta = \frac{v^2}{gr} = \frac{(73.5 \text{ ft/s})^2}{32 \text{ ft/s}^2 \times 1000 \text{ ft}} = 0.169 \qquad \theta = 9.6°$$

(b) From the solution to Problem 8.7, $r = v^2/\mu g$. Hence

$$\mu = \frac{v^2}{gr} = 0.169$$

8.9. A string 0.5 m long is used to whirl a 1-kg stone in a vertical circle at a uniform velocity of 5 m/s. (a) What is the tension in the string when the stone is at the top of the circle? (b) When the stone is at the bottom of the circle?

(a) The centripetal force needed to keep the stone moving at 5 m/s is

$$F_c = \frac{mv^2}{r} = \frac{1 \text{ kg} \times (5 \text{ m/s})^2}{0.5 \text{ m}} = 50 \text{ N}$$

At the top of the circle, the weight of the stone is a downward force on it acting toward the center of the circle:

$$w = mg = 1 \text{ kg} \times 9.8 \text{ m/s}^2 = 9.8 \text{ N}$$

From Fig. 8-1(a) it is seen that the tension in the string is

$$T = F_c - w = 50 \text{ N} - 9.8 \text{ N} = 40.2 \text{ N}$$

(b) At the bottom of the circle, the weight of the stone acts away from the center of the circle, so the string must provide a force equal to w plus F_c, as indicated in Fig. 8-1(b). Hence

$$T = F_c + w = 50 \text{ N} + 9.8 \text{ N} = 59.8 \text{ N}$$

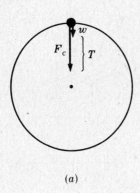

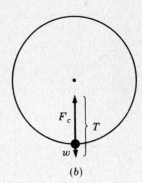

(a) (b)

Fig. 8-1

8.10. What is the gravitational force a 1-ton lead sphere exerts on another identical sphere 10 ft away?

The mass of each sphere is

$$m = \frac{w}{g} = \frac{2000 \text{ lb}}{32 \text{ ft/s}^2} = 62.5 \text{ slugs}$$

Hence

$$F = G \frac{m_1 m_2}{r^2} = 3.44 \times 10^{-8} \frac{\text{lb-ft}^2}{\text{slug}^2} \times \frac{62.5 \text{ slugs} \times 62.5 \text{ slugs}}{(10 \text{ ft})^2} = 1.3 \times 10^{-6} \text{ lb}$$

which is less than the force that would result from blowing gently on one of the spheres.

8.11. Find the velocity an artificial satellite must have in order to pursue a circular orbit around the earth just above the surface.

In a stable orbit, the gravitational force mg on the satellite must be equal to the centripetal force mv^2/r required. Hence

$$\frac{mv^2}{r} = mg$$

$$v^2 = rg$$

$$v = \sqrt{rg}$$

The mass of the satellite is irrelevant. To find v we use the radius of the earth r_e for r and the acceleration of gravity at the earth's surface for g. This gives

$$v = \sqrt{r_e g} = \sqrt{6.4 \times 10^6 \text{ m} \times 9.8 \text{ m/s}^2} = 7.9 \times 10^3 \text{ m/s}$$

which is about 18,000 mi/hr. With a smaller velocity than this, a space vehicle projected horizontally above the earth will fall to the surface; with a larger velocity, it will have an elliptical rather than a circular orbit.

8.12. A girl weighs 128 lb on the earth's surface. (*a*) What would she weigh at a height above the earth's surface of one earth radius? (*b*) What would her mass be there?

(*a*) Since the gravitational force the earth exerts on an object a distance r from its center varies as $1/r^2$, the gravitational force on the girl relative to its value mg at the earth's surface (where $r = r_e$) is

$$F = \left(\frac{r_e}{r} \right)^2 mg$$

When the girl is a distance r_e above the earth, $r = 2r_e$, and

$$F = \left(\frac{r_e}{r} \right)^2 mg = \left(\frac{r_e}{2r_e} \right)^2 w = \frac{1}{4} \times 128 \text{ lb} = 32 \text{ lb}$$

(b) Her mass of $m = \dfrac{w}{g} = \dfrac{128 \text{ lb}}{32 \text{ ft/s}^2} = 4$ slugs is the same everywhere.

Supplementary Problems

8.13. Are there any circumstances under which a body can move in a curved path without being accelerated?

8.14. In what way, if any, do g and G change with increasing height above the earth's surface?

8.15. The moon's mass is approximately 1% of the earth's mass. How does the gravitational pull of the earth on the moon compare with the gravitational pull of the moon on the earth?

8.16. A hole is drilled to the center of the earth and a stone whose mass is 1 kg at the earth's surface is dropped into it. What is the mass of the stone when it is at the earth's center? What is its weight?

8.17. A lasso is whirled so that the loop at its end describes a horizontal circle 5 ft in radius.   If each revolution of the lasso requires 4 s, find the centripetal acceleration of the loop.

8.18. A 0.02-kg ball is whirled in a horizontal circle at the end of a string 0.5 m long whose breaking strength is 1 N.   Neglecting gravity, what is the maximum velocity the ball can have?

8.19. How much centripetal force is needed to keep a 4-lb iron ball moving in a horizontal circle of radius 5 ft at a velocity of 15 ft/s?

8.20. A 5000-kg airplane makes a horizontal turn 1 km (1000 m) in radius at a velocity of 50 m/s. How much centripetal force is required?

8.21. A motorcycle begins to skid when it makes a turn of 50-ft radius at a velocity of 40 ft/s. What is the highest velocity at which it can make a turn of 100-ft radius?

8.22. What is the minimum coefficient of static friction required for a car traveling at 60 km/hr to make a level turn 40 m in radius?

8.23. The coefficient of static friction between a car's tires and a certain concrete road is 1.0 when the road is dry and 0.7 when the road is wet. If the car can safely make a certain turn at 25 mi/hr on a dry day, what is the maximum velocity on a rainy day?

8.24. An airplane whose velocity is 500 km/hr banks at an angle of 40°. If the rudder is not used, what is the radius of the turn the airplane makes?

8.25. A string 1 m long is used to whirl a ball of unknown mass in a vertical circle. What is the minimum velocity of the ball if the string is to be just taut when the ball is at the top of the circle?

8.26. A man swings a pail of water in a vertical circle 3.1 ft in radius. (a) If the water is not to spill, what is the minimum velocity the pail can have? (b) How much time per revolution is this equivalent to?

8.27. What is the gravitational attraction between an 80-kg man and a 50-kg woman who are 2 m apart?

8.28. What is the gravitational attraction between two 3200-lb elephants when they are 20 ft apart?

8.29. The acceleration of gravity on the surface of Mars is 0.4 g.   How much would a man weigh on the surface of Mars whose weight on the earth's surface is 180 lb?

8.30. A man weighs 180 lb on the earth's surface.   (a) What would he weigh at a distance from the center of the earth of three times the radius of the earth?   (b) What would his mass there be?

8.31. A 10-kg monkey is 10^6 m above the earth's surface.   (a) What is its weight there?   (b) What is its mass there?

8.32. Find the velocity in m/s an artificial earth satellite must have in order to pursue a circular orbit at an altitude of half an earth radius.

Answers to Supplementary Problems

8.13. No

8.14. The value of g decreases; the value of G is the same everywhere in the universe.

8.15.	They are equal.	**8.21.**	56 ft/s	**8.27.**	6.67×10^{-8} N
8.16.	1 kg; 0	**8.22.**	0.71	**8.28.**	8.6×10^{-7} lb
8.17.	12.3 ft/s^2	**8.23.**	21 mi/hr	**8.29.**	72 lb
8.18.	5 m/s	**8.24.**	2.35 km	**8.30.**	(a) 20 lb (b) 5.6 slugs
8.19.	5.6 lb	**8.25.**	3.1 m/s	**8.31.**	(a) 73 N (b) 10 kg
8.20.	1.25×10^4 N	**8.26.**	(a) 10 ft/s (b) 1.9 s	**8.32.**	6.5×10^3 m/s

Chapter 9

Energy

WORK

Work is a measure of the amount of change (in a general sense) a force produces when it acts upon a body. The change may be in the velocity of the body, in its position, in its size or shape, and so forth.

By definition, the work done by a force acting on a body is equal to the product of the force and the distance through which the force acts, provided that **F** and **s** are in the same direction. Thus

$$W = F \times s$$
$$\text{Work} = \text{force} \times \text{distance}$$

Work is a scalar quantity; no direction is associated with it.

If **F** and **s** are not parallel but are the angle θ apart, then

$$W = Fs \cos \theta$$

Since $\cos 0 = 1$, this formula becomes $W = Fs$ when **F** is parallel to **s**. When **F** is perpendicular to **s**, $\theta = 90°$ and $\cos 90° = 0$. No work is done in this case.

The unit of work is the product of a force unit and a length unit. In the SI system the unit of work is the *joule* (J).

SI units: 1 joule (J) = 1 newton-meter = 0.738 ft-lb

British units: 1 foot-pound (ft-lb) = 1.36 J

POWER

Power is the rate at which work is done by a force. Thus

$$P = \frac{W}{t}$$

$$\text{Power} = \frac{\text{work done}}{\text{time}}$$

The more power something has, the more work it can perform in a given time.

Two special units of power are in wide use, the *watt* and the *horsepower*, where

$$1 \text{ watt (W)} = 1 \text{ joule/second} = 1.34 \times 10^{-3} \text{ hp}$$

$$1 \text{ horsepower (hp)} = 550 \text{ ft-lb/s} = 746 \text{ W}$$

A *kilowatt* (kW) is equal to 10^3 W or 1.34 hp. A *kilowatt-hour* is the work done in 1 hr by an agency whose power output is 1 kW; hence 1 kw-hr = 3.6×10^6 J.

ENERGY

Energy is that property something has which enables it to do work. The more energy something has, the more work it can perform. Every kind of energy falls into one of three general categories: kinetic energy, potential energy, and rest energy.

The units of energy are the same as those of work, namely the joule and the foot-pound.

KINETIC ENERGY

The energy a body has by virtue of its motion is called *kinetic energy*. If the body's mass is m and its velocity is v, its kinetic energy is

$$\text{Kinetic energy} = \text{KE} = \frac{1}{2} mv^2$$

POTENTIAL ENERGY

The energy a body has by virtue of its position is called *potential energy*. A book held above the floor has gravitational potential energy because the book can do work on something else as it falls; a nail held near a magnet has magnetic potential energy because the nail can do work as it moves toward the magnet; the wound spring in a watch has elastic potential energy because the spring can do work as it unwinds.

The gravitational potential energy of a body of mass m that is at a height h above some reference level is

$$\text{Gravitational potential energy} = \text{PE} = mgh$$

where g is the acceleration of gravity. In terms of the weight w of the body,

$$\text{PE} = wh$$

REST ENERGY

Matter can be converted into energy, and energy into matter. The *rest energy* of a body is the energy it has by virtue of its mass alone. Thus mass can be regarded as a form of energy. The rest energy of a body is in addition to any KE or PE it might have.

If the mass of a body is m_0 when it is at rest, its rest energy is

$$\text{Rest energy} = E_0 = m_0 c^2$$

In this formula c is the velocity of light, whose value is

$$c = 3.00 \times 10^8 \text{ m/s} = 9.83 \times 10^8 \text{ ft/s} = 186{,}000 \text{ mi/s}$$

The rest mass m_0 is specified here because the mass of a moving body increases with its velocity; the increase is only significant at extremely high velocities, however.

CONSERVATION OF ENERGY

According to the law of *conservation of energy*, energy cannot be created or destroyed, although it can be transformed from one kind into another. The total amount of energy in the universe is constant. A falling stone provides a simple example: more and more of its initial potential energy turns into kinetic energy as its velocity increases, until finally all of its PE has become KE when it strikes the ground. The KE of the stone is then transferred to the ground by the impact.

Solved Problems

9.1. What is the nature of chemical energy?

Chemical energy is rest energy. When a chemical reaction occurs in which energy is given off, for instance, the products of the reaction always have less mass than the original substances did. Because even a minute quantity of matter is equivalent to a vast amount of energy (1 kg is equivalent to 9×10^{16} J), the mass changes in chemical reactions are too small to be detectable by ordinary means.

9.2. A force of 60 lb is used to push a 150-lb crate a distance of 10 ft across a level warehouse floor. (*a*) How much work is done?   (*b*) What is the change in the crate's potential energy?

(*a*) The weight of the crate does not matter here since its height does not change.   The work done is
$$W = Fs = 60 \text{ lb} \times 10 \text{ ft} = 600 \text{ ft-lb}$$

(*b*) The crate's height does not change, so its potential energy remains the same.

9.3. (*a*) How much work is done in raising a 2000-lb elevator cab to a height of 80 ft?   (*b*) How much potential energy does the cab have in its new position?

(*a*) The force needed is equal to the weight of the cab.   Hence
$$W = Fs = wh = 2000 \text{ lb} \times 80 \text{ ft} = 1.6 \times 10^5 \text{ ft-lb}$$
(*b*)
$$PE = wh = 1.6 \times 10^5 \text{ ft-lb}$$

9.4. (*a*) How much work is done in raising a 2-kg book from the ground to a height of 1.5 m?   (*b*) How much potential energy does the book have in its new position?

(*a*)
$$W = Fs = mgh = 2 \text{ kg} \times 9.8 \text{ m/s}^2 \times 1.5 \text{ m} = 29.4 \text{ J}$$
(*b*)
$$PE = mgh = 29.4 \text{ J}$$

9.5. (*a*) How much force must be applied to hold a 2-kg book 1.5 m above the ground?   (*b*) How much work is done when it is held there for 10 min?

(*a*) The force needed is the book's weight of
$$w = mg = 2 \text{ kg} \times 9.8 \text{ m/s}^2 = 19.6 \text{ N}$$

(*b*) No work is done since the book does not move.

9.6. Ten thousand joules of work is performed in raising a 200-kg bronze statue.   How high is it raised?

Since $W = Fs = mgh$,
$$h = \frac{W}{mg} = \frac{10^4 \text{ J}}{200 \text{ kg} \times 9.8 \text{ m/s}^2} = 5.1 \text{ m}$$

9.7. A 150-lb man runs up a staircase 10 ft high in 5 s.   Find his minimum power output in horsepower.

The minimum downward force the man's legs must exert is equal to his weight of 150 lb.   Hence
$$P = \frac{W}{t} = \frac{Fs}{t} = \frac{150 \text{ lb} \times 10 \text{ ft}}{5 \text{ s}} = 300 \text{ ft-lb/s}$$

Since 1 hp = 550 ft-lb/s,
$$P = \frac{300 \text{ ft-lb/s}}{550 \text{ (ft-lb/s)hp}} = 0.55 \text{ hp}$$

9.8. A man uses a horizontal force of 200 N to push a crate up a ramp 8 m long that is 20° above the horizontal.   (*a*) How much work does the man perform?   (*b*) If the man takes 12 s to push the crate up the ramp, what is his power output in watts and in hp?

(*a*)
$$W = Fs \cos \theta = 200 \text{ N} \times 8 \text{ m} \times \cos 20° = 1504 \text{ J}$$
(*b*)
$$P = \frac{W}{t} = \frac{1504 \text{ J}}{12 \text{ s}} = 125 \text{ W}$$

Since 1 hp = 746 W,

$$P = \frac{125 \text{ W}}{746 \text{ W/hp}} = 0.17 \text{ hp}$$

9.9. A hoist powered by a 20-hp motor is used to raise a 500-kg bucket of concrete to a height of 80 m. If the efficiency is 80%, find the time required.

The upward force F that is needed is equal to the bucket's weight of $mg = 500 \text{ kg} \times 9.8 \text{ m/s}^2 = 4900$ N. The power available is

$$P = 0.80 \times 20 \text{ hp} \times 746 \text{ W/hp} = 1.19 \times 10^4 \text{ W}$$

Since $P = W/t = Fs/t$,

$$t = \frac{Fs}{P} = \frac{4900 \text{ N} \times 80 \text{ m}}{1.19 \times 10^4 \text{ W}} = 32.8 \text{ s}$$

9.10. A motorboat requires 80 hp to move at the constant velocity of 10 mi/hr. How much resistive force does the water exert on the boat at this velocity?

Since $v = s/t$, $P = \dfrac{W}{t} = \dfrac{Fs}{t} = Fv$. Here

$$v = 10 \text{ mi/hr} \times 1.47 \frac{\text{ft/s}}{\text{mi/hr}} = 14.7 \text{ ft/s}$$

and

$$P = 80 \text{ hp} \times 550 \frac{\text{ft-lb/s}}{\text{hp}} = 44,000 \text{ ft-lb/s}$$

Hence

$$F = \frac{P}{v} = \frac{44,000 \text{ ft-lb/s}}{14.7 \text{ ft/s}} = 3000 \text{ lb}$$

9.11. A horse has a power output of 1 hp when it pulls a wagon with a force of 300 N. What is the wagon's velocity?

Since $P = 1 \text{ hp} = 746 \text{ W} = Fv$, $v = \dfrac{P}{F} = \dfrac{746 \text{ W}}{300 \text{ N}} = 2.5 \text{ m/s}$.

9.12. A 2400-lb car ascends a 10° slope at 30 mi/hr. If the overall efficiency is 70%, what is the power output of the car's engine?

From Fig. 9-1 the weight **w** of the car has a component **F** along the slope of magnitude

$$F = w \sin \theta = 2400 \text{ lb} \times \sin 10° = 417 \text{ lb}$$

The force supplied by the car's engine must equal this amount. The car's velocity is

$$v = 30 \text{ mi/hr} \times 1.47 \frac{\text{ft/s}}{\text{mi/hr}} = 44 \text{ ft/s}$$

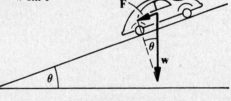

$F = w \sin \theta$

Fig. 9-1

and so, since $P = Fv$, the required power at 100% efficiency is

$$P = Fv = 417 \text{ lb} \times 44 \text{ ft/s} = 1.835 \times 10^4 \text{ ft-lb/s}$$

Since 1 hp = 550 ft-lb/s,

$$P = \frac{1.835 \times 10^4 \text{ ft-lb/s}}{550 \text{ (ft-lb/s)/hp}} = 33.4 \text{ hp}$$

At 70% efficiency,

$$P = \frac{33.4 \text{ hp}}{0.70} = 47.7 \text{ hp}$$

9.13. Find the kinetic energy of a 1000-kg car whose velocity is 20 m/s.

$$KE = \tfrac{1}{2}mv^2 = \tfrac{1}{2} \times 1000 \text{ kg} \times (20 \text{ m/s})^2 = 2 \times 10^5 \text{ J}$$

9.14. What velocity does a 1-kg object have when its kinetic energy is 1 J?

Since $KE = \tfrac{1}{2}mv^2$,

$$v = \sqrt{\frac{2\,KE}{m}} = \sqrt{\frac{2 \times 1 \text{ J}}{1 \text{ kg}}} = \sqrt{2} \text{ m/s} = 1.4 \text{ m/s}$$

9.15. A 128-lb girl skates at a velocity of 15 ft/s. What is her kinetic energy?

The girl's mass is $m = \dfrac{w}{g} = \dfrac{128 \text{ lb}}{32 \text{ ft/s}^2} = 4$ slugs, and so her kinetic energy is

$$KE = \tfrac{1}{2}mv^2 = \tfrac{1}{2} \times 4 \text{ slugs} \times (15 \text{ ft/s})^2 = 450 \text{ ft-lb}$$

9.16. What is the kinetic energy of a 2500-lb car whose velocity is 40 mi/hr?

The mass of the car is

$$m = \frac{w}{g} = \frac{2500 \text{ lb}}{32 \text{ ft/s}^2} = 78 \text{ slugs}$$

and its velocity is

$$v = 40 \text{ mi/hr} \times 1.47 \frac{\text{ft/s}}{\text{mi/hr}} = 59 \text{ ft/s}$$

Hence the car's kinetic energy is

$$KE = \tfrac{1}{2}mv^2 = \tfrac{1}{2} \times 78 \text{ slugs} \times (59 \text{ ft/s})^2 = 1.36 \times 10^5 \text{ ft-lb}$$

9.17. At her highest point, a girl on a swing is 7 ft above the ground, and at her lowest point she is 3 ft above the ground. What is her maximum velocity?

The girl's maximum velocity v occurs at the lowest point. Her kinetic energy there equals her loss of potential energy in descending through a height of $h = 7 \text{ ft} - 3 \text{ ft} = 4 \text{ ft}$. Hence

$$KE = PE$$
$$\tfrac{1}{2}mv^2 = mgh$$
$$v = \sqrt{2gh} = \sqrt{2 \times 32 \text{ ft/s}^2 \times 4 \text{ ft}} = \sqrt{256} \text{ ft/s} = 16 \text{ ft/s}$$

This result is independent of the girl's mass.

9.18. A man skis down a slope 200 m high. If his velocity at the bottom of the slope is 20 m/s, what percentage of his initial potential energy was lost due to friction and air resistance?

$$\frac{\text{Final KE}}{\text{Initial PE}} = \frac{\tfrac{1}{2}mv^2}{mgh} = \frac{v^2}{2gh} = \frac{(20 \text{ m/s})^2}{2 \times 9.8 \text{ m/s}^2 \times 200 \text{ m}} = 0.102 = 10.2\%$$

which means 89.8% of the initial PE was lost.

9.19. A hammer with a 3-lb head is used to drive a nail into a wooden board. If the hammer is moving at 15 ft/s when it strikes the nail and the nail moves $\tfrac{1}{2}$ in. into the board, find the average force the hammer exerts on the nail.

To find the force F exerted by the hammer on the nail we proceed as follows, with $s = \tfrac{1}{2} \text{in.} = \tfrac{1}{24} \text{ft}$:

$$KE \text{ of hammer} = \text{work done on nail}$$
$$\tfrac{1}{2}mv^2 = \tfrac{1}{2}\frac{w}{g}v^2 = Fs$$

$$F = \frac{wv^2}{2gs} = \frac{3 \text{ lb} \times (15 \text{ ft/s})^2}{2 \times 32 \text{ ft/s}^2 \times \tfrac{1}{24} \text{ ft}} = 253 \text{ lb}$$

9.20. Approximately 4×10^9 kg of matter is converted into energy in the sun each second. What is the power output of the sun?

The energy produced by the sun per second is

$$E_0 = m_0 c^2 = 4 \times 10^9 \text{ kg} \times (3 \times 10^8 \text{ m/s})^2 = 3.6 \times 10^{26} \text{ J}$$

Hence the power output is

$$P = \frac{E_0}{t} = \frac{3.6 \times 10^{26} \text{ J}}{1 \text{ s}} = 3.6 \times 10^{26} \text{ W}$$

9.21. How much mass is converted into energy per day in a nuclear power plant operated at a level of 100 megawatts (100×10^6 W)?

There are $60 \times 60 \times 24 = 86{,}400$ s/day, so the energy liberated per day is

$$E_0 = Pt = 10^8 \text{ W} \times 8.64 \times 10^4 \text{ s} = 8.64 \times 10^{12} \text{ J}$$

Since $E_0 = m_0 c^2$,

$$m_0 = \frac{E_0}{c^2} = \frac{8.64 \times 10^{12} \text{ J}}{(3 \times 10^8 \text{ m/s})^2} = 9.6 \times 10^{-5} \text{ kg}$$

Supplementary Problems

9.22. The earth exerts a gravitational force of 2×10^{20} N on the moon, and the moon travels 2.4×10^9 m each time it orbits the earth. How much work does the earth do on the moon in each orbit?

9.23. How much work must be done to raise a 1100-kg car 2 m above the ground?

9.24. A 20-lb object is raised to a height of 40 ft above the ground. (a) How much work was done? (b) What is the potential energy of the object? (c) If the object is dropped, what will its kinetic energy be just before it strikes the ground?

9.25. A boy pulls a wagon with a force of 15 lb by means of a rope that makes an angle of 40° with the ground. How much work does he do in moving the wagon 50 ft?

9.26. A horse exerts a force of 200 lb while pulling a sled for 3 miles. (a) How much work does the horse do? (b) If the trip takes 30 min, what is the power output of the horse in hp?

9.27. In 1970 the population of the world was about 3.5×10^9 and about 2×10^{20} J of work was performed under man's control. Find the average power consumption per person in watts and in hp. (1 year = 3.15×10^7 s)

9.28. A certain 80-kg mountain climber has an average power output of 0.1 hp. (a) How much work does he perform in climbing a mountain 2000 m high? (b) How long does he take to climb the mountain? (c) What is his potential energy at the top?

9.29. A man uses a rope and a system of pulleys to raise a 200-lb box to a height of 10 ft. He exerts a force of 60 lb on the rope and pulls a total of 40 ft of rope through the pulleys. (a) How much work does he perform? (b) By how much is the potential energy of the box increased? (c) If these answers are different, what do you think the reason is?

9.30. The four engines of a DC-8 airplane develop a total of 30,000 hp when its velocity is 240 m/s. How much force do the engines exert?

9.31. Neglecting friction and air resistance, is more work needed to accelerate a car from 10 mi/hr to 20 mi/hr or from 20 mi/hr to 30 mi/hr?

9.32. A 3000-lb car has an engine which can deliver 80 hp to the rear wheels. What is the maximum velocity at which the car can climb a 15° hill?

9.33. Find the kinetic energy of a 2-g (0.002-kg) insect when it is flying at 0.4 m/s.

9.34. The electrons in a television picture tube whose impacts on the screen produce the flashes of light that make up the image have masses of 9.1×10^{-31} kg and typical velocities of 3×10^7 m/s. What is the kinetic energy of such an electron?

9.35. (a) What is the kinetic energy of a 3200-lb car traveling at 100 ft/s (68 mi/hr)? (b) If the car can reach this velocity in 12 s starting from rest, what is the power output of its engine?

9.36. (a) What velocity does a 1-slug object have when its kinetic energy is 1 ft-lb? (b) What velocity does a 1-lb object have when its kinetic energy is 1 ft-lb?

9.37. A stone is dropped from a height of 100 ft. At what height is half of its energy potential and half kinetic?

9.38. From what height would a car have to fall to the ground in order to do as much work (that is, damage—largely to itself) as a car striking a wall at 88 ft/s (60 mi/hr)?

9.39. A 10-g bullet has a velocity of 600 m/s when it leaves the barrel of a rifle. If the barrel is 60 cm long, find the average force on the bullet while it is in the barrel.

9.40. A 16-lb shell has a velocity of 2000 ft/s when it leaves the barrel of a cannon. If the barrel is 10 ft long, find the average force on the shell while it is in the barrel.

9.41. A hammer with a 1-kg head is to be used to drive nails horizontally into a wall. A force of 1000 N is required to penetrate the wall, and it is desired that each blow of the hammer should force the nail 1 cm into the wall. What should the velocity of the hammer's head be when it strikes the nail?

9.42. This book weighs about 1.5 lb. What is its rest energy in ft-lb? In joules?

9.43. Approximately 4 million ft-lb of energy is liberated when 1 lb of dynamite explodes. How much matter is converted into energy in this process?

9.44. A sedentary person uses energy at an average rate of about 70 W. (a) How many joules of energy does he use per day? (b) All of this energy originates in the sun. How much matter is converted into energy per day to supply such a person?

Answers to Supplementary Problems

9.22. No work is done because the force on the moon is perpendicular to its direction of motion.

9.23. 2.16×10^4 J

9.24. (a) 800 ft-lb (b) 800 ft-lb (c) 800 ft-lb

9.25. 575 ft-lb

9.26. (a) 3.17×10^6 ft-lb (b) 3.2 hp

9.27. 1800 W; 2.4 hp

9.28. (a) 1.57×10^6 J (b) 5 hr 50 min (c) 1.57×10^6 J

9.29. (a) 2400 ft-lb (b) 2000 ft-lb (c) 400 ft-lb was expended in doing work against frictional forces in the pulleys

9.30. 9.3×10^4 N

9.31. From 20 mi/hr to 30 mi/hr

9.32. 56.7 ft/s = 38.6 mi/hr

9.33. 1.6×10^{-4} J

9.34. 4.1×10^{-16} J

9.35. (a) 5×10^5 ft-lb   (b) 4.17×10^4 ft-lb/s = 76 hp

9.36. (a) 1.4 ft/s   (b) 8 ft/s

9.37. 50 ft

9.38. 121 ft

9.39. 3000 N

9.40. 10^5 lb

9.41. 4.47 m/s

9.42. 4.5×10^{16} ft-lb; 6.1×10^{16} J

9.43. 1.32×10^{-10} lb

9.44. (a) 6.05×10^6 J   (b) 6.72×10^{-11} kg

Chapter 10

Momentum

LINEAR MOMENTUM

Work and energy are scalar quantities that have no directions associated with them. When two or more bodies interact with one another, or a single body breaks up into two or more others, the various directions of motion cannot be related by energy considerations alone. The vector quantities called *linear momentum* and *impulse* are important in analyzing such events.

The linear momentum (usually called simply *momentum*) of a body mass m and velocity $\mathbf{v}$ is the product of m and $\mathbf{v}$:

$$\text{Momentum} = m\mathbf{v}$$

The units of momentum are the kg-m/s and the slug-ft/s. The direction of the momentum of a body is the same as the direction in which it is moving.

The greater the momentum of a body, the greater its tendency to continue in motion. Thus a baseball struck by a bat (v large) is harder to stop than a baseball thrown by hand (v small), and an iron shot (m large) is harder to stop than a baseball (m small) of the same velocity.

IMPULSE

A force $\mathbf{F}$ that acts on a body during a time interval t provides the body with an *impulse* of $\mathbf{F}t$:

$$\text{Impulse} = \mathbf{F}t = \text{force} \times \text{time interval}$$

The units of impulse are the N-s and the lb-s.

When a force acts on a body to produce a change in its momentum, the momentum change $m(\mathbf{v}_2 - \mathbf{v}_1)$ is equal to the impulse provided by the force. Thus

$$\mathbf{F}t = m(\mathbf{v}_2 - \mathbf{v}_1)$$
$$\text{Impulse} = \text{momentum change}$$

CONSERVATION OF LINEAR MOMENTUM

According to the law of conservation of linear momentum, the total linear momentum of a system of bodies isolated from the rest of the universe remains constant regardless of what happens within the system. The total linear momentum of the system is the vector sum of the momenta of the various bodies included in the system; by "isolated" is meant that no net force of external origin acts on the system.

ROCKET PROPULSION

Momentum conservation underlies the operation of a rocket. The momentum of a rocket on the ground is zero. When the fuel is ignited, exhaust gases shoot downward and the rocket body rises to balance their momentum so that the total remains zero. A rocket does not "push" against the ground or the atmosphere, and indeed is most efficient in space where there is no air to cause friction.

The final velocity of a rocket depends upon the velocity of its exhaust gases and upon the amount of fuel it can carry. When extremely high velocities are needed, as in space exploration, it is more efficient to use two or more rocket stages. The first stage has as its payload another, smaller rocket. When the fuel of the first stage has been burnt up, the second stage is released from the shell of the first and is fired. Since it is already moving rapidly and does not have the burden of the motor and fuel tanks of the first stage, the second stage can reach a much higher final velocity. This process can be repeated a number of times, depending upon the required velocity.

COLLISIONS

Momentum is also conserved in collisions. If a billiard ball strikes a stationary one, the two move off in such a way that the vector sum of their momenta is the same as the initial momentum of the first ball (Fig. 10-1). This is true even if the balls move off in different directions.

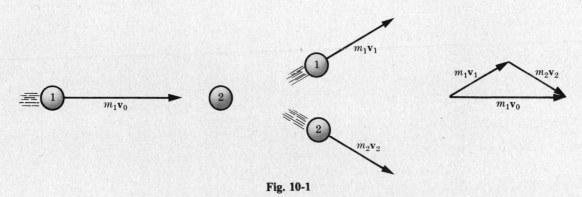

Fig. 10-1

A perfectly *elastic* collision is one in which the bodies involved move apart in such a way that kinetic energy as well as momentum is conserved. In a perfectly *inelastic* collision the bodies stick together and the kinetic energy loss is the maximum possible consistent with momentum conservation. Most collisions are intermediate between these two extremes.

The *coefficient of restitution e* is defined as the ratio between the relative velocity of recession $v_2 - v_1$ after a collision between two bodies and their relative velocity of approach $u_1 - u_2$:

$$\text{Coefficient of restitution} = e = \frac{v_2 - v_1}{u_1 - u_2}$$

Values of e range from 0 to 1. In a perfectly elastic collision, $e = 1$ and the relative velocity after the collision is the same as the relative velocity before it. In a perfectly inelastic collision, $e = 0$.

Solved Problems

10.1. A rocket explodes in midair. How does this affect (*a*) its total momentum and (*b*) its total kinetic energy?

   (*a*) The total momentum remains the same because no external forces acted on the rocket.

   (*b*) The total kinetic energy increases because the rocket fragments received additional KE from the explosion.

10.2. An airplane's velocity is doubled. (*a*) What happens to its momentum? Is the law of conservation of momentum obeyed? (*b*) What happens to its kinetic energy? Is the law of conservation of energy obeyed?

(a) The airplane's momentum of mv also doubles. Momentum is conserved because the increase in the airplane's velocity is accompanied by the backward motion of air through the action of its engines, so the total momentum of airplane + air remains the same when their opposite directions are taken into account.

(b) The airplane's kinetic energy of $\frac{1}{2}mv^2$ increases fourfold. Energy is conserved because the additional KE comes from chemical potential energy released in the airplane's engines.

10.3. Find the momentum of a 50-kg girl running at 6 m/s.

$$mv = 50 \text{ kg} \times 6 \text{ m/s} = 300 \text{ kg-m/s}$$

10.4. A 160-lb man runs a mile in 4 min. What is his average momentum?

The man's mass m and average velocity $\bar{v}$ are respectively

$$m = \frac{w}{g} = \frac{160 \text{ lb}}{32 \text{ ft/s}^2} = 5 \text{ slugs} \qquad \bar{v} = \frac{1 \text{ mile} \times 5280 \text{ ft/mi}}{4 \text{ min} \times 60 \text{ s/min}} = 22 \text{ ft/s}$$

Hence his average momentum is $m\bar{v} = 5 \text{ slugs} \times 22 \text{ ft/s} = 110 \text{ slug-ft/s}$.

10.5. A certain DC-9 airplane has a mass of 50,000 kg and a cruising velocity of 700 km/hr. Its engines develop a total thrust of 70,000 N. If air resistance, change in altitude, and fuel consumption are ignored, how long does it take the airplane to reach its cruising velocity starting from rest?

The airplane's initial velocity is $v_1 = 0$ and its final velocity is

$$v_2 = 700 \text{ km/hr} \times 0.278 \frac{\text{m/s}}{\text{km/hr}} = 195 \text{ m/s}$$

Hence

Impulse = momentum change

$$Ft = m(v_2 - v_1) = mv_2$$

$$t = \frac{mv_2}{F} = \frac{50,000 \text{ kg} \times 195 \text{ m/s}}{70,000 \text{ N}} = 139 \text{ s} = 2 \text{ min } 19 \text{ s}$$

10.6. A 2400-lb car strikes a fence at 30 ft/s (about 20 mi/hr) and comes to a stop in 1 s. What average force acted on the car?

The initial and final velocities of the car are 30 ft/s and 0. Hence

Impulse = momentum change

$$Ft = m(v_2 - v_1) = \frac{w}{g}(v_2 - v_1)$$

$$F = \frac{w(v_2 - v_1)}{gt} = \frac{3200 \text{ lb } (0 - 30 \text{ ft/s})}{32 \text{ ft/s}^2 \times 1 \text{ s}} = -3000 \text{ lb}$$

The minus sign means that the force that acted to stop the car is in the opposite direction to its initial velocity.

10.7. A 2000-lb car moving at 50 mi/hr collides head-on with a 3000-lb car moving at 20 mi/hr, and the two cars stick together. Which way does the wreckage move?

The 2000-lb car had the greater initial momentum, so the wreckage moves in the same direction it had.

10.8. A 5-kg rifle fires a 15-g (0.015-kg) bullet at a muzzle velocity of 600 m/s. Find the recoil velocity of the rifle.

From conservation of momentum, $m_r v_r = m_b v_b$, and so

$$v_r = \frac{m_b}{m_r} \times v_b = \frac{0.015 \text{ kg}}{5 \text{ kg}} \times 600 \text{ m/s} = 1.8 \text{ m/s}$$

10.9. An astronaut is in space at rest relative to an orbiting spacecraft. His total weight is 300 lb and he throws away a 1-lb wrench at a velocity of 15 ft/s relative to the spacecraft. How fast does he move off in the opposite direction?

From conservation of momentum,

$$m_a v_a = m_w v_w$$

$$\frac{w_a}{g} \times v_a = \frac{w_w}{g} \times v_w$$

$$v_a = \frac{w_w}{w_a} \times v_w = \frac{1 \text{ lb}}{300 \text{ lb}} \times 15 \text{ ft/s} = 0.05 \text{ ft/s}$$

We notice that the g's have canceled out, so that it was not necessary to find the mass values first; the ratio of two masses is always the same as the ratio of the corresponding weights.

10.10. A 0.5-kg snowball moving at 20 m/s strikes and sticks to a 70-kg man standing on the frictionless surface of a frozen pond. What is the man's final velocity?

Let v_1 = snowball's velocity and v_2 = final velocity of man + snowball. Then

Initial momentum of snowball = final momentum of man + snowball

$$m_s v_1 = (m_m + m_s) v_2$$

$$v_2 = \left(\frac{m_s}{m_m + m_s} \right) v_1 = \frac{0.5 \text{ kg}}{70.5 \text{ kg}} \times 20 \text{ m/s} = 0.14 \text{ m/s}$$

10.11. A 40-kg skater traveling at 4 m/s overtakes a 60-kg skater traveling at 2 m/s in the same direction and collides with him. (a) If the two skaters remain in contact, what is their final velocity? (b) How much kinetic energy is lost?

(a) Let v_1 = initial velocity of 40-kg skater, v_2 = initial velocity of 60-kg skater, and v_3 = final velocity of the two skaters. Then

Initial total momentum = final total momentum

$$m_1 v_1 + m_2 v_2 = (m_1 + m_2) v_3$$

$$v_3 = \frac{m_1 v_1 + m_2 v_2}{m_1 + m_2} = \frac{(40 \text{ kg} \times 4 \text{ m/s}) + (60 \text{ kg} \times 2 \text{ m/s})}{40 \text{ kg} + 60 \text{ kg}} = 2.8 \text{ m/s}$$

(b) $$\text{Initial KE} = \tfrac{1}{2} m_1 v_1^2 + \tfrac{1}{2} m_2 v_2^2$$

$$= \tfrac{1}{2} \times 40 \text{ kg} \times (4 \text{ m/s})^2 + \tfrac{1}{2} \times 60 \text{ kg} \times (2 \text{ m/s})^2 = 440 \text{ J}$$

$$\text{Final KE} = \tfrac{1}{2} (m_1 + m_2) v_3^2 = \tfrac{1}{2} \times 100 \text{ kg} \times (2.8 \text{ m/s})^2 = 392 \text{ J}$$

Therefore 48 J of energy is lost, 11% of the original amount.

10.12. The two skaters of Problem 10.11 are moving in opposite directions and collide head-on. (a) If they remain in contact, what is their final velocity? (b) How much kinetic energy is lost?

(a) We take into account the opposite directions of motion by letting $v_1 = +4$ m/s and $v_2 = -2$ m/s. Then

$$v_3 = \frac{m_1 v_1 + m_2 v_2}{m_1 + m_2} = \frac{(40 \text{ kg} \times 4 \text{ m/s}) - (60 \text{ kg} \times 2 \text{ m/s})}{40 \text{ kg} + 60 \text{ kg}} = +0.4 \text{ m/s}$$

Since v_3 is $+0.4$ m/s, the two skaters move off in the same direction the 40-kg skater had originally, which is to be expected since he had the greater initial momentum.

(b) Initial KE = 440 J (as in Problem 10.11)

$$\text{Final KE} = \tfrac{1}{2} (m_1 + m_2) v_3^2 = \tfrac{1}{2} \times 100 \text{ kg} \times (0.4 \text{ m/s})^2 = 8 \text{ J}$$

Therefore 432 J of energy is lost, 98% of the original amount. This is the reason why head-on collisions of automobiles produce much more damage than overtaking collisions.

10.13. A 2400-lb car moving north at 50 ft/s collides with a 3200-lb car moving west at 30 ft/s. If the cars stick together after the collision, at what velocity and in what direction does the wreckage begin to move?

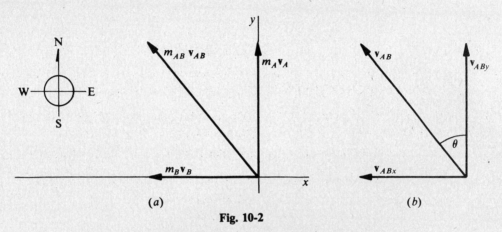

Fig. 10-2

Linear momentum must be conserved separately in both the north-south and east-west directions, which we shall call the y and x axes respectively as in Fig. 10-2(a). Thus we have

$$\text{Momentum before crash} = \text{momentum after crash}$$

x direction: $m_A v_{Ax} + m_B v_{Bx} = m_{AB} v_{ABx}$

y direction: $m_A v_{Ay} + m_B v_{By} = m_{AB} v_{ABy}$

Here

$$m_A = \frac{2400 \text{ lb}}{32 \text{ ft/s}^2} = 75 \text{ slugs} \qquad m_B = \frac{3200 \text{ lb}}{32 \text{ ft/s}^2} = 100 \text{ slugs}$$

$$m_{AB} = m_A + m_B = 175 \text{ slugs}$$

$$v_{Ax} = 0 \qquad\qquad v_{Bx} = -30 \text{ ft/s} \qquad v_{ABx} = ?$$

$$v_{Ay} = 50 \text{ ft/s} \qquad v_{By} = 0 \qquad\qquad v_{ABy} = ?$$

In the x direction we have

$$v_{ABx} = \frac{m_A v_{Ax} + m_B v_{Bx}}{m_{AB}} = \frac{0 + 100 \text{ slugs} \times (-30 \text{ ft/s})}{175 \text{ slugs}} = -17 \text{ ft/s}$$

and in the y direction we have

$$v_{ABy} = \frac{m_A v_{Ay} + m_B v_{By}}{m_{AB}} = \frac{75 \text{ slugs} \times 50 \text{ ft/s} + 0}{175 \text{ slugs}} = 21 \text{ ft/s}$$

The magnitude of the velocity v_{AB} is therefore

$$v_{AB} = \sqrt{v_{ABx}^2 + v_{ABy}^2} = \sqrt{(-17 \text{ ft/s})^2 + (21 \text{ ft/s})^2} = 27 \text{ ft/s}$$

The direction of $\mathbf{v}_{AB}$ may be specified by the angle θ between the $+y$ direction, which is north, and $\mathbf{v}_{AB}$. From Fig. 10-2(b), ignoring the sign of v_{ABx},

$$\tan \theta = \frac{v_{ABx}}{v_{ABy}} = \frac{17 \text{ ft/s}}{21 \text{ ft/s}} = 0.810 \qquad \theta = 39°$$

The wreckage begins to move at 27 ft/s in a direction 39° west of north.

10.14. A 1-kg ball moving at 5 m/s collides with a 2-kg ball moving in the opposite direction at 4 m/s. If the coefficient of restitution is 0.7, find the velocities of the two balls after the impact.

Here
$$m_1 = 1 \text{ kg} \quad u_1 = 5 \text{ m/s} \quad v_1 = ?$$
$$m_2 = 2 \text{ kg} \quad u_2 = -4 \text{ m/s} \quad v_2 = ?$$

First we make use of the fact that $e = 0.7$:

$$e = \frac{v_2 - v_1}{u_1 - u_2}$$

$$v_2 - v_1 = e(u_1 - u_2) = 0.7[5 \text{ m/s} - (-4 \text{ m/s})] = 6.3 \text{ m/s}$$

$$v_2 = v_1 + 6.3 \text{ m/s}$$

From conservation of momentum we have

Momentum before = momentum after
$$m_1 u_1 + m_2 u_2 = m_1 v_1 + m_2 v_2$$
$$1 \text{ kg} \times 5 \text{ m/s} - 2 \text{ kg} \times 4 \text{ m/s} = 1 \text{ kg} \times v_1 + 2 \text{ kg} \times v_2$$
$$v_2 = \frac{-3 \text{ kg-m/s} - v_1 \text{ kg}}{2 \text{ kg}}$$

We now set equal the two expressions for v_2 and solve for v_1:

$$v_1 + 6.3 \text{ m/s} = \frac{-3 \text{ kg-m/s} - v_1 \text{ kg}}{2 \text{ kg}}$$

$$v_1(1 + 0.5) = \frac{-3 \text{ kg-m/s}}{2 \text{ kg}} - 6.3 \text{ m/s} = -7.8 \text{ m/s}$$

$$v_1 = -5.2 \text{ m/s}$$

The minus sign means that the 1-kg ball moves off in the opposite direction from its initial one. To find v_2 it is simplest to use $v_2 = v_1 + 6.3 \text{ m/s}$:

$$v_2 = -5.2 \text{ m/s} + 6.3 \text{ m/s} = 1.1 \text{ m/s}$$

The 2-kg ball also reverses its direction as a result of the collision.

Supplementary Problems

10.15. Find the momentum of a 100-kg ostrich running at 15 m/s.

10.16. Find the momentum of a 3200-lb car moving at 60 mi/hr (88 ft/s).

10.17. An object at rest breaks up into two parts which fly off. Must they move in opposite directions?

10.18. A moving object strikes a stationary one. After the collision, must they move in the same direction? Must they move in opposite directions?

10.19. A 2500-kg truck crashes into a wall at 40 km/hr and comes to a stop in 0.5 s. What is the average force on the truck?

10.20. An 8-lb rifle and a 10-lb rifle fire identical bullets with the same muzzle velocities. Compare the recoil momenta and recoil velocities of the two rifles.

10.21. An empty dump truck is coasting with its engine off along a level road when rain starts to fall. Neglecting friction, what if anything happens to the velocity of the truck?

10.22. A 160-lb man dives horizontally from a 640-lb boat with a velocity of 6 ft/s. What is the recoil velocity of the boat?

10.23. Four 50-kg girls simultaneously dive horizontally at 2.5 m/s from the same side of a boat, whose recoil velocity is 0.1 m/s. What is the mass of the boat?

10.24. An unoccupied 2400-lb car has coasted down a hill and is moving along a level road at 40 ft/s. In order to stop the car, a 12,000-lb truck moving in the opposite direction collides head-on with it. What should the truck's velocity be in order that both vehicles come to a stop after the collision?

10.25. A 50-kg boy at rest on roller skates catches a 0.6-kg ball moving toward him at 30 m/s. How fast does he move backward as a result?

10.26. A 1200-kg car traveling at 10 m/s overtakes a 1000-kg car traveling at 8 m/s and collides with it. (*a*) If the two cars stick together, what is their final velocity? (*b*) How much kinetic energy is lost? What percentage of the original KE is this?

10.27. The cars of Problem 10.26 are moving in opposite directions and collide head-on. (*a*) If they stick together, what is their final velocity? (*b*) How much kinetic energy is lost? What percentage of the original KE is this?

10.28. A 1-lb stone moving south at 8 ft/s collides with a 5-lb lump of clay moving west at 3 ft/s and becomes embedded in the clay. Find the velocity (magnitude and direction) of the composite body.

10.29. A 5-lb ball moving to the right at 10 ft/s overtakes and collides with a 10-lb ball moving to the right at 8 m/s. Find the final velocities of the two balls if the coefficient of restitution is 0.8.

10.30. A rubber ball is dropped on the ground from a height of 5 ft. If the coefficient of restitution is 0.7, find the height to which the ball rebounds.

Answers to Supplementary Problems

10.15. 1500 kg-m/s

10.16. 8800 slug-ft/s

10.17. Yes

10.18. No; no; they can move in any direction provided that the vector sum of their momenta equals the initial momentum of the first object.

10.19. 18,500 N

10.20. The recoil momenta are the same, but the lighter rifle has a higher recoil velocity.

10.21. The truck's velocity decreases as rainwater accumulates in it, since the total momentum must remain constant despite the increase in mass.

10.22. 1.5 ft/s	**10.25.** 0.36 m/s	**10.28.** 2.83 ft/s at 28° south of west
10.23. 5000 kg	**10.26.** (*a*) 9.09 m/s (*b*) 1100 J; 1.2%	**10.29.** 7.6 ft/s; 9.2 ft/s
10.24. 8 ft/s	**10.27.** (*a*) 1.82 m/s (*b*) 88,356 J; 96%	**10.30.** 2.45 ft

Chapter 11

Rotational Motion

ANGULAR MEASURE

In everyday life, angles are measured in degrees, where 360° equals a full turn. A more suitable unit for technical purposes is the *radian* (rad). If a circle is drawn whose center is at the vertex of a particular angle (Fig. 11-1), the angle θ (Greek letter *theta*) in radians is equal to the ratio between the arc s cut by the angle and the radius r of the circle:

$$\theta = \frac{s}{r} \qquad\qquad \theta \text{ (rad)} = \frac{s}{r}$$

$$\text{Angle in radians} = \frac{\text{arc length}}{\text{radius}}$$

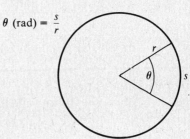

Because the circumference of a circle of radius r is $2\pi r$, there are 2π rad in a complete revolution. Hence

$$1 \text{ rev} = 360° = 2\pi \text{ rad}$$

and so

$$1° = 0.01745 \text{ rad} \qquad 1 \text{ rad} = 57.30°$$

Fig. 11-1

ANGULAR VELOCITY

The *angular velocity* of a body describes how fast it is turning about an axis. If the body turns through the angle θ in the time t, its angular velocity ω (Greek letter *omega*) is

$$\omega = \frac{\theta}{t}$$

$$\text{Angular velocity} = \frac{\text{angular displacement}}{\text{time}}$$

Angular velocity is usually expressed in radians/second (rad/s), revolutions/second (rev/s), and revolutions/minute (rpm), where

$$1 \text{ rev/s} = 2\pi \text{ rad/s} = 6.28 \text{ rad/s}$$
$$1 \text{ rpm} = 2\pi/60 \text{ rad/s} = 0.105 \text{ rad/s}$$

The linear velocity v of a particle that moves in a circle of radius r with the uniform angular velocity ω is given by

$$v = \omega r$$
$$\text{Linear velocity} = \text{angular velocity} \times \text{radius of circle}$$

This formula is valid only when ω is expressed in radian measure.

ANGULAR ACCELERATION

A rotating body whose angular velocity changes from ω_0 to ω_f in the time interval t has the *angular acceleration* α (Greek letter *alpha*) of

$$\alpha = \frac{\omega_f - \omega_0}{t}$$

$$\text{Angular acceleration} = \frac{\text{angular velocity change}}{\text{time}}$$

74

A positive value of α means that the angular velocity is increasing, a negative value means that it is decreasing. Only constant angular accelerations are considered here.

The formulas relating the angular displacement, velocity, and acceleration of a rotating body under constant acceleration are analogous to the formulas relating linear displacement, velocity, and acceleration given in Chapter 3. If a body has the initial angular velocity ω_0, its angular velocity ω_f after a time t during which its angular acceleration is α will be

$$\omega_f = \omega_0 + \alpha t$$

and in this time it will have turned through an angular displacement of

$$\theta = \omega_0 + \tfrac{1}{2}\alpha t^2$$

A relationship that does not involve the time t directly is sometimes useful:

$$\omega_t^2 = \omega_0^2 + 2\alpha\theta$$

MOMENT OF INERTIA

The rotational analog of mass is a quantity called *moment of inertia*. The greater the moment of inertia of a body, the greater its resistance to a change in its angular velocity. The value of the moment of inertia I of a body about a particular axis of rotation depends not only upon the body's mass but also upon how the mass is distributed about the axis.

Let us imagine a rigid body divided into a great many small particles whose masses are m_1, m_2, m_3, ... and whose distances from the axis of rotation are respectively r_1, r_2, r_3, ... (Fig. 11-2). The moment of inertia of this body is given by

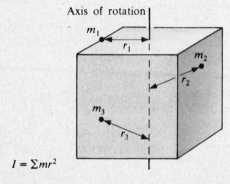

$I = \Sigma mr^2$

Fig. 11-2

$$I = m_1 r_1^2 + m_2 r_2^2 + m_3 r_3^2 + \cdots = \Sigma mr^2$$

where the symbol Σ (Greek capital letter *sigma*) means "sum of" as before. The farther a particle is

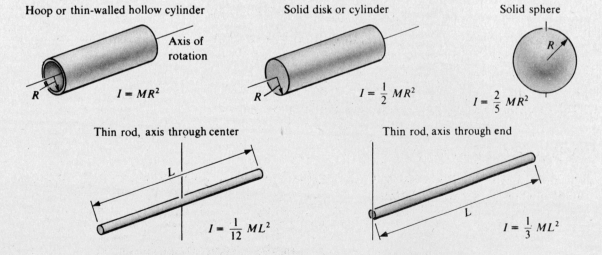

Hoop or thin-walled hollow cylinder
Axis of rotation
R
$I = MR^2$

Solid disk or cylinder
R
$I = \tfrac{1}{2}MR^2$

Solid sphere
R
$I = \tfrac{2}{5}MR^2$

Thin rod, axis through center
L
$I = \tfrac{1}{12}ML^2$

Thin rod, axis through end
L
$I = \tfrac{1}{3}ML^2$

Fig. 11-3

from the axis of rotation, the more it contributes to the moment of inertia. The units of I are the kg-m^2 and the slug-ft^2. Some examples of moments of inertia of bodies of mass M are shown in Fig. 11-3.

TORQUE

Torque plays the same role in rotational motion that force does in linear motion. A net force F acting on a body of mass m causes it to undergo the linear acceleration a in accordance with Newton's second law of motion, $F = ma$. Similarly a net torque τ acting on a body of moment of inertia I causes it to undergo the angular acceleration α (in rad/s^2) in accordance with the formula

$$\tau = I\alpha$$
Torque = moment of inertia $\times$ angular acceleration

ROTATIONAL ENERGY AND WORK

The kinetic energy of a body of moment of inertia I whose angular velocity is ω (in rad/s) is

$$KE = \tfrac{1}{2}I\omega^2$$

Kinetic energy = $\tfrac{1}{2} \times$ (moment of inertia) $\times$ (angular velocity)^2

The work done by a constant torque τ that acts on a body while it experiences the angular displacement θ (in rad) is

$$W = \tau\theta$$
Work = torque $\times$ angular displacement

The rate at which work is being done when a torque τ acts on a body that rotates at the constant angular velocity ω (in rad/s) is

$$P = \tau\omega$$
Power = torque $\times$ angular velocity

ANGULAR MOMENTUM

The equivalent of linear momentum in rotational motion is *angular momentum*. The angular momentum **L** of a rotating body has the magnitude

$$L = I\omega$$
Angular momentum = moment of inertia $\times$ angular velocity

The greater the angular momentum of a spinning object, such as a top, the greater its tendency to continue to spin.

Like linear momentum, angular momentum is a vector quantity with direction as well as magnitude. The direction of the angular momentum of a rotating body is given by the right-hand rule (Fig. 11-4): when the fingers of the right hand are curled in the direction of rotation, the thumb points in the direction of **L**.

Fig. 11-4

According to the principle of *conservation of angular momentum*, the total angular momentum of a system of bodies remains constant in the absence of a net torque regardless of what happens within the system. A skater performing a spin makes use of conservation of angular momentum. He begins to turn with his arms and one leg outstretched and then brings them in close to his body to distribute his mass nearer to the axis of rotation, thereby decreasing his moment of inertia I. Since L must remain constant, his angular velocity ω increases.

Because angular momentum is a vector quantity, its conservation implies that the direction of the axis of rotation tends to remain unchanged. For this reason a spinning top stays upright whereas a stationary one falls over immediately.

Solved Problems

11.1. Why do all helicopters have two propellers?

If a single propeller were used, the helicopter itself would have to rotate in the opposite direction in order to conserve angular momentum.

11.2. (a) Express 6 revolutions in radians. (b) How many revolutions are equivalent to 10 radians? (c) How many revolutions are equivalent to $\pi/2$ radians?

(a)
$$\theta = 6 \text{ rev} \times 2\pi \frac{\text{rad}}{\text{rev}} = 12\pi \text{ rad} = 37.7 \text{ rad}$$

(b)
$$\theta = \frac{10 \text{ rad}}{2\pi \text{ rad/rev}} = 1.6 \text{ rev}$$

(c)
$$\theta = \frac{\pi/2 \text{ rad}}{2\pi \text{ rad/rev}} = \tfrac{1}{4} \text{ rev}$$

11.3. (a) Express 8° in radians. (b) Express 2.5 rad in degrees. (c) Express π rad in degrees.

(a)
$$\theta = 8° \times 0.01745 \text{ rad/}° = 0.14 \text{ rad}$$
(b)
$$\theta = 2.5 \text{ rad} \times 57.3°/\text{rad} = 143 \text{ rad}$$
(c)
$$\theta = \pi \text{ rad} \times \frac{360°}{2\pi \text{ rad}} = 180°$$

11.4. A phonograph record 12 in. in diameter turns through an angle of 200°. How far does a point on the rim of the record travel?

The first step is to find the equivalent in radians of 200°, which is
$$\theta = 200° \times 0.01745 \text{ rad/}° = 3.49 \text{ rad}$$
The radius of the record is 6 in. Since $\theta = s/r$, a point on its rim travels
$$s = r\theta = 6 \text{ in.} \times 3.49 \text{ rad} = 20.9 \text{ in.}$$
when it turns through 200°.

11.5. A wheel 80 cm in diameter turns at 120 rpm. (a) What is the angular velocity of the wheel in rad/s? (b) What is the linear velocity of a point on the rim of the wheel in m/s?

(a) The angular velocity of the wheel is
$$\omega = 120 \text{ rpm} \times 0.105 \frac{\text{rad/s}}{\text{rpm}} = 12.6 \text{ rad/s}$$

(b) Since the radius of the wheel is $r = 40 \text{ cm} = 0.4 \text{ m}$, a point on its rim has the linear velocity
$$v = \omega r = 12.6 \text{ rad/s} \times 0.4 \text{ m} = 5.04 \text{ m/s}$$

11.6. A steel cylinder 5 in. in diameter is to be machined in a lathe. If the desired linear velocity of the cylinder's surface is to be 2.3 ft/s, at how many rpm should it rotate?

From the formula $v = \omega r$ we obtain, with $r = 2.5 \text{ in.} = 0.208 \text{ ft}$,
$$\omega = \frac{v}{r} = \frac{2.3 \text{ ft/s}}{0.208 \text{ ft}} = 11 \text{ rad/s}$$

and so, since 1 rpm = 0.105 rad/s,

$$\omega = \frac{11 \text{ rad/s}}{0.105 \text{ (rad/s)/rpm}} = 105 \text{ rpm}$$

11.7. The shaft of a motor rotates at 1800 rpm. Through how many radians does it turn in 18 s?

The angular velocity in rad/s that corresponds to 1800 rpm is

$$\omega = 1800 \text{ rpm} \times 0.105 \frac{\text{rad/s}}{\text{rpm}} = 189 \text{ rad/s}$$

Since $\omega = \theta/t$, in 18 s the shaft turns through

$$\theta = \omega t = 189 \text{ rad/s} \times 18 \text{ s} = 3402 \text{ rad}$$

11.8. An engine requires 5 s to go from its idling speed of 600 rpm to 1200 rpm. (*a*) What is its angular acceleration? (*b*) How many revolutions does it make in this period of time?

(*a*) The initial and final angular velocities of the engine are respectively

$$\omega_0 = 600 \text{ rpm} \times 0.105 \frac{\text{rad/s}}{\text{rpm}} = 63 \text{ rad/s}$$

$$\omega_f = 1200 \text{ rpm} \times 0.105 \frac{\text{rad/s}}{\text{rpm}} = 126 \text{ rad/s}$$

and so its angular acceleration is

$$\alpha = \frac{\omega_f - \omega_0}{t} = \frac{126 \text{ rad/s} - 63 \text{ rad/s}}{5 \text{ s}} = 12.6 \text{ rad/s}^2$$

(*b*) The angle through which the engine turns is

$$\theta = \omega_0 t + \tfrac{1}{2}\alpha t^2 = 63 \text{ rad/s} \times 5 \text{ s} + \tfrac{1}{2} \times 12.6 \text{ rad/s}^2 \times (5 \text{ s})^2$$

$$= 472.5 \text{ rad}$$

Since there are 2π rad in a revolution,

$$\theta = \frac{472.5 \text{ rad}}{2\pi \text{ rad/rev}} = 75.2 \text{ rev}$$

11.9. A phonograph turntable initially rotating at 3.5 rad/s makes three complete turns before coming to a stop. (*a*) What is its angular acceleration? (*b*) How much time does it take to come to a stop?

(*a*) The angle in radians that corresponds to 3 rev is

$$\theta = 3 \text{ rev} \times 2\pi \text{ rad/rev} = 6\pi \text{ rad}$$

From the formula $\omega_f^2 = \omega_0^2 + 2\alpha\theta$ we find that

$$\alpha = \frac{\omega_f^2 - \omega_0^2}{2\theta} = \frac{0 - (3.5 \text{ rad/s})^2}{2 \times 6\pi \text{ rad}} = -0.325 \text{ rad/s}^2$$

(*b*) Since $\omega_f = \omega_0 + \alpha t$ we have here

$$t = \frac{\omega_f - \omega_0}{\alpha} = \frac{0 - 3.5 \text{ rad/s}}{-0.325 \text{ rad/s}^2} = 10.8 \text{ s}$$

11.10. (*a*) Find the moment of inertia of a 1-kg phonograph turntable 34 cm in diameter. (*b*) What is its kinetic energy when it rotates at 45 rpm?

(*a*) The radius of the turntable is $R = 17 \text{ cm} = 0.17 \text{ m}$. Assuming that the turntable is a solid disk, its moment of inertia is

$$I = \tfrac{1}{2}MR^2 = \tfrac{1}{2} \times 1 \text{ kg} \times (0.17 \text{ m})^2 = 0.0145 \text{ kg-m}^2$$

(b) The angular velocity of the turntable is

$$\omega = 45 \text{ rpm} \times 0.105 \, \frac{\text{rad/s}}{\text{rpm}} = 4.73 \text{ rad/s}$$

and so its kinetic energy is

$$\text{KE} = \tfrac{1}{2} I \omega^2 = \tfrac{1}{2} \times 0.0145 \text{ kg-m}^2 \times (4.73 \text{ rad/s})^2 = 0.162 \text{ J}$$

(Note that 1 kg-m^2/s^2 = 1 N-m = 1 J.)

11.11. The *radius of gyration* of a body about a particular axis is the distance from that axis to a point at which the body's entire mass may be considered to be concentrated. Thus the moment of inertia of a body of mass M and radius of gyration k is $I = Mk^2$. (a) The radius of gyration of a hollow sphere of radius R and mass M is $k = \sqrt{2/3} \ R$. What is its moment of inertia? (b) Find the radius of gyration of a solid sphere.

(a)
$$I = Mk^2 = M \left(\sqrt{\frac{2}{3}} \ R \right)^2 = \frac{2}{3} MR^2$$

(b) From Fig. 11-3 the moment of inertia of a solid sphere is $I = \tfrac{2}{5} MR^2$. Hence its radius of gyration is

$$k = \sqrt{\frac{I}{M}} = \sqrt{\frac{2}{5} \frac{MR^2}{M}} = \sqrt{\frac{2}{5}} \ R$$

11.12. The radius of gyration of a 200-lb flywheel is 1 ft. Find its moment of inertia.

The mass of the flywheel is $M = w/g = 200 \text{ lb}/32 \text{ ft/s}^2 = 6.25 \text{ slugs}$, and so its moment of inertia is

$$I = Mk^2 = 6.25 \text{ slugs} \times (1 \text{ ft})^2 = 6.25 \text{ slug-ft}^2$$

11.13. A solid cylinder rolls from rest down an inclined plane 1.2 m high without slipping. What is its linear velocity at the foot of the plane?

The potential energy of the cylinder at the top of the plane is PE $= mgh$. At the foot of the plane it has both the kinetic energy of translation $\tfrac{1}{2} mv^2$ and the kinetic energy of rotation $\tfrac{1}{2} I \omega^2$, and its total KE then is equal to its initial PE:

$$\text{PE} = \text{KE}$$
$$mgh = \tfrac{1}{2} mv^2 + \tfrac{1}{2} I \omega^2$$

Because the cylinder rolls without slipping, its linear and angular velocities are related by $v = \omega R$, so that $\omega = v/R$. The moment of inertia of a solid cylinder is $I = \tfrac{1}{2} mR^2$. Hence

$$\tfrac{1}{2} I \omega^2 = \tfrac{1}{2} \times \left(\tfrac{1}{2} mR^2 \right) \times \left(\frac{v}{R} \right)^2 = \tfrac{1}{4} mv^2$$

and the energy equation becomes

$$mgh = \tfrac{1}{2} mv^2 + \tfrac{1}{4} mv^2 = \tfrac{3}{4} mv^2$$

Thus 2/3 of the cylinder's KE resides in its translational motion and 1/3 in its rotational motion. Solving for v yields

$$v = \sqrt{\frac{4}{3} \ gh} = \sqrt{\frac{4}{3} \times 9.8 \text{ m/s}^2 \times 1.2 \text{ m}} = 3.96 \text{ m/s}$$

11.14. The starting cord of an outboard motor is wound around a pulley 7 in. in diameter. How much torque is applied to the motor's crankshaft when the cord is pulled with a force of 12 lb?

Here the moment arm of the force is the pulley's radius of $r = 3.5$ in. $= 0.29$ ft. The torque is therefore

$$\tau = Fr = 12 \text{ lb} \times 0.29 \text{ ft} = 3.5 \text{ lb-ft}$$

11.15. A marine diesel engine develops 120 hp at 2000 rpm. (a) How much torque (in lb-ft and N-m) can it exert at this velocity? (b) The engine delivers its power through a spur gear 18 in. in diameter. If two teeth of the gear transmit torque to another gear at a certain instant, find the force in lb on each gear tooth.

(a) The angular velocity here is

$$\omega = 2000 \text{ rpm} \times 0.105 \frac{\text{rad/s}}{\text{rpm}} = 210 \text{ rad/s}$$

In British units,

$$P = 120 \text{ hp} \times 550 \frac{\text{ft-lb/s}}{\text{hp}} = 66,000 \text{ ft-lb/s}$$

and since $P = \tau\omega$,

$$\tau = \frac{P}{\omega} = \frac{66,000 \text{ ft-lb/s}}{210 \text{ rad/s}} = 314 \text{ lb-ft}$$

In SI (metric) units,

$$P = 120 \text{ hp} \times 746 \text{ W/hp} = 89,520 \text{ W}$$

and

$$\tau = \frac{P}{\omega} = \frac{89,520 \text{ W}}{210 \text{ rad/s}} = 426 \text{ N-m}$$

(b) Since $\tau = Fr$ and $r = 9$ in. $= 0.75$ ft here,

$$F = \frac{\tau}{r} = \frac{314 \text{ lb-ft}}{0.75 \text{ ft}} = 419 \text{ lb}$$

Each of the two gear teeth in contact with the driven gear thus exerts a force of half this amount, or 210 lb.

11.16. An air compressor is powered by an 1800-rpm electric motor using a V-belt drive. The motor pulley is 6 in. in diameter and the tension in the V-belt is 30 lb on one side and 10 lb on the other. Find the horsepower of the motor.

The torque exerted by the motor on the belt is, since $r = 3$ in. $= 0.25$ ft and the net force on the belt is the difference between the two tensions,

$$\tau = Fr = (30 \text{ lb} - 10 \text{ lb}) \times 0.25 \text{ ft} = 5 \text{ lb-ft}$$

The angular velocity of the motor is

$$\omega = 1800 \text{ rpm} \times 0.105 \frac{\text{rad/s}}{\text{rpm}} = 189 \text{ rad/s}$$

and so its power output is

$$P = \tau\omega = 5 \text{ lb-ft} \times 189 \text{ rad/s} = 945 \text{ ft-lb/s}$$

which is, in horsepower,

$$P = \frac{945 \text{ ft-lb/s}}{550 \text{ (ft-lb/s)/hp}} = 1.72 \text{ hp}$$

11.17. The winding drum of an elevator is 4 ft in diameter. (a) At how many rpm should the drum rotate in order to raise the cab at 500 ft/min? (b) If the total load is 2 tons, how much torque is required? (c) Neglecting friction, how many hp must the motor develop?

(a) Since

$$v = \frac{500 \text{ ft/min}}{60 \text{ s/min}} = 8.33 \text{ ft/s}$$

the angular velocity of the drum must be

$$\omega = \frac{v}{r} = \frac{8.33 \text{ ft/s}}{2 \text{ ft}} = 4.17 \text{ rad/s}$$

which is

$$\omega = \frac{41.7 \text{ rad/s}}{(0.105 \text{ rad/s})/\text{rpm}} = 39.7 \text{ rpm}$$

(b)   Since 1 ton = 2000 lb, the required torque is

$$\tau = Fr = 4000 \text{ lb} \times 2 \text{ ft} = 8000 \text{ lb-ft}$$

(c)   The required power can be calculated both from $P = Fv$ and from $P = \tau\omega$:

$$P = Fv = 4000 \text{ lb} \times 8.33 \text{ ft/s} = 3.34 \times 10^4 \text{ ft-lb/s}$$

$$P = \tau\omega = 8000 \text{ lb} \times 4.17 \text{ rad/s} = 3.34 \times 10^4 \text{ ft-lb/s}$$

Since 1 hp = 550 ft-lb/s,

$$P = \frac{3.34 \times 10^4 \text{ ft-lb/s}}{(550 \text{ ft-lb/s})/\text{hp}} = 60.6 \text{ hp}$$

11.18. A flywheel whose moment of inertia is 6 kg-m² is acted upon by a constant torque of 50 N-m.
(a) What is its angular acceleration?   (b) How long does it take to go from rest to a velocity of
90 rad/s?   (c) What is its kinetic energy at this velocity?

(a)
$$\alpha = \frac{\tau}{I} = \frac{50 \text{ N-m}}{6 \text{ kg-m}^2} = 8.33 \text{ rad/s}^2$$

(b)
$$t = \frac{\omega_f - \omega_0}{\alpha} = \frac{90 \text{ rad/s}}{8.33 \text{ rad/s}^2} = 10.8 \text{ s}$$

(c)
$$\text{KE} = \tfrac{1}{2}I\omega_f^2 = \tfrac{1}{2} \times 6 \text{ kg-m}^2 \times (90 \text{ rad/s})^2 = 2.43 \times 10^4 \text{ J}$$

11.19. A flywheel of moment of inertia 10 slug-ft² is rotating at 90 rad/s.   (a) What constant torque is
required to slow it down to 40 rad/s in 20 s?   (b) What is the angular displacement of the
flywheel while it is being slowed down?   (c) How much kinetic energy is lost by the flywheel?

(a)
$$\alpha = \frac{\omega_f - \omega_0}{t} = \frac{(40 - 90) \text{ rad/s}}{20 \text{ s}} = -2.5 \text{ rad/s}^2$$

The minus sign means that ω is decreasing.   The required torque is

$$\tau = I\alpha = 10 \text{ slug-ft}^2 \times (-2.5 \text{ rad/s}^2) = -25 \text{ lb-ft}$$

(b)   Either $\theta = \omega_0 t + \tfrac{1}{2}\alpha t^2$ or $\omega_f^2 = \omega_0^2 + 2\alpha\theta$ can be used to find θ.   From the first formula,

$$\theta = \omega_0 t + \tfrac{1}{2}\alpha t^2 = 90 \text{ rad/s} \times 20 \text{ s} - \tfrac{1}{2} \times 2.5 \text{ rad/s}^2 \times (20 \text{ s})^2 = 1300 \text{ rad}$$

(c)   The KE lost by the flywheel appears as work done against the applied torque.   Hence

$$\Delta \text{KE} = W = \tau\theta = -25 \text{ lb-ft} \times 1300 \text{ rad} = -3.25 \times 10^4 \text{ ft-lb}$$

As a check, we can find Δ KE directly:

$$\Delta \text{KE} = \text{KE}_f - \text{KE}_0 = \tfrac{1}{2}I\left(\omega_f^2 - \omega_0^2\right)$$

$$= \tfrac{1}{2} \times 10 \text{ slug-ft}^2 \times \left[(40 \text{ rad/s})^2 - (90 \text{ rad/s})^2\right] = -3.25 \times 10^4 \text{ ft-lb}$$

Supplementary Problems

11.20. (*a*) Express 12.5 revolutions in radians.   (*b*) Express 0.2 radian in revolutions.   (*c*) Express 3π radians in revolutions.

11.21. (*a*) Express 32° in radians.   (*b*) Express 4.8 rad in degrees.   (*c*) Express $\pi/8$ rad in degrees.

11.22. The minute hand of a clock is 40 cm long.   How many centimeters does its tip move in 25 min?

11.23. A drill bit 1/2 in. in diameter is turning at 400 rpm.   (*a*) What is its angular velocity in rad/s?   (*b*) What is the linear velocity in ft/s of a point on its circumference?

11.24. The teeth of a certain circular saw blade are supposed to move at 50 ft/s.   If the blade is 2 ft in radius, at how many rpm should it turn?

11.25. A car whose tires have radii of 50 cm travels at 20 km/hr.   What is the angular velocity of the tires?

11.26. The earth's radius is 6.4×10^6 m.   (*a*) Through how many radians does the earth turn in a year?   (*b*) How far does a point on the equator move in a year owing to this rotation?

11.27. A grindstone that is rotating at 2000 rpm requires 50 s to come to a stop when its motor is switched off.   (*a*) Find the angular acceleration of the grindstone.   (*b*) How many radians does it turn through before stopping?   (*c*) How many revolutions?

11.28. A wheel rotating at 20 rev/s is brought to rest by a constant torque in 12 s.   How many revolutions does it make in this period of time?

11.29. The propeller of a ship makes 300 revolutions while its speed increases from 200 rpm to 500 rpm.   (*a*) What is its angular acceleration?   (*b*) How much time did the increase in speed require?

11.30. The mass and radius of the earth are respectively 6×10^{24} kg and 6.4×10^6 m.   Assuming that it is a sphere of uniform density (which is not the case, since the earth's metallic core has a greater density than the mantle of rock around it), find the moment of inertia of the earth.

11.31. The moment of inertia of a 45-kg grindstone is 5 kg-m^2.   Find its radius of gyration.

11.32. A solid cylinder and a hollow cylinder of the same mass and diameter, both initially at rest, roll down the same inclined plane without stopping.   (*a*) Which reaches the bottom first?   (*b*) How do their kinetic energies at the bottom compare?

11.33. A bowling ball rolling at 8 m/s begins to move up an inclined plane.   What height does it reach?

11.34. A certain engine develops 40 hp at 1000 rpm and 65 hp at 1600 rpm.   At which angular velocity does it produce the greatest torque?

11.35. The alternator on a truck engine produces 350 W of electric power when it rotates at 4000 rpm.   If it is 95% efficient, what should the difference (in N) between the tensions in the tight and slack parts of its V-belt drive be?   The diameter of the alternator's pulley is 10 cm.

11.36. The anchor windlass of a boat is required to pull a load of 400 lb at 2 ft/s.   (*a*) How many horsepower are required?   (*b*) If the windlass drum is 8 in. in diameter, at how many rpm should it turn?   (*c*) What torque is developed by the drum?

11.37. A constant frictional torque of 200 N-m is applied to a turbine initially rotating at 120 rad/s, and it comes to a stop in 80 s.   What is the turbine's moment of inertia?

11.38. A wheel whose moment of inertia is 2 kg-m^2 has an initial angular velocity of 50 rad/s. (a) If a constant torque of 10 N-m acts on the wheel, how long does it take to be accelerated to 80 rad/s? (b) By how much does its kinetic energy increase?

11.39. A wheel whose moment of inertia is 0.4 slug-ft^2 is rotating at 1500 rpm. (a) What constant torque is required to increase its angular velocity to 2000 rpm? (b) How many turns does the wheel make while it is being accelerated? (c) How much work is done on the wheel?

Answers to Supplementary Problems

11.20. (a) 78.5 rad   (b) 0.032 rev   (c) 1.5 rev

11.21. (a) 0.558 rad   (b) 275°   (c) 22.5°

11.22. 105 cm

11.23. (a) 42 rad/s   (b) 0.875 ft/s

11.24. 238 rpm

11.25. 11.1 rad/s

11.26. (a) 2293 rad   (b) 1.47×10^{10} m

11.27. (a) -4.2 rad/s^2 (b) 5250 rad   (c) 836 rev

11.28. 120 rev

11.29. (a) 0.614 rad/s^2 (b) 51.3 s

11.30. 9.8×10^{37} kg-m^2

11.31. 33 cm

11.32. The solid cylinder reaches the bottom first because less of its total kinetic energy is KE of rotation. The total kinetic energies of the cylinders are the same since they had the same potential energies at the top of the plane.

11.33. 4.57 m

11.34. 1600 rpm

11.35. 17.5 N

11.36. (a) 1.45 hp   (b) 57 rpm   (c) 133 lb-ft

11.37. 133 kg-m^2

11.38. (a) 6 s   (b) 3900 J

11.39. (a) 2.63 lb-ft   (b) 234 turns   (c) 3.86×10^3 ft-lb

Chapter 12

Simple Machines

MACHINES

A *machine* is a device that changes the magnitude, direction, or mode of application of a force or torque while transmitting it for a particular purpose. There are only three basic machines, of which all others are developments: the lever, the inclined plane, and the hydraulic press. The hydraulic press is described in Chapter 16.

MECHANICAL ADVANTAGE

The *actual mechanical advantage* of a machine is the ratio between the output force F_{out} it exerts and the input force F_{in} that is applied to it:

$$AMA = \frac{F_{out}}{F_{in}}$$

$$\text{Actual mechanical advantage} = \frac{\text{output force}}{\text{input force}}$$

An AMA greater than 1 means that F_{out} exceeds F_{in}; an AMA less than 1 means that F_{out} is smaller than F_{in}.

The *theoretical mechanical advantage* of a machine is its mechanical advantage in the absence of friction. If the input force acts through the distance s_{in} when the output force acts through the distance s_{out}, then according to the principle of conservation of energy

$$\text{Work input} = \text{work output}$$

$$F_{in}s_{in} = F_{out}s_{out}$$

and so

$$\frac{F_{out}}{F_{in}} = \frac{s_{in}}{s_{out}}$$

when there is no friction present. Because friction acts to decrease the ratio F_{out}/F_{in} in an actual machine but does not change the ratio s_{in}/s_{out}, it is customary to define theoretical mechanical advantage in terms of the latter:

$$TMA = \frac{s_{in}}{s_{out}}$$

$$\text{Theoretical mechanical advantage} = \frac{\text{input distance}}{\text{output distance}}$$

EFFICIENCY

The *efficiency* of a machine equals the ratio between its actual and its theoretical mechanical advantages:

$$\text{Eff} = \frac{AMA}{TMA} = \frac{\text{work output}}{\text{work input}} = \frac{\text{power output}}{\text{power input}}$$

THE LEVER

The TMA of a *lever* is equal to the ratio between its moment arms L_{in} and L_{out} (Fig. 12-1):

$$\text{TMA} = \frac{L_{in}}{L_{out}}$$

The wheel and axle, belt and gear drives, and pulley systems are all developments of the lever.

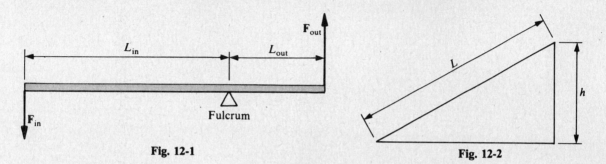

Fig. 12-1   Fig. 12-2

THE INCLINED PLANE

The TMA of an *inclined plane* is equal to the ratio between its length and its height (Fig. 12-2):

$$\text{TMA} = \frac{L}{h}$$

Wedges, cams, and screws are all developments of the inclined plane.

A *screw* is an inclined plane wrapped around a cylinder in the form of a helix. The *pitch p* of a screw is the distance from one thread to the next (Fig. 12-3). If the head of a screw is turned by a tangential force applied the distance L from its axis, the input force travels the distance $s_{in} = 2\pi L$ while the screw advances $s_{out} = p$. Hence the TMA of a screw is

$$\text{TMA} = \frac{s_{in}}{s_{out}} = \frac{2\pi L}{p}$$

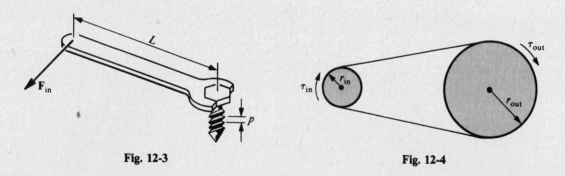

Fig. 12-3   Fig. 12-4

TORQUE TRANSMISSION

Belt and gear drives make possible the transmission of torques from one shaft to another (Fig. 12-4). The actual mechanical advantage of any such system is

$$\text{AMA} = \frac{\tau_{out}}{\tau_{in}} = \frac{\text{output torque}}{\text{input torque}}$$

The theoretical mechanical advantage of a pulley system equals the ratio between the radius of the driven pulley r_{out} and that of the driving pulley r_{in}, which is the same as the ratio of their diameters:

$$\text{TMA} = \frac{r_{out}}{r_{in}} = \frac{d_{out}}{d_{in}}$$

In the case of a gear drive, since the number of teeth on a gear is proportional to its radius,

$$\text{TMA} = \frac{N_{out}}{N_{in}} = \frac{\text{number of teeth on driven gear}}{\text{number of teeth on driving gear}}$$

In any torque transmission system, the angular velocity ratio is the inverse of the TMA:

$$\frac{\text{Output angular velocity}}{\text{Input angular velocity}} = \frac{\omega_{out}}{\omega_{in}} = \frac{d_{in}}{d_{out}} = \frac{1}{\text{TMA}}$$

An increase in torque is accompanied by a decrease in speed of rotation, and vice versa.

Solved Problems

[The efficiency is 100% in each case unless otherwise stated.]

12.1. One end of a 200-kg crate is to be lifted from the ground using a 3-m plank as a lever. If the maximum force that can be applied to the plank is 350 N, where should the fulcrum be placed? Assume that the crate's contents are uniformly distributed, so the mass to be raised is 100 kg.

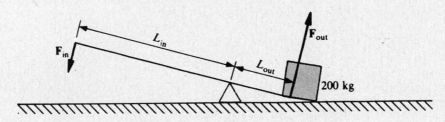

Fig. 12-5

The weight of a 100-kg mass is

$$w = mg = 100 \text{ kg} \times 9.8 \text{ m/s}^2 = 980 \text{ N}$$

Since the plank is 3 m long (Fig. 12-5),

$$L_{in} + L_{out} = 3 \text{ m} \qquad L_{in} = 3 \text{ m} - L_{out}$$

In the absence of friction,

$$\frac{F_{out}}{F_{in}} = \frac{L_{in}}{L_{out}}$$

$$\frac{980 \text{ N}}{350 \text{ N}} = \frac{3 \text{ m} - L_{out}}{L_{out}}$$

$$3.8 \, L_{out} = 3 \text{ m} \qquad L_{out} = 0.79 \text{ m}$$

The fulcrum should be located 79 cm from the end of the crate.

12.2. The *wheel and axle* (Fig. 12-6) is a development of the lever that permits continuous motion. (*a*) Where is the fulcrum?   (*b*) What is the TMA of a wheel and axle?

(a) The center of the axle acts as the fulcrum.

(b) The input force acts tangentially on the wheel rim, and the output force acts tangentially on the axle rim. When the combination makes one complete turn, a point on the wheel moves through $s_{in} = 2\pi R$ and a point on the axle moves through $s_{out} = 2\pi r$. Hence

$$TMA = \frac{s_{in}}{s_{out}} = \frac{2\pi R}{2\pi r} = \frac{R}{r}$$

The larger the wheel radius relative to that of the axle, the greater the mechanical advantage.

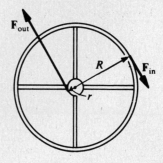

Fig. 12-6

12.3. Find the TMA of the pulley system shown in Fig. 12-7.

In order to raise the load through a height h, a length of rope equal to $4h$ must be pulled through the pulley system since the load is supported by four strands of rope. Hence the TMA of the pulley system is

$$TMA = \frac{s_{in}}{s_{out}} = \frac{4h}{h} = 4$$

As a general rule, the TMA of a pulley system is equal to the number of strands of rope that support the load. The rope leaving the upper pulley in Fig. 12-7 does not help to support the load and hence is not counted in determining the TMA of the system.

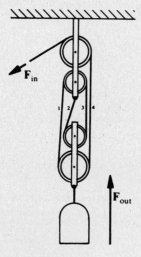

Fig. 12-7

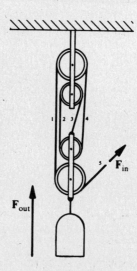

Fig. 12-8

12.4. Find the TMA of the pulley system shown in Fig. 12-8.

This pulley system is the same as that in Fig. 12-7 but is inverted. As a result five strands of rope now support the load, and the TMA is accordingly 5.

12.5. Find the TMA of the pulley system shown in Fig. 12-9.

There are actually two pulley systems here. The load is initially supported by the two strands of rope 1 and 2 that pass around pulley A, so $TMA_1 = 2$. The tension in rope 2 is then supported by strands 3 and 4 of another rope that passes around pulley B, so $TMA_2 = 2$. Pulley C serves merely to change the direction of the rope, and strand 5 does not help to support the load. Hence

$$TMA = TMA_1 \times TMA_2 = 2 \times 2 = 4$$

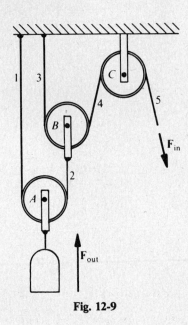

Fig. 12-9   **Fig. 12-10**

12.6. A man uses a force of 20 lb to raise a load of 70 lb by 2 ft with a pulley system. If he pulls a total of 8 ft of rope through the system, what is its efficiency?

The TMA and AMA of the system are respectively

$$\text{TMA} = \frac{s_{\text{in}}}{s_{\text{out}}} = \frac{8 \text{ ft}}{2 \text{ ft}} = 4$$

$$\text{AMA} = \frac{F_{\text{out}}}{F_{\text{in}}} = \frac{70 \text{ lb}}{20 \text{ lb}} = 3.5$$

Hence the efficiency is

$$\text{Eff} = \frac{\text{AMA}}{\text{TMA}} = \frac{3.5}{4} = 0.875 = 87.5\%$$

12.7. A *differential pulley* (often called a *chain hoist* because a chain running over notched pulleys is used to prevent slipping) is shown in Fig. 12-10. When the free loop of the chain is pulled as shown, strand 1 of the chain supporting the movable pulley is shortened while strand 2 is lengthened. Because strand 1 is shortened by more than strand 2 is lengthened, the load is raised. Pulling down on the other strand of the free loop lowers the load. (a) Find the TMA of the differential pulley. (b) If the upper pulleys of such a hoist are 11 in. and 12 in. in diameter and the hoist is 75% efficient, how heavy a load can be raised by a force of 40 lb? (c) If the load is to be raised by 4 ft, how much chain must be pulled through the pulleys?

(a) The radius of the large upper pulley is R and that of the small one is r. In a complete rotation of the upper pulleys the input force F_{in} acts through a distance equal to the circumference of the large pulley, so $s_{\text{in}} = 2\pi R$. In this rotation, strand 1 has been shortened by $2\pi R$ while strand 2 has been lengthened by $2\pi r$, so the total decrease in length is $2\pi R - 2\pi r = 2\pi(R - r)$. Since the decrease in length is shared by both strands 1 and 2, the movable pulley is raised by half this amount and $s_{\text{out}} = \pi(R - r)$. The TMA of the system is accordingly

$$\text{TMA} = \frac{s_{\text{in}}}{s_{\text{out}}} = \frac{2\pi R}{\pi(R - r)} = \frac{2R}{R - r}$$

(b) Since $R = 6$ in. and $r = 5.5$ in., the TMA of the hoist is

$$\text{TMA} = \frac{2R}{R - r} = \frac{2 \times 6 \text{ in.}}{6 \text{ in.} - 5.5 \text{ in.}} = 24$$

At 75% efficiency,

$$\text{AMA} = \text{Eff} \times \text{TMA} = 0.75 \times 24 = 18$$

and so an applied force of 40 lb can lift a load of

$$F_{\text{out}} = \text{AMA} \times F_{\text{in}} = 18 \times 40 \text{ lb} = 720 \text{ lb}$$

(c) Given that $s_{\text{out}} = 4$ ft and TMA = 24, we have

$$s_{\text{in}} = \text{TMA} \times s_{\text{out}} = 24 \times 4 \text{ ft} = 96 \text{ ft}$$

12.8. A ramp 80 ft long slopes down 5 ft to the edge of a lake. How much force is needed to pull out an 800-lb boat using a 200-lb trailer if friction in the trailer's wheels reduces the efficiency to 90%?

The TMA of the ramp is

$$\text{TMA} = \frac{L}{h} = \frac{80 \text{ ft}}{5 \text{ ft}} = 16$$

and its AMA is

$$\text{AMA} = \text{Eff} \times \text{TMA} = 0.90 \times 16 = 14.4$$

The output force is the combined weight of boat and trailer, so the required input force is

$$F_{\text{in}} = \frac{F_{\text{out}}}{\text{AMA}} = \frac{1000 \text{ lb}}{14.4} = 69 \text{ lb}$$

12.9. The screw of a machinist's vise has a pitch of 3/16 in. and a handle 8 in. long. If the efficiency is 40%, how much force is developed between the jaws of the vise when a force of 10 lb is applied to the end of the handle?

The TMA of the vise is

$$\text{TMA} = \frac{2\pi L}{P} = \frac{2\pi \times 8 \text{ in.}}{3/16 \text{ in.}} = 268$$

and its AMA is

$$\text{AMA} = \text{Eff} \times \text{TMA} = 0.40 \times 268 = 107$$

The output force is therefore

$$F_{\text{out}} = F_{\text{in}} \times \text{AMA} = 10 \text{ lb} \times 107 = 1070 \text{ lb}$$

12.10. A screw jack has a pitch of 5 mm and a handle 60 cm long. If a force of 50 N is needed to raise a load of 700 kg, find the efficiency of the jack.

The TMA and AMA of the jack are respectively

$$\text{TMA} = \frac{2\pi L}{p} = \frac{2\pi \times 60 \text{ cm}}{0.5 \text{ cm}} = 754$$

$$\text{AMA} = \frac{F_{\text{out}}}{F_{\text{in}}} = \frac{mg}{F_{\text{in}}} = \frac{700 \text{ kg} \times 9.8 \text{ m/s}^2}{50 \text{ N}} = 137$$

The efficiency of the jack is

$$\text{Eff} = \frac{\text{AMA}}{\text{TMA}} = \frac{137}{754} = 0.18 = 18\%$$

12.11. A 1200-rpm motor is connected to a 12-in. diameter circular saw blade with a V-belt and a pair of step pulleys (Fig. 12-11). (a) If the pulley diameters are 4 in., 5 in., and 6 in. in each set, find the possible angular velocities of the saw blade in rpm. (b) Find the corresponding linear velocities of the saw's teeth in ft/min.

(a) The three possible ratios of pulley diameters are 2/3, 1, and 3/2. Since

$$\omega_{out} = \omega_{in}\frac{d_{in}}{d_{out}}$$

the three possible angular velocities of the saw blade are

$$\omega_1 = 1200 \text{ rpm} \times \tfrac{2}{3} = 800 \text{ rpm}$$
$$\omega_2 = 1200 \text{ rpm} \times 1 = 1200 \text{ rpm}$$
$$\omega_3 = 1200 \text{ rpm} \times \tfrac{3}{2} = 1800 \text{ rpm}$$

(b) The linear velocity v of the saw's teeth is related to the radius r of the saw blade and its angular velocity ω by the formula $v = \omega r$, where ω must be expressed in radian measure. Since there are 2π radians in a revolution, the above angular velocities can also be expressed as

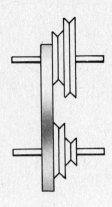

Fig. 12-11

$$\omega_1 = 2\pi\frac{\text{rad}}{\text{rev}} \times 800 \text{ rpm} = 5027 \text{ rad/min}$$

$$\omega_2 = 2\pi\frac{\text{rad}}{\text{rev}} \times 1200 \text{ rpm} = 7540 \text{ rad/min}$$

$$\omega_3 = 2\pi\frac{\text{rad}}{\text{rev}} \times 1800 \text{ rpm} = 11,310 \text{ rad/min}$$

The radius of the saw blade is $r = 6$ in. $= 0.5$ ft, and so the possible linear velocities of its teeth are

$$v_1 = \omega_1 r = 2514 \text{ ft/min} \qquad v_2 = \omega_2 r = 3770 \text{ ft/min} \qquad v_3 = \omega_3 r = 5655 \text{ ft/min}$$

12.12. An engine develops 8 hp at 4000 rpm. What gear ratio is needed if an output torque of 42 lb-ft is required?

Since

$$P = 8 \text{ hp} \times 550\frac{\text{ft-lb/s}}{\text{hp}} = 4400 \text{ ft-lb/s}$$

$$\omega = 4000 \text{ rpm} \times 0.105\frac{\text{rad/s}}{\text{rpm}} = 420 \text{ rad/s}$$

the engine itself develops a torque of

$$\tau = \frac{P}{\omega} = \frac{4400 \text{ ft-lb/s}}{420 \text{ rad/s}} = 10.5 \text{ lb-ft}$$

Hence the required gear ratio is

$$\frac{N_{out}}{N_{in}} = \frac{\tau_{out}}{\tau_{in}} = \frac{42 \text{ lb-ft}}{10.5 \text{ lb-ft}} = 4$$

Supplementary Problems

12.13. How great a force is needed to lift a 20-lb weight using the lever shown in Fig. 12-12?

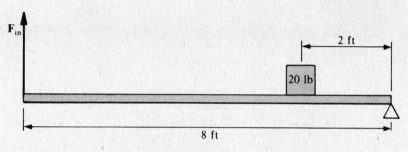

Fig. 12-12

12.14. A bolt is held in the jaws of a pair of pliers. If the bolt is 1 in. from the pivot and a force of 2 lb is applied to each handle 6 in. from the pivot, find the total compressive force on the bolt.

12.15. Find the TMA of the pulley system shown in Fig. 12-13.

12.16. Find the TMA of the pulley system shown in Fig. 12-14.

12.17. Find the TMA of the pulley system shown in Fig. 12-15.

12.18. What is the highest TMA that can be obtained with a system of two pulleys?

12.19. A chain hoist has upper pulleys 4 in. and 4.5 in. in radius. (*a*) If the hoist is 80% efficient, how much force is needed to raise a load of 200 lb? (*b*) How much chain must be pulled through the pulleys to raise the load by 2 ft?

12.20. A chain hoist is used to lift a 180-kg engine block by 1.2 m. To do this, a force of 100 N must be applied to the chain and 30 m of chain pulled through the system. What is the efficiency of the hoist?

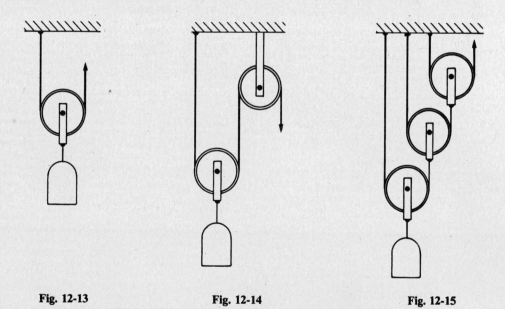

Fig. 12-13   Fig. 12-14   Fig. 12-15

12.21. A 400-lb piano on frictionless casters is to be pushed up a 20-ft ramp to a loading platform 3 ft above the ground. How much force is needed?

12.22. (*a*) What kind of simple machine does a screwdriver represent? (*b*) The handle of a screwdriver is 3 cm in diameter and its blade is 6 mm wide. What is its mechanical advantage?

12.23. A screwdriver whose handle is 1 in. in diameter is used to tighten a bolt which has 24 threads per inch. If the efficiency is 8%, find the force exerted on the nut when a force of 3 lb is applied to the handle.

12.24. A 600-rpm motor is used to operate an air compressor at 200 rpm through a V-belt drive. If the motor pulley is 15 cm in diameter, what should the diameter of the compressor pulley be?

12.25. The sprocket wheel on the rear axle of a certain bicycle is 4 in. in diameter and the sprocket wheel to which the pedals are attached is 8 in. in diameter. The wheels of the bicycle are 26 in. in diameter. At how many rpm must the pedals be turned in order to travel at 10 mi/hr?

Answers to Supplementary Problems

12.13.	5 lb	**12.18.**	3	**12.23.**	18 lb
12.14.	24 lb	**12.19.**	(*a*) 13.9 lb (*b*) 36 ft	**12.24.**	45 cm
12.15.	2	**12.20.**	71%	**12.25.**	64.8 rpm
12.16.	2	**12.21.**	60 lb		
12.17.	8	**12.22.**	(*a*) Wheel and axle (*b*) 5		

Chapter 13

Elasticity

STRESS AND STRAIN

The *stress* on a body acted upon by a deforming force is equal to the magnitude of the force F divided by the cross-sectional area A over which it acts. The unit of stress in the SI system is the N/m^2; in the British system it is customary to use the lb/in^2. The three categories of stress, *tension*, *compression*, and *shear*, are illustrated in Fig. 13-1.

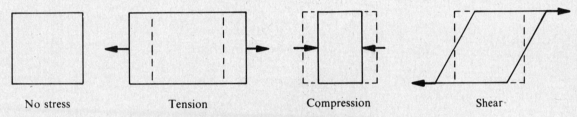

| No stress | Tension | Compression | Shear |

Fig. 13-1

The relative change in the size or shape of a body due to applied stress is called *strain*. Strain is a dimensionless quantity; for instance, the longitudinal strain that tension produces in a body is its change in length ΔL divided by its original length L_0, which is a pure number.

ELASTICITY

The *elastic limit* of a material is the maximum stress that can be applied to a body of it without causing a permanent deformation. For stresses below the elastic limit, the material exhibits *elastic* behavior: when the stress is removed, the body returns to its original size and shape.

Below the elastic limit, strain is found to be proportional to stress. This relationship is known as *Hooke's law*. In the case of tension, for example, doubling the applied force on a body will double the amount by which the body stretches. The *modulus of elasticity* of a material subjected to a particular kind of stress below its elastic limit is defined by the relationship

$$\text{Modulus of elasticity} = \frac{\text{stress}}{\text{strain}}$$

The *ultimate strength* of a material is the greatest stress it can withstand without rupture. In many materials the ultimate strength considerably exceeds the elastic limit. When a stress greater than its elastic limit but less than its ultimate strength is applied to such a material, the result is a permanent deformation. Bending a piece of metal is an example.

YOUNG'S MODULUS

When a tension or compression force F acts on an object of length L_0 and cross-sectional area A (Fig. 13-2), the result is a change in length ΔL. Below the elastic limit, the ratio between stress and strain in this situation is called *Young's modulus*:

$$Y = \frac{F/A}{\Delta L/L_0}$$

$$\text{Young's modulus} = \frac{\text{longitudinal stress}}{\text{longitudinal strain}}$$

The value of Young's modulus depends only on the composition of the object, not upon its size or shape. The usual units of Y are N/m^2 and lb/in^2.

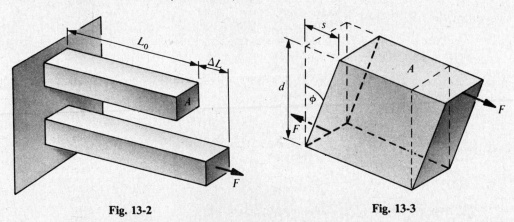

Fig. 13-2   Fig. 13-3

SHEAR MODULUS

A shear stress changes the shape of an object, not its volume. Figure 13-3 shows a rectangular block acted upon by shear forces F. The shearing stress is equal to F/A and the shearing strain is equal to the *angle of shear* ϕ, expressed in radians. Because ϕ is always small, it is very nearly the same as the ratio s/d between the displacement s of the block's faces and the distance d between these faces. Below the elastic limit, then, there are two equivalent expressions for the *shear modulus* (or *modulus of rigidity*):

$$s = \frac{F/A}{\phi} = \frac{F/A}{s/d}$$

$$\text{Shear modulus} = \frac{\text{shear stress}}{\text{shear strain}}$$

BULK MODULUS

When compressive forces act over the entire surface of a body, its volume decreases. If the compressive force per unit area F/A is uniform, the *bulk modulus* is given by

$$B = -\frac{F/A}{\Delta V/V_0}$$

$$\text{Bulk modulus} = -\frac{\text{volume stress}}{\text{volume strain}}$$

The minus sign is included because an increase in the volume stress leads to a decrease in the volume.

Volume stresses occur when objects are immersed in liquids, since a liquid exerts a uniform force perpendicular to any surface in its interior. As discussed in Chapter 16, the stress F/A exerted by a liquid is called *pressure p*, so that we can also write

$$B = -\frac{p}{\Delta V/V_0}$$

Solved Problems

13.1. A braided nylon rope 1 in. in diameter has a breaking strength of 28,500 lb.   Find the breaking strength of similar ropes (*a*) $\frac{1}{2}$ in. and (*b*) 2 in. in diameter.

Since the breaking stress F/A is the same for all the ropes, their breaking strengths F are in proportion to their cross-sectional areas A.   The cross-sectional area of a cylinder of diameter d is $A = \pi r^2 = \pi d^2/4$, and so in each case F varies directly with d^2.

(*a*)   A rope $\frac{1}{2}$ in. in diameter has an area $(\frac{1}{2})^2 = \frac{1}{4}$ that of a rope 1 in. in diameter, hence its breaking strength is $\frac{1}{4}$ as much, or 7125 lb.

(*b*)   A rope 2 in. in diameter has an area $2^2 = 4$ times that of a rope 1 in. in diameter, hence its breaking strength is 4 times as much, or 114,000 lb.

13.2. A wire 8 ft long with a cross-sectional area of 0.01 in^2 stretches by 0.05 in. when a weight of 100 lb is suspended from it.   Find the stress on the wire, the resulting strain, and the value of Young's modulus for the wire's material.

$$\text{Stress} = \frac{F}{A} = \frac{100 \text{ lb}}{0.01 \text{ in}^2} = 10^4 \text{ lb/in}^2$$

$$\text{Strain} = \frac{\Delta L}{L_0} = \frac{0.05 \text{ in.}}{96 \text{ in.}} = 5.2 \times 10^{-4}$$

$$Y = \frac{F/A}{\Delta L/L_0} = \frac{10^4 \text{ lb/in}^2}{5.2 \times 10^{-4}} = 1.92 \times 10^7 \text{ lb/in}^2$$

13.3. An aluminum wire 3 mm in diameter and 4 m long is used to support a mass of 50 kg.   What is the elongation of the wire?   Young's modulus for aluminum is 7×10^{10} N/m^2.

The cross-sectional area of a wire of radius $r = 1.5$ mm $= 1.5 \times 10^{-3}$ m is

$$A = \pi r^2 = \pi \times (1.5 \times 10^{-3} \text{ m})^2 = 7.07 \times 10^{-6} \text{ m}^2$$

The applied force is

$$F = mg = 50 \text{ kg} \times 9.8 \text{ m/s}^2 = 490 \text{ N}$$

and so the elongation of the wire is

$$\Delta L = \frac{L_0}{Y} \frac{F}{A} = \frac{4 \text{ m} \times 490 \text{ N}}{7 \times 10^{10} \text{ N/m}^2 \times 7.07 \times 10^{-6} \text{ m}^2} = 3.96 \times 10^{-3} \text{ m} = 3.96 \text{ mm}$$

13.4. The elastic limit of aluminum is 1.3×10^8 N/m^2.   What is the maximum mass the wire of Problem 13.3 can support without exceeding its elastic limit?

Since the cross-sectional area of the wire is $A = 7.07 \times 10^{-6}$ m^2 and

$$\left(\frac{F}{A}\right)_{\text{max}} = 1.3 \times 10^8 \text{ N/m}^2$$

the maximum force is

$$F_{\text{max}} = 1.3 \times 10^8 \text{ N/m}^2 \times 7.07 \times 10^{-6} \text{ m}^2 = 919 \text{ N}$$

which corresponds to a mass of

$$m = \frac{w}{g} = \frac{F_{\text{max}}}{g} = \frac{919 \text{ N}}{9.8 \text{ m/s}^2} = 94 \text{ kg}$$

13.5. A steel tube 12 ft long is used to support a sagging floor.   The inside diameter of the tube is 3 in., its outside diameter is 4 in., and $Y = 3 \times 10^7$ lb/in^2.   A sensitive strain gage indicates that

the tube's length decreases by 0.004 in.   What is the magnitude of the load the tube supports?

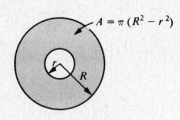

$A = \pi(R^2 - r^2)$

The area of the tube (Fig. 13-4) is

$$A = \pi(R^2 - r^2) = \pi\left[(2 \text{ in.})^2 - (1.5 \text{ in.})^2\right] = 5.5 \text{ in}^2$$

Since $L_0 = 12 \text{ ft} = 144 \text{ in.}$,

$$F = YA\frac{\Delta L}{L_0} = 3 \times 10^7 \frac{\text{lb}}{\text{in}^2} \times 5.5 \text{ in}^2 \times \frac{0.004 \text{ in.}}{144 \text{ in.}} = 4580 \text{ lb}$$

Fig. 13-4

13.6. By how much can a steel wire 3 m long and 2 mm in diameter be stretched before the elastic
limit is exceeded?   Young's modulus for the wire is 2×10^{11} N/m^2 and its elastic limit is
2.5×10^8 N/m^2.

The cross-sectional area of a wire of radius $r = 1 \text{ mm} = 10^{-3}$ m is

$$A = \pi r^2 = 3.14 \times 10^{-6} \text{ m}^2$$

The maximum force that can be applied without exceeding the elastic limit is therefore

$$F = \left(\frac{F}{A}\right)_{\text{max}} \times A = 2.5 \times 10^8 \text{ N/m}^2 \times 3.14 \times 10^{-6} \text{ m}^2 = 785 \text{ N}$$

When this force is applied, the wire will stretch by

$$\Delta L = \frac{L_0}{Y}\frac{F}{A} = \frac{3 \text{ m} \times 785 \text{ N}}{2 \times 10^{11} \text{ N/m}^2 \times 3.14 \times 10^{-6} \text{ m}^2} = 3.75 \times 10^{-3} \text{ m} = 3.75 \text{ mm}$$

13.7. A steel cable whose cross-sectional area is 1 in^2 is used to support an elevator cab weighing 5000
lb.   If the stress in the cable is not to exceed 20% of the cable's elastic limit of 40,000 lb/in^2,
find the maximum permissible upward acceleration.

The force that corresponds to a maximum stress in the cable of $0.20 \times 40,000 \text{ lb} = 8000 \text{ lb}$ is

$$F = \left(\frac{F}{A}\right)_{\text{max}} \times A = 8000 \text{ lb/in}^2 \times 1 \text{ in}^2 = 8000 \text{ lb}$$

This force is to equal the weight w of the cab plus the force ma that provides it with an upward
acceleration, so that

$$F = w + ma \qquad a = \frac{F - w}{m}$$

Since $m = w/g$,

$$a = \frac{g(F - w)}{w} = \frac{32 \text{ ft/s}^2(8000 \text{ lb} - 5000 \text{ lb})}{5000 \text{ lb}} = 19.2 \text{ ft/s}^2$$

13.8. How much force is required to punch a hole $\frac{1}{2}$ in. in diameter in a steel sheet $\frac{1}{8}$ in. thick whose
shearing strength is 4×10^4 lb/in^2?

The shear stress is exerted over the cylindrical surface that is
the boundary of the hole (Fig. 13-5).   The area of this surface is

$$A = 2\pi rh = 2\pi \times 0.25 \text{ in.} \times 0.125 \text{ in.} = 0.196 \text{ in}^2$$

Since the minimum shear stress needed to rupture the steel is

$$\left(\frac{F}{A}\right)_{\text{min}} = 4 \times 10^4 \frac{\text{lb}}{\text{in}^2}$$

the required force is

$$F = \left(\frac{F}{A}\right)_{\text{min}} \times A = 4 \times 10^4 \frac{\text{lb}}{\text{in}^2} \times 0.196 \text{ in}^2 = 7840 \text{ lb}$$

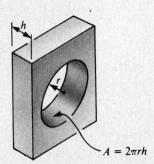

$A = 2\pi rh$

Fig. 13-5

13.9. The shearing strength of a certain steel alloy is 3.5×10^4 lb/in². Two 3/16-in.-diameter bolts of this alloy are used to fasten a bracket to a wall. What is the maximum load the bracket can support without shearing off the bolts?

The shearing stress here is exerted perpendicular to each bolt, so for each bolt $A = \pi r^2$ and

$$F = \left(\frac{F}{A}\right)_{max} \times A = 3.5 \times 10^4 \frac{\text{lb}}{\text{in}^2} \times \pi \left(\frac{3}{32} \text{ in.}\right)^2 = 966 \text{ lb}$$

The combined load is twice this, or 1932 lb.

13.10. The pressure at a depth of 1000 ft in the ocean exceeds sea-level atmospheric pressure by 444 lb/in². By how much does the volume of a 50-in³ aluminum object contract when it is lowered to this depth? The bulk modulus of aluminum is 10^7 lb/in².

$$\Delta V = -\frac{pV_0}{B} = \frac{-444 \text{ lb/in}^2 \times 50 \text{ in}^3}{10^7 \text{ lb/in}^2} = -2.22 \times 10^{-3} \text{ in}^3$$

The minus sign means that the volume of the object decreases.

13.11. The reciprocal of the bulk modulus B of a liquid is called its *compressibility* k, so that $k = 1/B$. The bulk modulus of water is 2.3×10^9 N/m². (*a*) Find its compressibility per atmosphere of pressure, where 1 atm $= 1.013 \times 10^5$ N/m² is the pressure exerted by the earth's atmosphere at sea level. (*b*) How much pressure in atm is needed to compress a sample of water by 0.1%?

(*a*) In terms of atm, the bulk modulus of water is

$$B = \frac{2.3 \times 10^9 \text{ N/m}^2}{1.013 \times 10^5 \text{ (N/m}^2)/\text{atm}} = 2.27 \times 10^4 \text{ atm}$$

and so its compressibility is

$$k = \frac{1}{B} = \frac{1}{2.27 \times 10^4 \text{ atm}} = 4.4 \times 10^{-5} \text{ atm}^{-1}$$

(*b*) Here $\Delta V/V_0 = -0.1\% = -0.001$, hence the required pressure is

$$p = -\frac{1}{k} \frac{\Delta V}{V_0} = \frac{0.001}{4.4 \times 10^{-5} \text{ atm}^{-1}} = 23 \text{ atm}$$

Supplementary Problems

13.12. A cable is shortened to half its original length. (*a*) How does this affect its elongation under a given load? (*b*) How does this affect the maximum load it can support without exceeding its elastic limit?

13.13. A cable is replaced by another one of the same length and material but of twice the diameter. (*a*) How does this affect its elongation under a given load? (*b*) How does this affect the maximum load it can support without exceeding its elastic limit?

13.14. A wire 5 m long and 4 mm in diameter supports a load of 80 kg. If the wire stretches by 2.6 mm, find the value of Young's modulus for its material.

13.15. A steel wire ($Y = 2.9 \times 10^7$ lb/in²) 6 ft long and 0.002 in² in cross section supports a load of 15 lb. (*a*) What is its elongation? (*b*) What would its elongation be if the load were doubled to 30 lb?

13.16. A brass cube 1 in. on each edge is held in the jaws of a vise with a force of 800 lb. By how much is the cube compressed? Young's modulus for brass is 1.3×10^7 lb/in².

13.17. The ultimate strength in tension of a certain type of steel is 7×10^4 lb/in^2. What is the maximum tension a rod made of this steel and 1 in. in diameter can withstand?

13.18. An elevator cab weighing 3000 lb is designed for a maximum upward acceleration of 12 ft/s^2. If the stress in its cable is not to exceed 6000 lb/in^2, what should the cable diameter be?

13.19. Two steel beams are riveted together to form a single longer beam. Eight rivets 0.4 in. in diameter are used. If a tension force of 6 tons is applied to the new beam, what is the shearing stress on the rivets? How does this compare with their shearing strength of 4×10^4 lb/in^2?

13.20. Find the force needed to punch a hole 1 in. square in a steel sheet 0.05 in. thick whose shearing strength is 5×10^4 lb/in^2.

13.21. When a pressure of 300 lb/in^2 is applied to a mercury sample, it contracts by 0.008%. Find the bulk modulus of mercury.

Answers to Supplementary Problems

13.12. (*a*) Its elongation will be half the former amount.   (*b*) No change.

13.13. (*a*) Its elongation will be one-quarter the former amount.   (*b*) The maximum load will be four times greater.

13.14. 1.2×10^{11} N/m^2

13.15.   (*a*) 0.0186 in.   (*b*) 0.0372 in.

13.16. 6.15×10^{-5} in.

13.17.   55,000 lb

13.18.   0.936 in^2

13.19. 1.2×10^4 lb/in^2; 30%

13.20.   10,000 lb

13.21. 3.8×10^6 lb/in^2

Chapter 14

Simple Harmonic Motion

RESTORING FORCE

When an elastic object such as a spring is stretched or compressed, a *restoring force* appears that tries to return the object to its normal length. It is this restoring force that must be overcome by the applied force in order to deform the object. From Hooke's law, the restoring force F is proportional to the displacement s provided the elastic limit is not exceeded. Hence

$$F_r = -ks$$

Restoring force = $-$(force constant $\times$ displacement)

The minus sign is required because the restoring force acts in the opposite direction to the displacement. The greater the value of the *force constant k*, the greater the restoring force for a given displacement and the greater the applied force $F = ks$ needed to produce the displacement.

ELASTIC POTENTIAL ENERGY

Because work must be done by an applied force to stretch or compress an object, it has *elastic potential energy* as a result, where

$$PE = \tfrac{1}{2}ks^2$$

When a deformed elastic object is released, its elastic potential energy turns into kinetic energy or into work done on something else.

SIMPLE HARMONIC MOTION

In *periodic motion*, a body repeats a certain motion indefinitely, always returning to its starting point after a constant time interval and then starting a new cycle. *Simple harmonic motion* is periodic motion that occurs when the restoring force on a body displaced from an equilibrium position is proportional to the displacement and in the opposite direction. A mass m attached to a spring executes simple harmonic motion when the spring is pulled out and released. The spring's PE becomes KE as the mass begins to move, and the KE of the mass becomes PE again as its momentum causes the spring to overshoot the equilibrium position and become compressed (Fig. 14-1).

The *amplitude A* of a body undergoing simple harmonic motion is the maximum value of its displacement on either side of the equilibrium position.

PERIOD AND FREQUENCY

The *period T* of a body undergoing simple harmonic motion is the time involved in each complete cycle; T is independent of the amplitude A. If the acceleration of the body is a when its displacement is s,

$$T = 2\pi\sqrt{-\frac{s}{a}}$$

$$\text{Period} = 2\pi\sqrt{-\frac{\text{displacement}}{\text{acceleration}}}$$

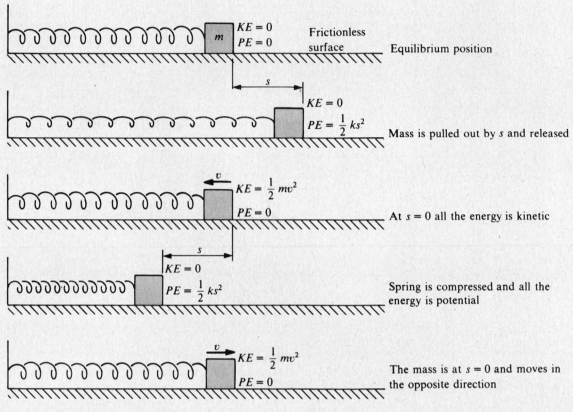

Fig. 14-1

In the case of a body of mass m attached to a spring of force constant k, $F_r = -ks = ma$ and so $-s/a = m/k$. Hence

$$T = 2\pi \sqrt{\frac{m}{k}} \qquad \text{(stretched spring)}$$

The *frequency* f of a body undergoing simple harmonic motion is the number of cycles per second it executes, so that

$$f = \frac{1}{T}$$

$$\text{Frequency} = \frac{1}{\text{period}}$$

The unit of frequency is the *hertz* (Hz), where 1 Hz = 1 cycle/s.

DISPLACEMENT, VELOCITY, ACCELERATION

If $t = 0$ when a body undergoing simple harmonic motion is in its equilibrium position of $s = 0$ and is moving in the direction of increasing s, then at any time t thereafter its displacement is

$$s = A \sin 2\pi ft$$

Often this formula is written

$$s = A \sin \omega t$$

where $\omega = 2\pi f$ is the *angular frequency* of the motion in rad/s.   (If instead the body is at $s = +A$ when $t = 0$, then $s = A \cos 2\pi ft = A \cos \omega t$.) Figure 14-2 is a graph of s vs. t.

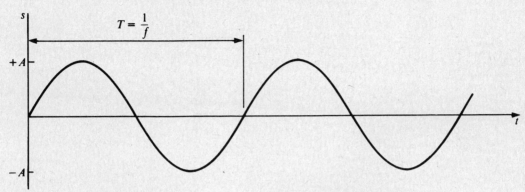

Fig. 14-2

The velocity of the body at the time t is

$$v = 2\pi fA \cos 2\pi ft = \omega A \cos \omega t$$

When v is $+$, the body is moving in the direction of increasing s; when v is $-$, it is moving in the direction of decreasing s.   In terms of the displacement s the magnitude of the velocity is

$$v = 2\pi f \sqrt{A^2 - s^2}$$

The acceleration of the body at the time t is

$$a = -4\pi^2 f^2 A \sin 2\pi ft = -\omega^2 A \sin \omega t$$

In terms of the displacement s the acceleration is

$$a = 4\pi^2 f^2 s$$

PENDULUMS

A *simple pendulum* has its entire mass concentrated at the end of a string, as in Fig. 14-3(*a*), and undergoes simple harmonic motion provided that the arc through which it travels is only a few degrees. The period of a simple pendulum of length L is

$$T = 2\pi \sqrt{\frac{L}{g}} \qquad \text{(simple pendulum)}$$

The *physical pendulum* of Fig. 14-3(*b*) is an object of any kind that is pivoted so that it can oscillate freely.   If the moment of inertia of the object about the pivot O is I, its mass is m, and the distance

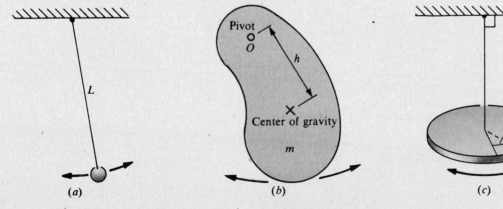

Fig. 14-3

from its center of gravity to the pivot is h, then its period is

$$T = 2\pi\sqrt{\frac{I}{mgh}} \qquad \text{(physical pendulum)}$$

A *torsion pendulum* consists of an object suspended by a wire or thin rod, as in Fig. 14-3(c), which undergoes rotational simple harmonic oscillations. Form Hooke's law, the torque τ needed to twist the object through an angle θ is given by

$$\tau = K\theta$$

provided the elastic limit is not exceeded, where K is a constant that depends upon the material and dimensions of the wire. If I is the moment of inertia of the object about its point of suspension, the period of the oscillations is

$$T = 2\pi\sqrt{\frac{I}{K}} \qquad \text{(torsion pendulum)}$$

Solved Problems

14.1. The amplitude of a simple harmonic oscillator is doubled. How does this affect (a) the period, (b) the total energy, and (c) the maximum velocity of the oscillator?

(a) The period of such an oscillator does not depend on the amplitude of its motion, hence T is unchanged.

(b) The total energy of the oscillator is equal to the elastic potential energy $\frac{1}{2}kA^2$ at either extreme of its motion, when $v = 0$. Hence doubling A means that the total energy increases fourfold.

(c) The maximum velocity occurs at $s = 0$, the equilibrium position, when the entire energy of the oscillator is KE. Since $KE = \frac{1}{2}mv^2$ and the total energy is four times greater than originally, $v_{\max}$ must double.

14.2. (a) A pendulum clock is in an elevator that descends at a constant velocity. Does it keep correct time? (b) The same clock is in an elevator in free fall. Does it keep correct time?

(a) The motion of the pendulum bob is not affected by motion of its support at constant velocity, so the clock keeps correct time.

(b) In free fall the pendulum's support has the same downward acceleration of g as the bob, so no oscillations occur and the clock does not operate at all.

14.3. A force of 1 lb compresses a spring by 2 in. (a) Find the force constant of the spring. (b) Find the elastic potential energy of the compressed spring.

(a)
$$k = \frac{F}{s} = \frac{1\ \text{lb}}{2\ \text{in.}} \times 12\ \frac{\text{in.}}{\text{ft}} = 6\ \frac{\text{lb}}{\text{ft}}$$

(b)
$$\text{PE} = \tfrac{1}{2}ks^2 = \tfrac{1}{2} \times 6\ \frac{\text{lb}}{\text{ft}} \times \left(\frac{2\ \text{in.}}{12\ \text{in/ft}}\right)^2 = 0.083\ \text{ft-lb}$$

14.4. A spring with a force constant of 12 lb/ft is compressed by 3 in. A 0.1-lb ball is placed against the end of the spring, which is then released. What is the ball's velocity when it leaves the spring?

The kinetic energy of the ball equals the elastic potential energy of the compressed spring.   Hence

$$KE = PE$$

$$\tfrac{1}{2}mv^2 = \tfrac{1}{2}\,\frac{w}{g}\,v^2 = \tfrac{1}{2}ks^2$$

$$v^2 = \frac{kgs^2}{w}$$

$$v = \sqrt{\frac{kg}{w}}\; s = \sqrt{\frac{12\ \text{lb/ft} \times 32\ \text{ft/s}^2}{0.1\ \text{lb}}} \times 0.25\ \text{ft} = 15.5\ \text{ft/s}$$

14.5. A 100-g object is suspended from a spring whose force constant is 50 N/m. (a) By how much does the spring stretch? (b) What is the period of oscillation of the system? (c) What is its frequency?

(a)   Here $F = mg = 0.1\ \text{kg} \times 9.8\ \text{m/s}^2 = 0.98\ \text{N}$, and so the spring stretches by

$$s = \frac{F}{k} = \frac{0.98\ \text{N}}{50\ \text{N/m}} = 0.0196\ \text{m} = 1.96\ \text{cm}$$

(b)
$$T = 2\pi\sqrt{\frac{m}{k}} = 2\pi\sqrt{\frac{0.1\ \text{kg}}{50\ \text{N/m}}} = 0.281\ \text{s}$$

(c)
$$f = \frac{1}{T} = \frac{1}{0.281\ \text{s}} = 3.56\ \text{s}^{-1} = 3.56\ \text{Hz}$$

14.6. An object of unknown mass is suspended from a spring, which stretches by 10 cm as a result. If the system is set in oscillation, what will its frequency be?

The force F that causes the spring to stretch by $s = 10\ \text{cm} = 0.1\ \text{m}$ is the weight mg of the unknown mass, so that the force constant of the spring is

$$k = \frac{F}{s} = \frac{mg}{0.1\ \text{m}} = 10mg\ \text{m}^{-1}$$

The period of oscillation of the system is therefore

$$T = 2\pi\sqrt{\frac{m}{k}} = 2\pi\sqrt{\frac{m}{10mg\ \text{m}^{-1}}} = 2\pi\sqrt{\frac{1}{10\ \text{m}^{-1} \times 9.8\ \text{m/s}^2}} = 0.635\ \text{s}$$

and the frequency is $f = 1/T = 1.58\ \text{Hz}$.

14.7. An object is oscillating in simple harmonic motion with an amplitude of 6 in. and a period of 2 s. Find the magnitudes of its velocity and acceleration when its displacement from the equilibrium position is (a) $s = 0$, (b) $s = +3$ in., and (c) $s = -6$ in.

(a)   The frequency of the oscillations is $f = 1/T = 0.5\ \text{s}^{-1}$. At $s = 0$,

$$v = 2\pi f\sqrt{A^2 - s^2} = 2\pi fA = 2\pi \times 0.5\ \text{s}^{-1} \times 0.5\ \text{ft} = 1.57\ \text{ft/s}$$

$$a = -4\pi^2 f^2 s = 0$$

(b)   At $s = +3$ in. $= +0.25$ ft,

$$v = 2\pi f\sqrt{A^2 - s^2} = 2\pi \times 0.5\ \text{s}^{-1} \times \sqrt{(0.5\ \text{ft})^2 - (0.25\ \text{ft})^2} = 1.36\ \text{ft/s}$$

$$a = -4\pi^2 f^2 s = -4\pi^2 \times (0.5\ \text{s}^{-1})^2 \times 0.25\ \text{ft} = -2.47\ \text{ft/s}^2$$

(c)   At $s = -6$ in. $= -0.5$ ft $= -A$,

$$v = 2\pi f\sqrt{A^2 - s^2} = 0$$

$$a = -4\pi^2 f^2 s = -4\pi^2 \times (0.5\ \text{s}^{-1})^2 \times (-0.5\ \text{ft}) = +4.93\ \text{ft/s}^2$$

14.8. What is the maximum velocity of a body undergoing simple harmonic motion? At what displacement does this velocity occur?

For a body in simple harmonic motion, $v = 2\pi f \sqrt{A^2 - s^2}$, which is a maximum when $s = 0$. Hence

$$v_{max} = 2\pi f \sqrt{A^2 - 0} = 2\pi f A$$

This formula only gives the magnitude of v_{max}.

14.9. What is the maximum acceleration of a body undergoing simple harmonic motion? At what displacement does this acceleration occur?

For a body in simple harmonic motion, $a = -4\pi^2 f^2 s$, which is a maximum when $s = \pm A$, where A is the amplitude. Hence

$$a_{max} = \pm 4\pi^2 f^2 A$$

The acceleration is negative when $s = +A$ and the acceleration is positive when $s = -A$.

14.10. Each piston of a certain car engine weighs 2 lb and has a "stroke" (total travel distance) of 4 in. When the engine is operating at 3000 rpm, find (a) the maximum velocity of each piston, (b) its maximum acceleration, and (c) the force on it.

(a) The frequency of oscillation of each piston is 3000 cycles/min, which is

$$f = \frac{3000 \text{ cycles/min}}{60 \text{ s/min}} = 50 \text{ Hz}$$

The amplitude of the motion is half the total travel, so $A = 2$ in. $= 0.167$ ft. Hence the maximum velocity is

$$v_{max} = 2\pi f A = 2\pi \times 50 \text{ s}^{-1} \times 0.167 \text{ ft} = 52.4 \text{ ft/s}$$

(b) The maximum acceleration is

$$a_{max} = 4\pi^2 f^2 A = 4\pi^2 \times (50 \text{ s}^{-1})^2 \times 0.167 \text{ ft} = 1.65 \times 10^4 \text{ ft/s}^2$$

(c) The mass of each piston is $m = w/g = 2 \text{ lb}/(32 \text{ ft/s}^2) = 0.0625$ slug, and the force on it is therefore

$$F = ma = 0.0625 \text{ slug} \times 1.65 \times 10^4 \text{ ft/s}^2 = 1030 \text{ lb}$$

14.11. A lamp is suspended from a high ceiling with a cord 12 ft long. Find its period of oscillation.

$$T = 2\pi \sqrt{\frac{L}{g}} = 2\pi \sqrt{\frac{12 \text{ ft}}{32 \text{ ft/s}^2}} = 3.85 \text{ s}$$

14.12. Find the length in meters of a simple pendulum whose period is 2 s.

The first step is to solve the formula $T = 2\pi\sqrt{L/g}$ for L. We proceed as follows:

$$T^2 = \frac{4\pi^2 L}{g} \qquad L = \frac{gT^2}{4\pi^2}$$

Now we substitute $g = 9.8 \text{ m/s}^2$ and $T = 2$ s, and obtain

$$L = \frac{9.8 \text{ m/s}^2 \times (2 \text{ s})^2}{4\pi^2} = 0.993 \text{ m}$$

14.13. A broomstick 1.5 m long is suspended from one end and set in oscillation. (a) What is the period of oscillation? (The moment of inertia of a thin rod pivoted at one end is $I = \frac{1}{3}mL^2$.) (b) What would be the length of a simple pendulum with the same period?

(a) The distance h from the pivot to the center of gravity of the broomstick is $L/2$. Hence

$$T = 2\pi\sqrt{\frac{I}{mgh}} = 2\pi\sqrt{\frac{mL^2/3}{mgL/2}} = 2\pi\sqrt{\frac{2L}{3g}} = 2\pi\sqrt{\frac{2 \times 1.5 \text{ m}}{3 \times 9.8 \text{ m/s}^2}} = 2.01 \text{ s}$$

(b) From the solution to Problem 14.12,

$$L = \frac{gT^2}{4\pi^2} = \frac{9.8 \text{ m/s}^2 \times (2.01 \text{ s})^2}{4\pi^2} = 1.0 \text{ m}$$

14.14. A certain torsion pendulum consists of a 6-lb aluminum disk 1 ft in diameter that is suspended from its center by a wire. When a torque of 1 lb-ft is applied to the disk, it rotates through 15°. Find the frequency of oscillation of the disk.

Since 1° = 0.01745 rad, 15° = 0.262 rad, and the torque constant of the suspension wire is

$$K = \frac{\tau}{\theta} = \frac{1 \text{ lb-ft}}{0.262 \text{ rad}} = 3.82 \text{ lb-ft/rad}$$

From Fig. 11-3 the moment of inertia of the disk is

$$I = \tfrac{1}{2}MR^2 = \frac{1}{2}\frac{w}{g} R^2 = \frac{6 \text{ lb} \times (0.5 \text{ ft})^2}{2 \times 32 \text{ ft/s}^2} = 0.0234 \text{ slug-ft}^2$$

The period of oscillation is therefore

$$T = 2\pi\sqrt{\frac{I}{K}} = 2\pi\sqrt{\frac{0.0234 \text{ slug-ft}^2}{3.82 \text{ lb-ft/rad}}} = 0.492 \text{ s}$$

The frequency is $f = 1/T = 1/0.492 \text{ s} = 2.03$ Hz.

14.15. In order to determine its moment of inertia about a diameter, a brass hoop is suspended by a wire whose torsion constant is $K = 25$ N-m/rad. The hoop executes 6 oscillations per second. What is the moment of inertia?

From the formula $T = 2\pi\sqrt{I/K}$ for the period of a torsion pendulum we obtain $I = T^2K/4\pi^2$. Since $T = 1/f$, the moment of inertia of the hoop is

$$I = \frac{K}{4\pi^2 f^2} = \frac{25 \text{ N-m/rad}}{4\pi^2 \times (6 \text{ s}^{-1})^2} = 0.0176 \text{ kg-m}^2$$

Supplementary Problems

14.16. What change in mass is required to double the frequency of a harmonic oscillator?

14.17. A 1-lb object is dropped on a vertical spring from a height of 7 ft. If the maximum compression of the spring is 4 in., find its force constant.

14.18. A toy gun uses a spring whose force constant is 100 N/m to propel an 8-g rubber pellet. If the spring is compressed by 5 cm when the trigger is pulled, what is the pellet's initial velocity?

14.19. A 5-lb object is suspended from a spring whose force constant is 40 lb/ft. Find (a) the amount by which the spring is stretched, (b) the elastic potential energy of the stretched spring, and (c) the period of oscillation of the system.

14.20. A 30-g mass is suspended from a spring whose force constant is 20 N/m. Find (a) the amount by which the spring is stretched, (b) the elastic potential energy of the stretched spring, and (c) the period of oscillation of the system.

14.21. A 200-lb portable gasoline-powered generating set is placed on four springs, which are depressed by 0.5 in. What is the natural frequency of vibration of this system?

14.22. An object is oscillating in simple harmonic motion with an amplitude of 1 cm and a period of 0.2 s. Find the magnitudes of its velocity and acceleration when its displacement from the equilibrium position is 0.3 cm.

14.23. The piston of a refrigeration compressor weighs 1 lb and has a stroke of 3 in. When the compressor is operated at 600 rpm, find the piston's maximum velocity and acceleration and the maximum force acting on it.

14.24. Find the frequency of a simple pendulum 20 cm long.

14.25. What is the length in inches of a simple pendulum whose period is 1 s?

14.26. A pendulum 1.00 m long oscillates 30.0 times per minute in a certain location. What is the value of g there?

14.27. The prongs of a tuning fork vibrate at 440 Hz (the musical note A) with an amplitude of 0.5 mm at the tips. Find the maximum velocity of the tips.

14.28. A uniform steel girder 20 ft long is suspended from one end. What is the period of its oscillations?

14.29. A 24-kg wooden sphere whose diameter is 40 cm is suspended by a wire. A torque of 0.5 N-m is found to rotate the sphere through 10°. Find the period of oscillation of the sphere.

14.30. A disk whose moment of inertia is 0.5 slug-ft^2 is suspended by a thin rod. When the disk is turned through a few degrees and then released, it oscillates back and forth twice per second. Find the torsion constant of the rod.

Answers to Supplementary Problems

14.16. The mass must be reduced to $\frac{1}{4}$ its original value.

14.17. 126 lb/ft

14.18. 5.6 m/s

14.19. (a) 0.125 ft   (b) 0.3125 ft-lb   (c) 0.393 s

14.20. (a) 1.47 cm   (b) 2.16×10^{-3} J   (c) 0.24 s

14.21. 2.21 Hz

14.22. 0.300 m/s; 2.96 m/s^2

14.23. 7.85 ft/s; 493 ft/s^2; 15.4 lb

14.24. 1.11 Hz

14.25. 9.73 in.

14.26. 9.87 m/s^2

14.27. 0.69 m/s

14.28. 7.33 s

14.29. 2.30 s

14.30. 79 N-m/rad

Chapter 15

Waves and Sound

WAVES

A *wave* is, in general, a disturbance that moves through a medium. A wave carries energy, but there is no transport of matter. In a *periodic wave*, pulses of the same kind follow one another in regular succession.

In a *transverse wave*, the particles of the medium move back and forth perpendicular to the direction of the wave. Waves that travel down a stretched string when one end is shaken are transverse (Fig. 15-1).

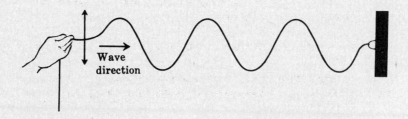

Fig. 15-1

In a *longitudinal wave*, the particles of the medium move back and forth in the same direction as the wave. Waves that travel down a coil spring when one end is pulled out and released are longitudinal (Fig. 15-2).

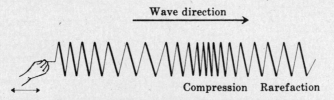

Fig. 15-2

Water waves are a combination of longitudinal and transverse waves. Each particle near the surface moves in a circular orbit, as shown in Fig. 15-3, so that a succession of crests and troughs occurs. At a crest, the surface water moves in the direction of the wave; at a trough, it moves in the opposite direction. As in all types of wave motion, there is no net movement of matter from one place to another.

FREQUENCY AND WAVELENGTH

The *frequency f* of a periodic wave is the number of waves that pass a given point per second. The unit of frequency is the *hertz* (Hz), where

$$1 \text{ Hz} = 1 \text{ wave/s}$$

Multiples of the hertz are the kilohertz (kHz) and the megahertz (MHz), equal respectively to 10^3 Hz and 10^6 Hz.

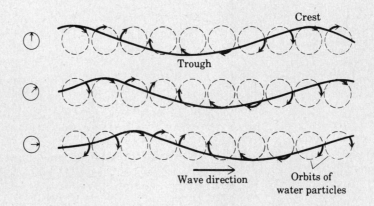

Fig. 15-3

The *wavelength* λ (Greek letter *lambda*) of a periodic wave is the distance between adjacent wave crests (Fig. 15-4). Frequency and wavelength are related to wave velocity by the formula

$$v = f\lambda$$

Wave velocity = frequency × wavelength

AMPLITUDE AND INTENSITY

The *amplitude A* of a wave is the maximum displacement of the particles of the medium through which the wave passes on either side of their equilibrium positions. In a transverse wave, the amplitude is half the distance between the top of a crest and the bottom of a trough (Fig. 15-4).

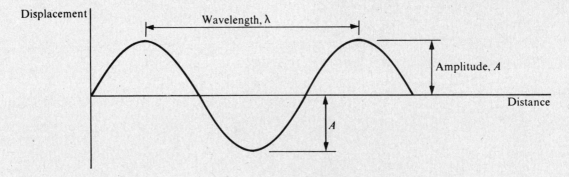

Fig. 15-4

The *intensity I* of a wave is the rate at which it transports energy per unit area perpendicular to its direction of motion. The intensity of a mechanical wave (one that involves moving matter, in contrast to, say, an electromagnetic wave) is proportional to f^2, the square of its frequency, and to A^2, the square of its amplitude.

SOUND

Sound waves are longitudinal waves in which alternate regions of compression and rarefaction move away from a source. Sound waves can travel through solids, liquids, and gases. The velocity of sound is a constant for a given material at a given pressure and temperature; in air at 1 atm pressure and 0 °C it is 331 m/s = 1086 ft/s.

When sound waves spread out uniformly in space, their intensity decreases inversely with the square of the distance R from their source. Thus if the intensity of a certain sound is I_1 at the distance R_1, its intensity I_2 at the distance R_2 can be found from the formula

$$\frac{I_2}{I_1} = \frac{R_1^2}{R_2^2}$$

The response of the human ear to sound intensity is not proportional to the intensity, so that doubling the actual intensity of a certain sound does not lead to the sensation of a sound twice as loud but only of one that is slightly louder than the original. For this reason the *decibel* (db) scale is used for sound intensity. An intensity of 10^{-12} W/m^2, which is just audible, is given the value 0 db; a sound 10 times more intense is given the value 10 db; a sound 10^2 times more intense than 0 db is given the value 20 db; a sound 10^3 times more intense than 0 db is given the value 30 db; and so forth. More formally, the intensity I(db) of a sound wave whose intensity is I in W/m^2 is given by

$$I(\text{db}) = 10 \log \frac{I}{I_0}$$

where $I_0 = 10^{-12}$ W/m^2. Normal conversation might be 60 db, city traffic noise might be 90 db, and a jet aircraft might produce as much as 140 db (which produces damage to the ear) at a distance of 100 ft.

Solved Problems

15.1. As a phonograph record turns, a certain groove passes the needle at 25 cm/s. If the wiggles in the groove are 0.1 mm apart, what is the frequency of the sound that results?

Here the wavelength of the wiggles is $\lambda = 0.1$ mm $= 10^{-4}$ m, so they pass the needle at the rate of

$$f = \frac{v}{\lambda} = \frac{0.25 \text{ m/s}}{10^{-4} \text{ m}} = 2500 \text{ Hz}$$

This is therefore the frequency of the sound waves that are produced by the electronic system of the record player.

15.2. The velocity of sound in seawater is 5020 ft/s. Find the wavelength in seawater of a sound wave whose frequency is 256 Hz.

$$\lambda = \frac{v}{f} = \frac{5020 \text{ ft/s}}{256 \text{ Hz}} = 19.6 \text{ ft}$$

15.3. An anchored boat is observed to rise and fall through a total range of 2 m once every 4 s as waves whose crests are 30 m apart pass by it. Find (*a*) the frequency of the waves, (*b*) their velocity, (*c*) their amplitude, and (*d*) the velocity of an individual water particle at the surface.

(*a*)
$$f = \frac{1}{T} = \frac{1}{4 \text{ s}} = 0.25 \text{ Hz}$$

(*b*)
$$v = f\lambda = 0.25 \text{ Hz} \times 30 \text{ m} = 7.5 \text{ m/s}$$

(*c*) The amplitude is half the total range, hence $A = 1$ m.

(d) As each wave passes by, the water particles at the surface move in circular orbits of radius $r = A = 1$ m (see Fig. 15-3). The circumference of such an orbit is

$$s = 2\pi r = 2\pi \times 1 \text{ m} = 6.28 \text{ m}$$

The waves have the period 4 s, which means that each surface water particle must move through its 6.28-m orbit in 4 s. The velocity of such a water particle is therefore

$$V = \frac{s}{T} = \frac{6.28 \text{ m}}{4 \text{ s}} = 1.57 \text{ m/s}$$

It is worth noting that the *wave* velocity here is 7.5 m/s, nearly 5 times greater. This signifies that the motion of a wave can be much faster than the motions of the individual particles of the medium in which the wave travels.

15.4. A tuning fork vibrating at 300 Hz is placed in a tank of water. (a) Find the frequency and wavelength of the sound waves in the water. (b) Find the frequency and wavelength of the sound waves produced in the air above the tank by the vibrations of the water surface. The velocity of sound is 4913 ft/s in water and 1086 ft/s in air.

(a) In the water, the frequency of the sound waves is the 300 Hz of their source and their wavelength is

$$\lambda_1 = \frac{v_1}{f} = \frac{4913 \text{ ft/s}}{300 \text{ Hz}} = 16.4 \text{ ft}$$

(b) In the air, the frequency of the sound waves is the same as the frequency of their source, which is the vibrating water surface. Hence $f = 300$ Hz. The wavelength is different, however:

$$\lambda_2 = \frac{v_2}{f} = \frac{1086 \text{ ft/s}}{300 \text{ Hz}} = 3.6 \text{ ft}$$

15.5. How many times more intense is a 50-db sound than a 40-db sound? Than a 20-db sound?

Each interval of 10 db represents a change in sound intensity by a factor of 10. Hence a 50-db sound is 10 times more intense than a 40-db sound and $10 \times 10 \times 10 = 1000$ times more intense than a 20-db sound.

15.6. What is the intensity in W/m² of the 70-db noise of a truck passing by?

An intensity of 0 db is equivalent to 10^{-12} W/m². Since a sound of 70 db is 10^7 times more intense, it is equivalent to a rate of energy flow of

$$I = 10^7 \times 10^{-12} \text{ W/m}^2 = 10^{-5} \text{ W/m}^2$$

15.7. A certain person speaking normally produces a sound intensity of 40 db at a distance of 3 ft. If the threshold intensity for reasonable audibility is 20 db, how far away can he be heard clearly?

A change of 20 db in sound intensity is equivalent to a ratio of $10 \times 10 = 100$. Hence

$$\frac{I_2}{I_1} = \frac{R_1^2}{R_2^2}$$

$$R_2 = R_1 \sqrt{\frac{I_1}{I_2}} = 3 \text{ ft} \times \sqrt{100} = 30 \text{ ft}$$

Supplementary Problems

15.8. What quantity is carried by all types of waves from their source to the place where they are eventually absorbed?

15.9. Why do submarines submerge in storms?

15.10. The velocity of sound waves in air at sea level is 331 m/s. Find the wavelength in air of a sound wave whose frequency is 440 Hz.

15.11. A tuning fork vibrating at 600 Hz is immersed in a tank of water, and the resulting sound waves in the water are found to have a wavelength of 8.2 ft. What is the velocity of sound in the water?

15.12. A wave of frequency f_1 and wavelength λ_1 goes from a medium in which its velocity is v to another medium in which its velocity is $2v$. Find the frequency and wavelength of the wave in the second medium.

15.13. How many times more intense than the 60-db sound of a person talking loudly is the 100-db sound of a power lawnmower?

15.14. Sound waves whose intensities exceed about 1 W/m² cause damage to the ear. How many db is this equivalent to?

15.15. A certain jet aircraft produces sound of 140 db intensity at a distance of 100 ft. How many miles away is the intensity 90 db?

Answers to Supplementary Problems

15.8. Energy

15.9. Below a depth of about half the wavelength of an ocean wave, there is almost no disturbance of the water, hence a submarine at such depths is in no danger.

15.10. 0.75 m

15.11. 4920 ft/s

15.12. $f_2 = f_1, \lambda_2 = 2\lambda_1$

15.13. $10^4 = 10,000$ times more intense

15.14. 120 db

15.15. 6 mi

Chapter 16

Fluids at Rest

DENSITY

The *density* (d) of a substance is its mass per unit volume. The SI unit of density is the kilogram per cubic meter (kg/m^3); the density of aluminum, for instance, is 2700 kg/m^3. Another common unit of density is the gram per cubic centimeter (g/cm^3). Since 1 kg = 1000 g and 1 m^3 = (100 cm)^3 = 10^6 cm^3,

$$1 \text{ g/cm}^3 = 10^3 \text{ kg/m}^3$$

Hence the density of aluminum can also be given as 2.7 g/cm^3.

In British units density is properly expressed in slugs/ft^3. The density of aluminum in these units is 5.3 slugs/ft^3. Because weight rather than mass is normally specified in this system, the quantity *weight density* is customarily used. The weight density of a substance is its weight per unit volume. Thus the weight density of aluminum is 170 lb/ft^3. There is no special symbol for weight density, and either w/V or dg can be used for it.

SPECIFIC GRAVITY

The *specific gravity* of a substance is its density relative to that of pure water, which is

$$d(\text{water}) = 1000 \text{ kg/m}^3 = 1.00 \text{ g/cm}^3 = 1.94 \text{ slugs/ft}^3$$

The weight density of water is

$$dg(\text{water}) = 62 \text{ lb/ft}^3$$

Since the density of water is 1 g/cm^3, the specific gravity of a substance is the same as the numerical value of its density when given in g/cm^3. Thus the specific gravity of aluminum is 2.7.

PRESSURE

When a force acts perpendicular to a surface, the *pressure* exerted is the ratio between the magnitude of the force and the area of the surface:

$$p = \frac{F}{A}$$

$$\text{Pressure} = \frac{\text{force}}{\text{area}}$$

Pressures are properly expressed in N/m^2 or in lb/ft^2, but other units are often used:

$$1 \text{ lb/in}^2 = 144 \text{ lb/ft}^2$$

1 *atmosphere* (atm) = average pressure exerted by the earth's atmosphere at sea level

$$= 1.013 \times 10^5 \text{ N/m}^2 = 14.7 \text{ lb/in}^2$$

1 *millibar* (mb) = 100 N/m^2 (widely used in meteorology)

PRESSURE IN A FLUID

Pressure is a useful quantity where fluids (gases and liquids) are concerned because of the following properties of fluids.

1. The forces a fluid exerts on the walls of its container, and those the walls exert on the fluid, always act perpendicular to the walls.

2. The force exerted by the pressure in a fluid is the same in all directions at a given depth.

3. An external pressure exerted on a fluid is transmitted uniformly throughout the fluid. This does not mean that pressures in a fluid are the same everywhere, because the weight of the fluid itself exerts pressures that increase with increasing depth. The pressure at a depth h in a fluid of density d due to the weight of fluid above is

$$p = dgh$$

Hence the total pressure at that depth is

$$p = p_{\text{external}} + dgh$$

When a body of fluid is in an open container, the atmosphere exerts an external pressure on it.

GAUGE PRESSURE

Pressure gauges measure the difference between an unknown pressure and atmospheric pressure. What they measure is known as *gauge pressure*, and the true pressure is known as *absolute pressure*:

$$p = p_{\text{gauge}} + p_{\text{atm}}$$

Absolute pressure = gauge pressure + atmospheric pressure

A tire whose gauge pressure is 30 lb/in^2 contains air at an absolute pressure of 45 lb/in^2, since sea-level atmospheric pressure is about 15 lb/in^2.

ARCHIMEDES' PRINCIPLE

An object immersed in a fluid is acted upon by an upward force that arises because pressures in a fluid increase with depth. Hence the upward force on the bottom of the object is more than the downward force on its top. The difference between the two, called the *buoyant force*, is equal to the weight of a body of the fluid whose volume is the same as that of the object. This is *Archimedes' principle*: The buoyant force on a submerged object is equal to the weight of fluid the object displaces.

If the buoyant force is less than the weight of the object itself, the object sinks; if the buoyant force equals the weight of the object, the object floats in equilibrium at any depth in the fluid; if the buoyant force is more than the weight of the object, the object floats with part of its volume above the surface.

HYDRAULIC PRESS

The *hydraulic press* is a basic machine which makes use of the fact that an external pressure exerted on a fluid is transmitted uniformly throughout the fluid. In a hydraulic press, a piston whose cross-sectional area is A_{in} is moved through a distance L_{in} by an applied force, and fluid in the cylinder transmits the applied pressure to a piston of area A_{out} which moves the distance L_{out}. The theoretical mechanical advantage of the system is

$$\text{TMA} = \frac{L_{\text{in}}}{L_{\text{out}}} = \frac{A_{\text{out}}}{A_{\text{in}}}$$

Since the area of a piston is proportional to the square of its diameter, a ratio of piston diameters of only 5, for instance, will yield a TMA of 25.

Solved Problems

16.1. A wooden block is on the bottom of a tank when water is poured in. The contact between the block and the tank is so good that no water gets between them. Is there a buoyant force on the block?

There is no buoyant force since there is no water under the block to exert an upward force on it.

16.2. The specific gravity of gold is 19. (*a*) What is the mass of a cubic centimeter of gold? (*b*) What is the weight of a cubic inch of gold?

(*a*) Since the density of water is 1 g/cm³, the density of gold is 19 g/cm³ and 1 cm³ has a mass of 19 g.

(*b*) Since the weight density of water is 62 lb/ft³, the weight density of gold is $dg = 19 \times 62$ lb/ft³ $= 1200$ lb/ft³. Because 1 ft³ $= 12$ in.$\times 12$ in.$\times 12$ in.$= 1728$ in³, a cubic inch of gold weighs

$$w = (dg)V = 1200\,\frac{\text{lb}}{\text{ft}^3} \times \frac{1\text{ ft}^3}{1728\text{ in}^3} = 0.7\text{ lb}$$

16.3. An oak beam 10 cm by 20 cm by 4 m has a mass of 58 kg. (*a*) Find the density and specific gravity of oak. (*b*) Does oak float in water?

(*a*) The volume of the beam is $V = 0.1$ m $\times 0.2$ m $\times 4$ m $= 0.08$ m³ and so its density is

$$d = \frac{m}{V} = \frac{58\text{ kg}}{0.08\text{ m}^3} = 725\text{ kg/m}^3$$

Since the density of water is 1000 kg/m³, the specifc gravity of oak is

$$\text{sp gr} = \frac{d_{\text{oak}}}{d_{\text{water}}} = \frac{725}{1000} = 0.725$$

(*b*) Any material whose specific gravity is less than 1 floats in water, so oak does.

16.4. How much does the air in a room 12 ft square and 10 ft high weigh? The weight density of air is 0.08 lb/ft³ at sea level.

The volume of the room is $V = 12$ ft $\times 12$ ft $\times 10$ ft $= 1440$ ft³. Hence the weight of the air is

$$w = (dg)V = 0.08\text{ lb/ft}^3 \times 1440\text{ ft}^3 = 115\text{ lb}$$

16.5. The density of mammals is roughly the same as that of water. Find the volume of a 500-lb lion.

$$V = \frac{w}{dg} = \frac{500\text{ lb}}{62\text{ lb/ft}^3} = 8.06\text{ ft}^3$$

16.6. A 130-lb woman balances on the heel of her right shoe, which is 1 in. in radius. How much pressure does she exert on the ground? How does this compare with atmospheric pressure?

The area of the heel is $A = \pi r^2 = 3.14$ in², so the pressure is

$$p = \frac{F}{A} = \frac{130\text{ lb}}{3.14\text{ in}^2} = 41.4\text{ lb/in}^2$$

Since $p_{\text{atm}} = 14.7$ lb/in², this pressure is 2.8 times greater.

16.7. An airplane whose mass is 20,000 kg and whose wing area is 60 m² is in level flight. What is the average difference in pressure between the upper and lower surfaces of its wings? Express the answer in N/m² and in atm.

The upward force on an airplane in level flight is equal to its weight, which here is

$$F = w = mg = 20,000\text{ kg} \times 9.8\text{ m/s}^2 = 1.96 \times 10^5\text{ N}$$

The pressure difference p is therefore

$$p = \frac{F}{A} = \frac{1.96 \times 10^5 \text{ N}}{60 \text{ m}^2} = 3267 \text{ N/m}^2$$

Since 1 atm $= 1.013 \times 10^5$ N/m^2,

$$p = \frac{3.267 \times 10^3 \text{ N/m}^2}{(1.013 \times 10^5 \text{ N/m}^2)/\text{atm}} = 0.0322 \text{ atm}$$

16.8. What is the pressure at the bottom of a swimming pool 6 ft deep that is filled with fresh water? Express the answer in lb/in^2.

$$p = p_{\text{atm}} + (dg)h$$

$$= 14.7 \frac{\text{lb}}{\text{in}^2} + 62 \frac{\text{lb}}{\text{ft}^3} \times 6 \text{ ft} \times \frac{1 \text{ ft}^2}{144 \text{ in}^2}$$

$$= 17.3 \text{ lb/in}^2$$

Note the use of the conversion factor 1 ft^2/144 in^2.

16.9. The interior of a submarine located at a depth of 50 m in seawater is maintained at sea-level atmospheric pressure. Find the force acting on a window 20 cm square. The density of seawater is 1.03×10^3 kg/m^3.

The pressure outside the submarine is $p = p_{\text{atm}} + dgh$ and the pressure inside it is p_{atm}. Hence the net pressure p' acting on the window is

$$p' = dgh = 1.03 \times 10^3 \frac{\text{kg}}{\text{m}^3} \times 9.8 \frac{\text{m}}{\text{s}^2} \times 50 \text{ m} = 5.05 \times 10^5 \frac{\text{N}}{\text{m}^2}$$

Since the area of the window is $A = 0.2 \text{ m} \times 0.2 \text{ m} = 0.04 \text{ m}^2$, the force acting on it is

$$F = p'A = 5.05 \times 10^5 \text{ N/m}^2 \times 4 \times 10^{-2} \text{ m}^2 = 2.02 \times 10^4 \text{ N}$$

16.10. An iron anchor weighs 200 lb in air. How much force is required to support the anchor when it is immersed in seawater? The weight density of iron is 480 lb/ft^3 and that of seawater is 64 lb/ft^3.

Since $dg = w/V$, the volume of the anchor is

$$V = \frac{w}{dg} = \frac{200 \text{ lb}}{480 \text{ lb/ft}^3} = 0.417 \text{ ft}^3$$

The weight of seawater displaced by the anchor is

$$w = (dg)V = 64 \text{ lb/ft}^3 \times 0.417 \text{ ft}^3 = 27 \text{ lb}$$

Thus the buoyant force on the anchor is 27 lb and the net force needed to support it is

$$200 \text{ lb} - 27 \text{ lb} = 173 \text{ lb}$$

16.11. A 70-kg man jumps off a raft 2 m square moored in a freshwater lake. By how much does the raft rise?

The volume of water that must be displaced by the raft to support the man is

$$V = \frac{m}{d} = \frac{70 \text{ kg}}{10^3 \text{ kg/m}^3} = 0.07 \text{ m}^3$$

The area of the raft is $A = 2 \text{ m} \times 2 \text{ m} = 4 \text{ m}^2$. Since Volume $=$ height $\times$ area, the raft rises by

$$h = \frac{V}{A} = \frac{0.07 \text{ m}^3}{4 \text{ m}^2} = 0.018 \text{ m} = 1.8 \text{ cm}$$

16.12. The density of ice is 920 kg/m^3 and that of seawater is 1030 kg/m^3. What percentage of the volume of an iceberg is submerged?

When an iceberg of volume V floats, its weight of $d_{ice}gV$ is balanced by the buoyant force on it, which is equal to the weight of water displaced. If V_{sub} is the volume of the iceberg that is submerged, the weight of water displaced is $d_{water}gV_{sub}$. Hence

$$\text{Weight of iceberg} = \text{weight of displaced water}$$

$$d_{ice}gV = d_{water}gV_{sub}$$

$$\frac{V_{sub}}{V} = \frac{d_{ice}}{d_{water}} = \frac{920 \text{ kg/m}^3}{1030 \text{ kg/m}^3} = 0.89 = 89\%$$

Eighty-nine percent of the volume of an iceberg is under the water's surface.

16.13. A 100-gallon steel tank weighs 50 lb when empty. Will it float in seawater when filled with gasoline? The weight density of gasoline is 42 lb/ft³, that of seawater is 64 lb/ft³, and 1 gallon = 0.134 ft³.

    The volume of the tank is $V = 100$ gal $\times 0.134$ ft³/gal $= 13.4$ ft³. The total weight of the tank when filled with gasoline is

$$w = 50 \text{ lb} + (dg)_{gasoline}V = 50 \text{ lb} \times 42 \text{ lb/ft}^3 \times 13.4 \text{ ft}^3$$

$$= 50 \text{ lb} + 563 \text{ lb} = 613 \text{ lb}$$

The maximum buoyant force on the tank is exerted when the tank is completely submerged. Thus

$$F_{max} = (dg)_{water}V = 64 \text{ lb/ft}^3 \times 13.4 \text{ ft}^3 = 858 \text{ lb}$$

Since the weight of the filled tank is less than 858 lb, it will float.

16.14. A hydraulic press has an input cylinder 1 in. in diameter and an output cylinder 6 in. in diameter. (*a*) Assuming 100% efficiency, find the force exerted by the output piston when a force of 10 lb is applied to the input piston. (*b*) If the input piston is moved through 4 in., how much is the output piston moved?

(*a*)   At 100% efficiency the actual mechanical advantage equals the TMA, and

$$F_{out}/F_{in} = A_{out}/A_{in}$$

Since $A = \pi r^2 = \pi d^2/4$,

$$F_{out} = F_{in}\left(\frac{d_{out}^2}{d_{in}^2}\right) = 10 \text{ lb} \times \frac{(6 \text{ in.})^2}{(1 \text{ in.})^2} = 360 \text{ lb}$$

(*b*) $$L_{out} = L_{in}\left(\frac{A_{in}}{A_{out}}\right) = L_{in}\left(\frac{d_{in}^2}{d_{out}^2}\right) = 4 \text{ in.} \times \frac{(1 \text{ in.})^2}{(6 \text{ in.})^2} = 0.11 \text{ in.}$$

Supplementary Problems

16.15. A sailboat has a lead keel to help keep it upright despite the pressure wind exerts on its sails. What difference, if any, is there between the stability of a sailboat in fresh water and in seawater?

16.16. Dam A and dam B are identical in size and shape, and the water levels at both are the same height above their bases. Dam A holds back a lake that contains 1 cubic mile of water, and dam B holds back a lake that contains 2 cubic miles of water. What is the ratio between the total force exerted on dam A and that exerted on dam B?

16.17. A 50-g gold bracelet is dropped into a full glass of water and 2.6 cm³ of water overflows. What is the density of gold? What is its specific gravity?

16.18. The weight density of ice is 58 lb/ft^3. What is its specific gravity?

16.19. How much does the water in a swimming pool 20 ft long, 10 ft wide, and 6 ft deep weigh?

16.20. The density of iron is 7.8×10^3 kg/m^3. (a) What is the specific gravity of iron? (b) How many cubic meters does a metric ton (1000 kg) of iron occupy?

16.21. An airplane whose weight is 50,000 lb and whose wing area is 600 ft^2 is in level flight. What is the average difference in pressure between the upper and lower surfaces of its wings?

16.22. A 70-kg man wears shoes whose area is 200 cm^2 each. How much pressure does he exert on the ground?

16.23. A phonograph needle whose point is 0.1 mm (10^{-4} m) in radius exerts a downward force of 0.02 N. What is the pressure on the record groove? How many atm is this?

16.24. What is the pressure at a depth of 100 m in the ocean? How many atm is this? The density of seawater is 1.03×10^3 kg/m^3.

16.25. What pressure is experienced by a skin diver 20 ft below the surface of a freshwater lake?

16.26. (a) How much force is required to raise a 1000-kg block of concrete to the surface of a freshwater lake? (b) How much force is needed to lift it out of the water? The density of concrete is 2.3×10^3 kg/m^3.

16.27. An aluminum bar weighs 17 lb in air. How much force is required to support the bar when it is immersed in gasoline? The weight density of aluminum is 170 lb/ft^3 and that of gasoline is 42 lb/ft^3.

16.28. A raft 8 ft wide, 12 ft long, and 2 ft high is made from solid balsa wood ($dg = 8$ lb/ft^3). How much weight can it support in seawater ($dg = 64$ lb/ft^3)?

16.29. People have roughly the same density as fresh water. Find the buoyant force exerted by the atmosphere on a 50-kg woman at sea level where the density of air is 1.3 kg/m^3.

16.30. A balloon weighing 100 kg has a capacity of 1000 m^3. If it is filled with hydrogen, how great a payload in kg can it support? At sea level the density of hydrogen is 0.09 kg/m^3 and that of air is 1.3 kg/m^3.

16.31. A lever with a mechanical advantage of 10 is used to apply force to the input piston of a hydraulic jack whose input piston is 1 in. in diameter and whose output piston is 4 in. in diameter. (a) If the jack is 90% efficient, how much weight can it lift when a force of 50 lb is applied to the lever? (b) If each stroke of the lever moves the input piston 3 in., how many strokes are needed to raise the output piston 1 ft?

Answers to Supplementary Problems

16.15. The boat is more stable in fresh water because the buoyancy of the lead keel is less there.

16.16. The forces are the same.

16.17. 19 g/cm^3; 19

16.18. 0.93

16.19. 74,400 lb

16.20. (a) 7.8   (b) 0.128 m^3

16.21. 83.3 lb/ft^2

16.22. 1.72×10^4 N/m^2

16.23. 6.37×10^5 N/m^2; 6.3 atm

16.24. 1.11×10^6 N/m^2; 11.0 atm

16.25. 23.3 lb/in^2

16.26. (a) 5539 N   (b) 9800 N

16.27. 12.8 lb

16.28. 10,752 lb

16.29. 0.64 N

16.30. 1110 kg

16.31. (a) 7200 lb   (b) 64 strokes

Chapter 17

Fluids in Motion

FLUID FLOW

In the *streamline flow* of a fluid, the direction of motion of the individual fluid particles is the same as that of the fluid as a whole. Each particle of the fluid that passes any point follows the same path as those particles that passed that point before. *Turbulent flow*, on the other hand, is characterized by the presence of irregular whirls and eddies; it occurs at high velocities and when the fluid's path changes direction sharply, for instance near an obstruction.

The rate at which a fluid whose velocity is v flows through a pipe or channel of cross-sectional area A is

$$R = vA$$

Rate of flow = velocity × cross-sectional area

It is common for R to be expressed in such units as gal/min and liters/s instead of the proper units of ft^3/s and m^3/s. (1 U.S. gal = 0.134 ft^3 and 1 liter = 10^{-3} m^3 = 10^3 cm^3.)

When a fluid is incompressible, which is approximately true for most liquids, its rate of flow R is constant even though the size of the pipe or channel varies. Thus if a liquid's velocity is v_1 when the cross-sectional area is A_1 and v_2 when it is A_2,

$$v_1 A_1 = v_2 A_2$$

BERNOULLI'S EQUATION

Bernoulli's equation applies to the streamline flow of an incompressible fluid of density d with negligible viscosity (internal friction). According to this equation, which is derived from the law of conservation of energy, the quantity $p + dgh + \frac{1}{2}dv^2$ has the same value at all points in the motion of such a fluid, where p is the absolute pressure, h is the height above an arbitrary reference level, and v is the fluid velocity. Thus at the two locations 1 and 2

$$p_1 + dgh_1 + \tfrac{1}{2}dv_1^2 = p_2 + dgh_2 + \tfrac{1}{2}dv_2^2$$

The quantity dgh is the potential energy of the fluid per unit volume and the quantity $\frac{1}{2}dv^2$ is its kinetic energy per unit volume. Each term of this equation has the units of pressure.

Another form of Bernoulli's equation is obtained by dividing each term of the above equation by dg, which gives

$$\frac{p_1}{dg} + \frac{v_1^2}{2g} + h_1 = \frac{p_2}{dg} + \frac{v_2^2}{2g} + h_2$$

Each term of this equation has the dimensions of a length and is called a *head*, where

$$\frac{p}{dg} = \text{pressure head} \qquad \frac{v^2}{2g} = \text{velocity head} \qquad h = \text{elevation head}$$

TORRICELLI'S THEOREM

In the case of a liquid flowing out of an open tank through a small orifice a distance h below the surface, as in Fig. 17-1, the pressure on the liquid is the same at the surface and at the orifice, and the downward velocity of the fluid surface is negligible compared with the outward velocity of the fluid. Here Bernoulli's equation reduces to

$$v = \sqrt{2gh}$$

which is *Torricelli's theorem*. This is the same velocity an object would have if dropped from rest from a height h.

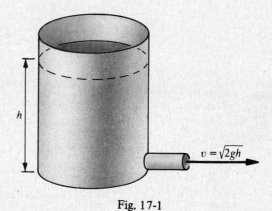

Fig. 17-1

PRESSURE AND VELOCITY

When the flow is horizontal, so $h_1 = h_2$, Bernoulli's equation becomes

$$p_1 + \tfrac{1}{2} d v_1^2 = p_2 + \tfrac{1}{2} d v_2^2$$

The pressure is greatest where the fluid velocity is least, and the pressure is least where the fluid velocity is greatest. The lift developed by an airplane wing is an example of this effect: the air moving across the curved upper surface of the wing travels faster than that moving across the lower surface because it must cover a greater distance, hence the pressure is less on the upper surface and the net result is an upward force on the wing.

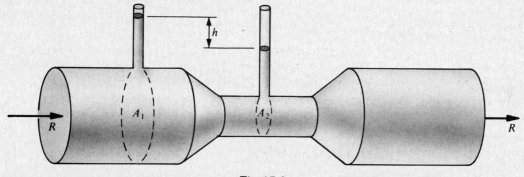

Fig. 17-2

The device shown in Fig. 17-2, known as a *venturi meter*, provides a convenient method for determining the rate of flow R of a liquid through a pipe in terms of the difference in height h between the levels of the liquid in the vertical tubes (called *manometers*). From the above equation plus the fact that $p_1 - p_2 = dgh$ is the difference in pressure between locations 1 and 2 in the pipe, it is possible to show that

$$R = A_1 \sqrt{\frac{2gh}{(A_1/A_2)^2 - 1}}$$

where A_1 and A_2 are the cross-sectional areas of the pipe at the two locations.

Solved Problems

17.1. Bernoulli's equation holds for incompressible, nonviscous fluids. How is this relationship changed when the viscosity of a fluid is not negligible?

The effect of viscosity, which is a kind of friction, is to dissipate mechanical energy into heat. Hence the quantity $p + dgh + \frac{1}{2}dv^2$ decreases along the direction of flow of a nonviscous fluid.

17.2. An airplane whose mass is 40,000 kg and whose total wing area is 120 m² is in level flight. What is the difference in pressure between the upper and lower surfaces of its wings?

The lift developed by the wings in level flight is equal to the airplane's weight. Hence

$$F = w = mg = 4 \times 10^4 \text{ kg} \times 9.8 \text{ m/s}^2 = 3.92 \times 10^5 \text{ N}$$

The pressure difference is therefore

$$\Delta p = \frac{F}{A} = \frac{3.92 \times 10^5 \text{ N}}{120 \text{ m}^2} = 3.27 \times 10^3 \text{ N/m}^2$$

which is equal to 0.032 atm or 0.47 lb/in².

17.3. A barrel 80 cm high is filled with kerosene. When a tap at the bottom of the barrel is opened, with what velocity does the kerosene emerge?

$$v = \sqrt{2gh} = \sqrt{2 \times 9.8 \text{ m/s}^2 \times 0.8 \text{ m}} = 3.96 \text{ m/s}$$

17.4. A boat strikes an underwater rock which punctures a hole 2 in. in diameter in its hull 5 ft below the waterline. At what rate in gal/min does water enter the hull?

From Torricelli's theorem, the velocity with which water enters the hull is $v = \sqrt{2gh}$. Since the rate of flow through an orifice of area A is $R = vA$ when the fluid velocity is v, $R = A\sqrt{2gh}$. Here the radius of the hole is $r = 1$ in. $= 0.0833$ ft, and so

$$A = \pi r^2 = \pi \times (0.0833 \text{ ft})^2 = 0.0218 \text{ ft}^2$$

Therefore water enters the hull at the rate

$$R = A\sqrt{2gh} = 0.0218 \text{ ft}^2 \times \sqrt{2 \times 32 \text{ ft/s}^2 \times 5 \text{ ft}}$$
$$= 0.39 \text{ ft}^3/\text{s}$$

To convert this figure into gal/min, we note that

$$1 \text{ gal} = 0.134 \text{ ft}^3 \qquad \text{and} \qquad 1 \text{ min} = 60 \text{ s}$$

so that

$$R = \frac{0.39 \text{ ft}^3/\text{s} \times 60 \text{ s/min}}{0.134 \text{ ft}^3/\text{gal}} = 175 \text{ gal/min}$$

17.5. A garden hose has an inside diameter of $\frac{1}{2}$ in. and water flows through it at 8 ft/s. (a) What nozzle diameter is required for the water to emerge at 30 ft/s? (b) At what rate does water leave the nozzle?

(a) The cross-sectional areas of hose and nozzle are in the same ratio as the squares of their diameters, since $A = \pi r^2 = \pi d^2/4$. From $v_1 A_1 = v_2 A_2$ we obtain

$$v_1 d_1^2 = v_2 d_2^2$$

$$d_2 = d_1 \sqrt{\frac{v_1}{v_2}} = 0.5 \text{ in.} \times \sqrt{\frac{8 \text{ ft/s}}{30 \text{ ft/s}}} = 0.26 \text{ in.}$$

(b) Since $r_1 = \frac{1}{4}$ in./12 (in./ft) $= 0.021$ ft,

$$R = v_1 A_1 = v_1 \pi r_1^2 = 8 \text{ ft/s} \times \pi \times (0.021 \text{ ft})^2 = 0.011 \text{ ft}^3/\text{s}$$

The same result, of course, would be obtained from $R = v_2 A_2$.

17.6. Water emerges horizontally at 30 ft/s from the nozzle of a hose held 4 ft above the ground. How far away does the water stream reach?

From Chapter 4 the time required for the water to strike the ground from a height h is $t = \sqrt{2h/g}$. During this time the water will travel horizontally the distance

$$s = v_0 t = v_0 \sqrt{\frac{2h}{g}} = 30 \text{ ft/s} \times \sqrt{\frac{2 \times 4 \text{ ft}}{32 \text{ ft/s}^2}} = 15 \text{ ft}$$

17.7. At what velocity should water emerge from the nozzle of a fire hose if it is to reach a height of 80 ft when the hose is aimed vertically upward?

The velocity needed to reach a height h is the same as the velocity $v = \sqrt{2gh}$ that would be acquired in free fall from that height. Hence

$$v = \sqrt{2gh} = \sqrt{2 \times 32 \text{ ft/s}^2 \times 80 \text{ ft}} = 72 \text{ ft/s}$$

17.8. The bilge pump on a boat is required to lift 0.4 m³ of seawater per minute through a height of 1.5 m. If the overall efficiency is 50%, what should the power rating of the pump motor be? The density of seawater is 1.03×10^3 kg/m³.

The work done in lifting $V = 0.4$ m³ of seawater through $h = 1.5$ m is

$$W = mgh = dVgh = 1.03 \times 10^3 \text{ kg/m}^3 \times 0.4 \text{ m}^3 \times 9.8 \text{ m/s}^2 \times 1.5 \text{ m}$$
$$= 6056 \text{ J}$$

Since $t = 1$ min $= 60$ s here, the required power at 50% efficiency is

$$P = \frac{1}{\text{Eff}} \times \frac{W}{t} = \frac{6056 \text{ J}}{0.5 \times 60 \text{ s}} = 202 \text{ W}$$

To convert this result to hp we divide by 746 W/hp:

$$P = \frac{202 \text{ W}}{746 \text{ W/hp}} = 0.27 \text{ hp}$$

17.9. A 100-hp motor is used to power the pump of a fire engine. At 60% efficiency, how many tons of water per minute can be delivered to a height of 100 ft?

The available power is

$$P = 0.6 \times 100 \text{ hp} \times 550 \frac{\text{ft-lb/s}}{\text{hp}} = 3.3 \times 10^4 \text{ ft-lb/s}$$

The power needed to raise a weight w to a height h in the time t is

$$P = \frac{W}{t} = \frac{wh}{t}$$

and so, with $t = 1$ min $= 60$ s,

$$w = \frac{Pt}{h} = \frac{3.3 \times 10^4 \text{ ft-lb/s} \times 60 \text{ s}}{100 \text{ ft}} = 1.98 \times 10^4 \text{ lb}$$

Since 1 ton = 2000 lb,

$$w = \frac{1.98 \times 10^4 \text{ lb}}{2000 \text{ lb/ton}} = 9.9 \text{ tons}$$

17.10. Water emerges from the 2-in.-diameter nozzle of a hose at the rate of 200 gal/min. (*a*) Find the force with which the nozzle must be held. (*b*) The water strikes a window of a house and moves off parallel to the window. Find the force exerted on the window.

(*a*) The rate of flow is

$$R = 200 \frac{\text{gal}}{\text{min}} \times 0.134 \frac{\text{ft}^3}{\text{gal}} \times \frac{1}{60 \text{ s/min}} = 0.447 \text{ ft}^3/\text{s}$$

Hence the mass of water that flows through the nozzle per second is

$$\frac{m}{t} = dR = 1.94 \text{ slugs/ft}^3 \times 0.447 \text{ ft}^3/\text{s} = 0.867 \text{ slug/s}$$

Since $R = vA$ and $A = \pi r^2$ the velocity of the stream is

$$v = \frac{R}{A} = \frac{R}{\pi r^2} = \frac{0.447 \text{ ft}^3/\text{s}}{\pi (1/12 \text{ ft})^2} = 20.5 \text{ ft/s}$$

Equating the impulse Ft with the momentum change mv of the water yields

$$Ft = mv$$

$$F = \frac{mv}{t} = 0.867 \text{ slug/s} \times 20.5 \text{ ft/s} = 17.8 \text{ lb}$$

(b) The change in momentum of the water stream per second is the same as in (a), hence the force on the window is also 17.8 lb.

17.11. (a) At what velocity does water emerge from an orifice in a tank in which the gauge pressure is 3×10^5 N/m²? . (b) In which the gauge pressure is 40 lb/in²? ($d_{\text{water}} = 10^3$ kg/m³ = 1.94 slugs/ft³)

(a) Let point 1 be at the orifice and point 2 be inside the tank at the same level. In this situation $h_1 = h_2$ and $v_2 = 0$ (very nearly). Substituting in Bernoulli's equation yields

$$p_1 + \tfrac{1}{2}dv_1^2 = p_2 \qquad v_1 = \sqrt{\frac{2(p_2 - p_1)}{d}}$$

The quantity $p_2 - p_1$ is the gauge pressure of 3×10^5 N/m² since p_1 is atmospheric pressure. Hence

$$v_1 = \sqrt{\frac{2 \times 3 \times 10^5 \text{ N/m}^2}{10^3 \text{ kg/m}^3}} = 24.5 \text{ m/s}$$

(b) The same procedure is followed here, but first it is necessary to convert 40 lb/in² to its equivalent in lb/ft²:

$$p_2 - p_1 = 40 \frac{\text{lb}}{\text{in}^2} \times 144 \frac{\text{in}^2}{\text{ft}^2} = 5760 \text{ lb/ft}^2$$

Hence

$$v_1 = \sqrt{\frac{2 \times 5760 \text{ lb/ft}^2}{1.94 \text{ slugs/ft}^3}} = 77 \text{ ft/s}$$

17.12. A horizontal pipe 1 in. in radius is joined to a pipe 4 in. in radius, as in Fig. 17-3. (a) If the velocity of seawater ($d = 2.00$ slugs/ft³) in the small pipe is 20 ft/s and the pressure there is 30 lb/in², find the velocity and pressure in the large pipe. (b) What is the rate of flow through the pipe in lb/min?

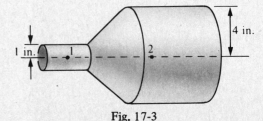

Fig. 17-3

(a) The cross-sectional areas of the pipes are in the same ratio as the squares of their radii, since $A = \pi r^2$. From $v_1 A_1 = v_2 A_2$ we obtain

$$v_2 = v_1 \frac{A_1}{A_2} = v_1 \frac{r_1^2}{r_2^2} = 20 \text{ ft/s} \times \frac{(1 \text{ in.})^2}{(4 \text{ in.})^2} = 1.25 \text{ ft/s}$$

Because both pipes are horizontal, $h_1 = h_2$, and Bernoulli's equation becomes

$$p_1 + \tfrac{1}{2}dv_1^2 = p_2 + \tfrac{1}{2}dv_2^2$$

Hence

$$p_2 = p_1 + \tfrac{1}{2}d(v_1^2 - v_2^2) = 30\,\frac{\text{lb}}{\text{in}^2} \times 144\,\frac{\text{in}^2}{\text{ft}^2} + \frac{1}{2} \times 2\,\frac{\text{slugs}}{\text{ft}^3} \times \left[(20\,\text{ft/s})^2 - (1.25\,\text{ft/s})^2\right]$$

$$= 4718\,\text{lb/ft}^2$$

which is

$$p_2 = \frac{4718\,\text{lb/ft}^2}{144\,\text{in}^2/\text{ft}^2} = 33\,\text{lb/in}^2$$

(b) $$R = v_1 A_1 = v_1 \times \pi r_1^2 = 20\,\text{ft/s} \times \pi \times \left(\tfrac{1}{12}\,\text{ft}\right)^2 = 0.436\,\text{ft}^3/\text{s}$$

Since $dg = 64\,\text{lb/ft}^3$ and 1 min = 60 s, the rate of flow in the required units is

$$0.436\,\text{ft}^3/\text{s} \times 64\,\text{lb/ft}^3 \times 60\,\text{s/min} = 1674\,\text{lb/min}$$

17.13. Water flows through the pipe shown in Fig. 17-4 at the rate of 3 ft^3/s. If the pressure at point 1 is 25 lb/in^2, find (a) the velocity at point 1, (b) the velocity at point 2, and (c) the pressure at point 2.

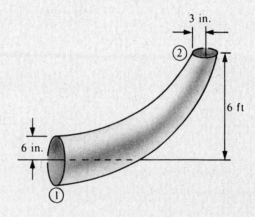

Fig. 17-4

(a) Since $A = \pi r^2$ and $R = v_1 A_1$,

$$v_1 = \frac{R}{A_1} = \frac{R}{\pi r_1^2} = \frac{3\,\text{ft}^3/\text{s}}{\pi \times (0.5\,\text{ft})^2} = 3.82\,\text{ft/s}$$

(b) From $v_1 A_1 = v_2 A_2$ we have

$$v_2 = v_1 \frac{A_1}{A_2} = v_1 \frac{r_1^2}{r_2^2} = 3.82\,\text{ft/s} \times \frac{(0.5\,\text{ft})^2}{(0.25\,\text{ft})^2}$$

$$= 15.3\,\text{ft/s}$$

(c) We now substitute the known quantities p_1, v_1, and v_2 with $h_1 = 0$ and $h_2 = 6$ ft into Bernoulli's equation,

$$p_1 + dgh_1 + \tfrac{1}{2}dv_1^2 = p_2 + dgh_2 + \tfrac{1}{2}dv_2^2$$

This yields

$$p_2 = p_1 + \tfrac{1}{2}d(v_1^2 - v_2^2) - dgh_2$$

$$= 25\,\frac{\text{lb}}{\text{in}^2} \times 144\,\frac{\text{in}^2}{\text{ft}^2} + \tfrac{1}{2} \times 1.94\,\frac{\text{slugs}}{\text{ft}^3}\left[(3.82\,\text{ft/s})^2 - (15.3\,\text{ft/s})^2\right]$$

$$- 1.94\,\frac{\text{slugs}}{\text{ft}^3} \times 32\,\frac{\text{ft}}{\text{s}^2} \times 6\,\text{ft}$$

$$= 3015\,\text{lb/ft}^2$$

which is

$$p_2 = \frac{3015\,\text{lb/ft}^2}{144\,\text{in}^2/\text{ft}^2} = 21\,\text{lb/in}^2$$

17.14. Oil flows through a venturi meter whose cross-sectional areas are 0.6 and 0.2 ft². Find the rate of flow when the difference in manometer heights is 3 in.

Here $A_1 = 0.6$ ft², $A_2 = 0.2$ ft², and $h = 3$ in. $= 0.25$ ft. The rate of flow is therefore

$$R = A_1 \sqrt{\frac{2gh}{(A_1/A_2)^2 - 1}} = 0.6 \text{ ft}^2 \sqrt{\frac{2 \times 32 \text{ ft/s}^2 \times 0.25 \text{ ft}}{(0.6 \text{ ft}^2/0.2 \text{ ft}^2)^2 - 1}} = 0.85 \text{ ft}^3/\text{s}$$

Supplementary Problems

17.15. An airplane's wings are designed to have a pressure difference of 0.4 lb/in² between their upper and lower surfaces. If the area of the wings is 300 ft², what is the design weight of the airplane when fully loaded?

17.16. A water tank is on the roof of an apartment house. If a faucet is opened on the ground floor, 140 ft below the water level in the tank, and there is no frictional resistance in the plumbing system, with what velocity will water emerge from the faucet?

17.17. (a) What gauge pressure is required in a fountain if the jet of water is to be 20 ft high? (b) 20 m high?

17.18. Water leaves the nozzle of a hose at a velocity of 40 ft/s. (a) If the nozzle is held vertically, how high does the water reach? (b) If the nozzle is held horizontally 5 ft above the ground, how far does the water stream go before striking the ground?

17.19. Water emerges from a faucet 5 mm in diameter at 2 m/s. (a) If the pipe leading to the faucet is 3 cm in diameter, find the water velocity in the pipe. (b) At what rate does water leave the faucet?

17.20. Water passes through a turbine at a rate of 100 ft³/s. If the water enters the turbine at 20 ft/s and leaves at 5 ft/s, find the average force exerted on the turbine blades.

17.21. A pressure gauge in a water-supply system reads 40 lb/in² when no water flows and 34 lb/in² when a tap is opened. What is the flow velocity?

17.22. A tank of gasoline has a crack 1 mm wide and 5 cm long at its base. If the liquid level is 2 m above the base, find the rate at which gasoline leaks out in m³/s and in liters/s.

17.23. How many kilograms of milk can a $\frac{1}{3}$-hp pump in a dairy transfer per hour from one tank to another in which the liquid level is 10 m higher? Assume 70% efficiency.

17.24. A pump in a fuel station is required to deliver gasoline ($dg = 42$ lb/ft³) at the rate of 30 gal/min. If the gasoline must be raised through 6 ft and the efficiency is 75%, find the minimum power rating of the pump motor.

17.25. Water leaves the safety valve of a boiler at a velocity of 100 ft/s. What is the gauge pressure in the boiler?

17.26. A horizontal pipe 6 in. in diameter has a constriction 3 in. in diameter. If the velocity of water in the constriction is 30 ft/s, find the difference between the pressure there and that in the rest of the pipe. Which pressure is the greater?

17.27. Water enters the pipe shown in Fig. 17-5 from the left at a velocity of 10 ft/s and a pressure of 22 lb/in². Find (a) the velocity with which water leaves the pipe at the right, (b) the pressure there, and (c) the rate of flow of water through the pipe in lb/min.

17.28. Water flows through a venturi meter whose main tube is 4 cm in diameter and whose constriction is 1 cm in diameter. Find the rate of flow in liters/min when the difference in manometer heights is 8 cm.

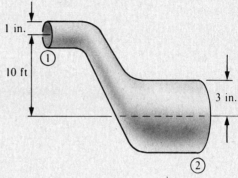

Fig. 17-5

Answers to Supplementary Problems

17.15. 17,280 lb

17.16. 95 ft/s

17.17. (a) 1240 lb/ft² = 8.6 lb/in²   (b) 196 × 10⁵ N/m²

17.18. (a) 25 ft   (b) 22 ft

17.19. (a) 0.0556 m/s   (b) 3.93 × 10⁻⁵ m³/s

17.20. 2910 lb

17.21. 5.28 ft/s

17.22. 3.13 × 10⁻⁴ m³/s; 0.313 liter/s

17.23. 6394 kg

17.24. 0.041 hp

17.25. 9700 lb/ft² = 67 lb/in²

17.26. 818 lb/ft² = 5.7 lb/in²; the pressure in the large part of the pipe is greater

17.27. (a) 1.11 ft/s   (b) 27 lb/in²   (c) 812 lb/min

17.28. 5.9 liters/min

Chapter 18

Heat

INTERNAL ENERGY

Every body of matter, whether solid, liquid, or gas, consists of atoms or molecules which are in rapid motion. The kinetic energies of these particles constitute the *internal energy* of the body of matter. The *temperature* of the body is a measure of the average kinetic energy of its particles. *Heat* may be thought of as internal energy in transit. When heat is added to a body, its internal energy increases and its temperature rises; when heat is removed from a body, its internal energy decreases and its temperature falls.

TEMPERATURE

Temperature is familiar as the property of a body of matter responsible for sensations of hot or cold when it is touched. Temperature provides an indicator of the direction of internal energy flow: when two objects are in contact, internal energy goes from the one at the higher temperature to the one at the lower temperature, regardless of the total amounts of internal energy in each one. Thus if hot coffee is poured into a cold cup, the coffee becomes cooler and the cup becomes warmer.

A *thermometer* is a device for measuring temperature. Matter usually expands when heated and contracts when cooled, the relative amount of change being different for different substances. This behavior is the basis of most thermometers, which make use of the different rates of expansion of mercury and glass, or of two metal strips joined together, to indicate temperature.

TEMPERATURE SCALES

The *Celsius* (or *centigrade*) temperature scale assigns 0° to the freezing point of water and 100° to its boiling point. On the *Fahrenheit* scale these points are respectively 32° and 212°. A Fahrenheit degree is therefore 5/9 as large as a Celsius degree. The following formulas give the procedure for converting a temperature expressed in one scale to the corresponding value in the other.

$$T_F = \frac{9}{5} T_C + 32°$$

$$T_C = \frac{5}{9} (T_F - 32°)$$

HEAT

Heat is a form of energy which, when added to a body of matter, increases its internal energy content and thereby causes its temperature to rise. The customary symbol for heat is Q.

Because heat is a form of energy, the proper SI unit of heat is the joule. However, the *kilocalorie* is still widely used with SI units: 1 kilocalorie (kcal) is the amount of heat needed to raise the temperature of 1 kg of water by 1 °C. The calorie itself is the amount of heat needed to raise the temperature of 1 g of water by 1 °C; hence 1 kcal = 1000 calories. (The calorie used by dieticians to measure the energy content of foods is the same as the kilocalorie.)

The British unit of heat is the *British thermal unit* (Btu): 1 Btu is the amount of heat needed to raise the temperature of 1 lb of water by 1 °F. To convert heat figures from one system to the other we note that

$$1 \text{ kcal} = 3.97 \text{ Btu}$$

$$1 \text{ Btu} = 0.252 \text{ kcal}$$

Although weight rather than mass is specified in the British system when dealing with heat, in practice this makes no difference in the various calculations. Whenever m appears in the equations of heat, it is understood to refer to mass in kg when metric units are used and to weight in lb when British units are used.

SPECIFIC HEAT CAPACITY

Different substances respond differently to the addition or removal of heat. For instance, 1 kg of water increases in temperature by 1 °C when 1 kcal of heat is added, but 1 kg of aluminum increases in temperature by 4.5 °C when this is done. The *specific heat capacity* of a substance is the amount of heat needed to change the temperature of a unit quantity of it by 1°. The symbol of specific heat capacity is c; its metric unit is the kcal/kg-°C and its British unit is the Btu/lb-°F. The numerical value of c for a substance is the same in both systems of units.

Among common materials, water has the highest specific heat capacity, namely 1.00 kcal/kg-°C (or Btu/lb-°F). Ice and steam have lower specific heat capacities than water, respectively 0.50 and 0.48 kcal/kg-°C (or Btu/lb-°F). Metals usually have low specific heat capacities; thus lead and iron have $c = 0.03$ and 0.11 kcal/kg-°C (or Btu/lb-°F) respectively.

When an amount of heat Q is transferred to or from a mass m of a substance whose specific heat capacity is c, the resulting temperature change ΔT is related to Q, m, and c by the formula

$$Q = mc \, \Delta T$$

Heat transferred = mass × specific heat capacity × temperature change

CHANGE OF STATE

When heat is continuously added to a solid, it grows hotter and hotter and finally begins to melt. While it is melting, the material remains at the same temperature and the absorbed heat goes into changing its state from solid to liquid. After all the solid is melted, the temperature of the resulting liquid then increases as more heat is supplied until it begins to boil. Now the material again stays at a constant temperature until all of it has become a gas, after which the gas temperature rises.

The amount of heat that must be added to a unit quantity (1 kg or 1 lb) of a substance at its melting point to change it from a solid to a liquid is called its *heat of fusion* (L_f). The same amount of heat must be removed from a unit quantity of the substance when it is a liquid at its melting point to change it to a solid.

The amount of heat that must be added to a unit quantity of a substance at its boiling point to change it from a liquid to a gas is called its *heat of vaporization* (L_v). The same amount of heat must be removed from a unit quantity of the substance when it is a gas at its boiling point to change it to a liquid.

The heat of fusion of water is $L_f = 80$ kcal/kg = 144 Btu/lb and its heat of vaporization is $L_v = 540$ kcal/kg = 972 Btu/lb.

PRESSURE AND BOILING POINT

The boiling point of a liquid depends upon the pressure applied to it: the higher the pressure, the higher the boiling point. Thus water under a pressure of 2 atm boils at 121 °C instead of at 100 °C as it does at sea-level atmospheric pressure. At high altitudes, where the atmospheric pressure is less than at sea level, water boils at a lower temperature than 100 °C. At an elevation of 2000 m, for instance, atmospheric pressure is about three-quarters of its sea-level value and water boils at 93 °C there.

Solved Problems

18.1. A person is dissatisfied with the rate at which eggs cook in a pan of boiling water. Would they cook faster if he (a) turns up the gas flame; (b) uses a pressure cooker?

(a) No. The maximum temperature that water can have while in the liquid state is its boiling point. Increasing the rate at which heat is supplied to a pan of water increases the rate at which steam is produced, but does not raise the temperature of the water beyond 100 °C (212 °F).

(b) Yes. In a pressure cooker, the pressure is greater than normal atmospheric pressure, which elevates the boiling point and so causes the eggs to cook faster.

18.2. What is the Celsius equivalent of 80 °F?

$$T_C = \frac{5}{9}(T_F - 32°) = \frac{5}{9}(80° - 32°) = 26.7 \text{ °C}$$

18.3. What is the Fahrenheit equivalent of 80 °C?

$$T_F = \frac{9}{5}T_C + 32° = \frac{9}{5} \times 80° + 32° = 176 \text{ °F}$$

18.4. Oxygen freezes at −362 °F. What is the Celsius equivalent of this temperature?

$$T_C = \frac{5}{9}(T_F - 32°) = \frac{5}{9}(-362° - 32°) = -219 \text{ °C}$$

18.5. Nitrogen freezes at −210 °C. What is the Fahrenheit equivalent of this temperature?

$$T_F = \frac{9}{5}T_C + 32° = \frac{9}{5}(-210°) + 32° = -346 \text{ °F}$$

18.6. How much heat must be added to 3 kg of water to raise its temperature from 20 °C to 80 °C?

The temperature change is $\Delta T = 80 \text{ °C} - 20 \text{ °C} = 60 \text{ °C}$. Hence

$$Q = mc \, \Delta T = 3 \text{ kg} \times 1 \text{ kcal/kg-°C} \times 60 \text{ °C} = 180 \text{ kcal}$$

18.7. Two hundred Btu of heat is removed from a 50-lb block of ice initially at 25 °F. What is its final temperature? ($c_{ice} = 0.5$ Btu/lb-°F)

$$Q = mc \, \Delta T$$

$$\Delta T = \frac{Q}{mc} = \frac{200 \text{ Btu}}{50 \text{ lb} \times 0.5 \text{ Btu/lb-°F}} = 8 \text{ °F}$$

The final temperature is therefore 25 °F − 8 °F = 17 °F.

18.8. Ten kcal of heat is added to a 1-kg sample of wood and its temperature is found to rise from 20 °C to 44 °C. What is the specific heat capacity of the wood?

$$Q = mc \, \Delta T$$

$$c = \frac{Q}{m \, \Delta T} = \frac{10 \text{ kcal}}{1 \text{ kg} \times 24 \text{ °C}} = 0.42 \text{ kcal/kg-°C}$$

18.9. Three lb of water at 100 °F is added to 5 lb of water at 40 °F. What is the final temperature of the mixture?

If T is the final temperature, then the 5 lb of water initially at 40 °F undergoes a temperature change of $\Delta T_1 = T - 40$ °F and the 3 lb of water initially at 100 °F undergoes a temperature change of $\Delta T_2 = 100$ °F $- T$. We proceed as follows:

$$\text{Heat gained} = \text{heat lost}$$
$$m_1 c_1 \Delta T_1 = m_2 c_2 \Delta T_2$$
$$5 \text{ lb} \times 1 \text{ Btu/lb-°F} \times (T - 40 \text{ °F}) = 3 \text{ lb} \times 1 \text{ Btu/lb-°F} \times (100 \text{ °F} - T)$$
$$(5T - 200) \text{ Btu} = (300 - 3T) \text{ Btu}$$
$$8T = 500$$
$$T = 62.5 \text{ °F}$$

18.10. In preparing tea, 600 g of water at 90 °C is poured into a 200-g china pot ($c_{pot} = 0.2$ kcal/kg-°C) at 20 °C. What is the final temperature of the water?

$$\text{Heat gained by pot} = \text{heat lost by water}$$
$$m_{pot} c_{pot} \Delta T_{pot} = m_{water} c_{water} \Delta T_{water}$$
$$0.2 \text{ kg} \times 0.2 \text{ kcal/kg-°C} \times (T - 20 \text{ °C}) = 0.6 \text{ kg} \times 1 \text{ kcal/kg-°C} \times (90 \text{ °C} - T)$$
$$(0.04T - 0.8) \text{ kcal} = (54 - 0.6T) \text{ kcal}$$
$$0.64T = 54.8$$
$$T = 86 \text{ °C}$$

18.11. In order to raise the temperature of 5 kg of water from 20 °C to 30 °C a 2-kg iron bar is heated and then dropped into the water. What should the temperature of the bar be? ($c_{iron} = 0.11$ kcal/kg-°C)

Let the temperature of the iron bar be T. Then the change in the water's temperature is $\Delta T_w = 30$ °C $- 20$ °C $= 10$ °C and the change in the bar's temperature is $\Delta T_{iron} = T - 30$ °C. We proceed in the usual way:

$$\text{Heat gained by water} = \text{heat lost by bar}$$
$$m_w c_w \Delta T_w = m_{iron} c_{iron} \Delta T_{iron}$$
$$5 \text{ kg} \times 1 \text{ kcal/kg-°C} \times 10 \text{ °C} = 2 \text{ kg} \times 0.11 \text{ kcal/kg-°C} \times (T - 30 \text{ °C})$$
$$50 \text{ kcal} = (0.22T - 6.6) \text{ kcal}$$
$$0.22T = 56.6$$
$$T = 257 \text{ °C}$$

18.12. How much heat must be added to 200 lb of lead at 70 °F to cause it to melt? The specific heat capacity of lead is 0.03 Btu/lb-°F, it melts at 626 °F, and its heat of fusion is 10.6 Btu/lb.

Here $\Delta T = 626$ °F $- 70$ °F $= 556$ °F. Hence
$$Q = mc \, \Delta T + m L_f$$
$$= 200 \text{ lb} \times 0.03 \text{ Btu/lb-°F} \times 556 \text{ °F} + 200 \text{ lb} \times 10.6 \text{ Btu/lb}$$
$$= 3336 \text{ Btu} + 2120 \text{ Btu} = 5456 \text{ Btu}$$

18.13. Five kg of water at 40 °C is poured on a large block of ice at 0 °C. How much ice melts?

$$\text{Heat lost by water} = \text{heat gained by ice}$$
$$m_w c \, \Delta T = m_{ice} L_f$$
$$5 \text{ kg} \times 1 \text{ kcal/kg-°C} \times 40 \text{ °C} = m_{ice} \times 80 \text{ kcal/kg}$$
$$m_{ice} = \left(\frac{200}{80} \right) \text{kg} = 2.5 \text{ kg}$$

18.14. Five hundred kcal of heat is added to 2 kg of water at 80 °C. How much steam is produced?

The heat needed to raise the temperature of the water from 80 °C to the boiling point of 100 °C is

$$Q_1 = m_1 c \, \Delta T = 2 \text{ kg} \times 1 \text{ kcal/kg-°C} \times 20 \text{ °C} = 40 \text{ kcal}$$

Hence $Q_2 = 500 \text{ kcal} - 40 \text{ kcal} = 460 \text{ kcal}$ of heat is available to convert water at 100 °C to steam at the same temperature. Since $Q_2 = m_{\text{steam}} L_v$, the steam produced is

$$m_{\text{steam}} = \frac{Q_2}{L_v} = \frac{460 \text{ kcal}}{540 \text{ kcal/kg}} = 0.85 \text{ kg}$$

18.15. A 30-g ice cube at 0 °C is dropped into 200 g of water at 30 °C. What is the final temperature?

If T is the final temperature, then $\Delta T_{\text{ice}} = T - 0 \text{ °C}$ and $\Delta T_{\text{water}} = 30 \text{ °C} - T$. Therefore

$$\text{Heat gained by ice} = \text{heat lost by water}$$

$$m_{\text{ice}} L_f + m_{\text{ice}} c_{\text{water}} \Delta T_{\text{ice}} = m_{\text{water}} c_{\text{water}} \Delta T_{\text{water}}$$

$$0.03 \text{ kg} \times 80 \frac{\text{kcal}}{\text{kg}} + 0.03 \text{ kg} \times 1 \frac{\text{kcal}}{\text{kg-°C}} \times (T - 0 \text{ °C}) = 0.2 \text{ kg} \times 1 \frac{\text{kcal}}{\text{kg-°C}} \times (30 \text{ °C} - T)$$

$$(2.4 + 0.03 \, T) \text{ kcal} = (6 - 0.2T) \text{ kcal}$$

$$0.23T = 3.6$$

$$T = 15.7 \text{ °C}$$

18.16. How much steam at 292 °F is required to melt 1 lb of ice at 32 °F?

Here $\Delta T_1 = 292 \text{ °F} - 212 \text{ °F} = 80 \text{ °F}$ and $\Delta T_2 = 212 \text{ °F} - 32 \text{ °F} = 180 \text{ °F}$. Therefore, if m_s is the mass of steam,

$$\text{Heat gained by ice} = \text{heat lost by steam}$$

$$m_{\text{ice}} L_f = m_s c_s \Delta T_1 + m_s L_v + m_s c_{\text{water}} \Delta T_2$$

$$1 \text{ lb} \times 144 \text{ Btu/lb} = m_s \times 0.48 \text{ Btu/lb-°F} \times 80 \text{ °F} + m_s \times 972 \text{ Btu/lb}$$

$$+ m_s \times 1 \text{ Btu/lb-°F} \times 180 \text{ °F}$$

$$144 \text{ Btu} = (38.4 + 972 + 180)m_s \text{ Btu} = 1190 m_s \text{ Btu}$$

$$m_s = \left(\frac{144}{1190} \right) \text{lb} = 0.12 \text{ lb}$$

Supplementary Problems

18.17. A glass of water is stirred and then allowed to stand until the water stops moving. What has happened to the kinetic energy of the moving water?

18.18. Why is an ice cube at 0 °C more effective in cooling a drink than the same mass of water at 0 °C?

18.19. Ethyl alcohol melts at −114 °C and boils at 78 °C. What are the Fahrenheit equivalents of these temperatures?

18.20. Bromine melts at 19 °F and boils at 140 °F. What are the Celsius equivalents of these temperatures?

18.21. How much heat must be removed from 4 lb of water at 200 °F to reduce its temperature to 50 °F?

18.22. How much heat must be added to a 20-kg block of ice to raise its temperature from −20 °C to −5° C? ($c_{\text{ice}} = 0.50 \text{ kcal/kg-°C}$)

18.23. How much heat is lost by a 50-g silver spoon when it is cooled from 20 °C to 0 °C? ($c_{silver} = 0.056$ kcal/kg-°C)

18.24. Seven hundred kcal of heat is added to a 100-kg marble statue of Isaac Newton initially at 18 °C. What is its final temperature? ($c_{marble} = 0.21$ kcal/kg-°C)

18.25. Two Btu of heat is added to a $\frac{1}{2}$-lb glass vessel at 70 °F and its temperature is found to rise to 90 °F. What is the specific heat capacity of the glass?

18.26. Ten kg of water at 5 °C is added to 100 kg of water at 80 °C. What is the final temperature of the mixture?

18.27. A 600-g copper dish contains 1500 g of water at 20 °C. A 100-g iron bar at 120 °C is dropped into the water. What is the final temperature of the water? ($c_{copper} = 0.093$ kcal/kg-°C; $c_{iron} = 0.11$ kcal/kg-°C)

18.28. Four pounds of soup at 140 °F is poured into a 4-lb china serving dish at 70 °F. What is the final temperature of the soup? ($c_{soup} = 0.9$ Btu/lb-°F; $c_{dish} = 0.2$ Btu/lb-°F)

18.29. How much water at 35 °F must be added to 10 lb of punch at 70 °C that is in a 3-lb silver punch bowl to lower its temperature to 60 °C? ($c_{punch} = 0.7$ Btu/lb-°F; $c_{silver} = 0.056$ Btu/lb-°F)

18.30. How much heat must be removed from 200 g of water at 30 °C to convert it to ice at 0 °C?

18.31. Three lb of water at 100 °F is poured on a large block of ice at 32 °F. How much ice melts?

18.32. Five hundred kcal of heat is added to 10 kg of zinc at 70 °F. How much zinc melts? The specific heat capacity of zinc is 0.092 kcal/kg-°C, it melts at 420 °C, and its heat of fusion is 24 kcal/kg.

18.33. How much ice at 0 °C must be added to 200 g of water at 30 °C to lower its temperature to 20 °C?

18.34. Four thousand Btu of heat is added to 3 lb of water at 100 °F. How much steam is produced? What is the temperature of the steam?

18.35. River water at 50 °F is used to condense spent steam at 250 °F in an electric generating plant to water at 120 °F. If the cooling water leaves the condenser at 90 °F, how many pounds of river water are needed per pound of spent steam?

Answers to Supplementary Problems

18.17. The kinetic energy of the moving water becomes dissipated into internal energy, and the water therefore has a higher temperature than before it was stirred.

18.18. The ice absorbs heat from the drink in order to melt to water at 0 °C.

18.19. -173 °F; 172 °F	**18.25.** 0.2 Btu/lb-°F	**18.31.** 1.4 lb
18.20. -7 °C; 60 °C	**18.26.** 73 °C	**18.32.** 7.4 kg
18.21. 600 Btu	**18.27.** 20.7 °C	**18.33.** 20 g
18.22. 150 kcal	**18.28.** 127 °F	**18.34.** 3 lb; 731 °F
18.23. 0.056 kcal	**18.29.** 2.9 lb	**18.35.** 27 lb
18.24. 51 °C	**18.30.** 22 kcal	

Chapter 19

Expansion of Solids, Liquids, and Gases

LINEAR EXPANSION

A change in temperature ΔT causes most solids to change in length (or other linear dimension) by an amount ΔL that is proportional to both the original length L_0 and ΔT:

$$\Delta L = aL_0\Delta T$$

Change in length = a × original length × temperature change

The quantity a is a constant that depends upon the nature of the material; it is called the *coefficient of linear expansion*.

VOLUME EXPANSION

The change in volume ΔV of a solid or liquid whose original volume is V_0 when its temperature is changed by ΔT is given by

$$\Delta V = bV_0\,\Delta T$$

Change in volume = b × original volume × temperature change

The quantity b is the *coefficient of volume expansion*. Generally $b = 3a$ for a given material.

BOYLE'S LAW

At constant temperature, the volume of a sample of gas is inversely proportional to the absolute pressure applied to the gas. The greater the pressure, the smaller the volume. This relationship is known as *Boyle's law*. If p_1 is the gas pressure when its volume is V_1 and p_2 is its pressure when its volume is V_2, then Boyle's law states that

$$p_1V_1 = p_2V_2 \qquad (T = \text{constant})$$

ABSOLUTE TEMPERATURE SCALES

When the temperature of a sample of gas is changed while the pressure on it is held constant, its volume changes by $1/273$ of its volume at 0 °C for each temperature change of 1 °C. If it were possible to cool a gas sample to -273 °C, its volume would diminish to zero. Since all gases condense into liquids at temperatures above -273 °C, this experiment cannot be carried out; nevertheless, -273 °C is a significant temperature.

On the *absolute temperature scale*, the zero point is set at -273 °C. Temperatures in this scale are expressed in *kelvins* (K); these units are equal to Celsius degrees. Thus

$$T_K = T_C + 273$$

The freezing point of water on the absolute scale is 273 K and its boiling point is 373 K.

The *Rankine scale* is an absolute temperature scale based upon the Fahrenheit scale. Absolute zero in the Rankine scale is -460 °F, and

$$T_R = T_F + 460°$$

The freezing point of water on the Rankine scale is 460 °R and its boiling point is 672 °R.

CHARLES'S LAW

Because of the way the absolute temperature scales are defined, the relationship between the temperature and volume of a gas sample at constant pressure can be expressed as

$$\frac{V_1}{T_1} = \frac{V_2}{T_2} \qquad (p = \text{constant})$$

In this formula, which is called *Charles's law*, V_1 is the volume of the sample at the absolute temperature T_1 and V_2 is its volume at the absolute temperature T_2; the formula only holds when the temperatures are expressed in an absolute scale.

IDEAL GAS LAW

Boyle's law and Charles's law can be combined to form the *ideal gas law*:

$$\frac{p_1 V_1}{T_1} = \frac{p_2 V_2}{T_2}$$

This law is obeyed fairly well by all gases through a wide range of pressures and temperatures. An *ideal gas* is one for which $pV/T = \text{constant}$ under all circumstances; though no such gas actually exists, the fact that a real gas behaves approximately like an ideal one provides a specific target for theories of the gaseous state.

The ideal gas law is further discussed in Chapter 20.

Solved Problems

19.1. A surveyor's steel tape measure is calibrated at 65 °F. A reading of 200 ft is found when the tape measure is used to determine the width of a building lot when the temperature is 10 °F. How great an error does the temperature difference introduce? The coefficient of linear expansion of steel is $6.7 \times 10^{-6}/°F$.

Since $\Delta T = 65\ °F - 10\ °F = 55\ °F$, the 200-ft length of steel tape shrinks by

$$\Delta L = a L_0\, \Delta T = 6.7 \times 10^{-6}/°F \times 200\ \text{ft} \times 55\ °F = 0.0737\ \text{ft} = 0.88\ \text{in.}$$

19.2. A wooden wagon wheel has an outside diameter of 120 cm. The iron tire for this wheel is deliberately made smaller so that it can be shrunk in place to be a tight fit. If the tire's inside diameter is 119.6 cm at 20 °C, find the temperature to which it must be heated in order to fit over the wheel. The coefficient of linear expansion of steel is $1.2 \times 10^{-5}/°C$.

Here $L_0 = 119.6$ cm and $\Delta L = 0.4$ cm. Hence

$$\Delta T = \frac{\Delta L}{a L_0} = \frac{0.4\ \text{cm}}{1.2 \times 10^{-5}/°C \times 119.6\ \text{cm}} = 279\ °C$$

The required temperature is therefore 20 °C + 279 °C = 299 °C.

19.3. Lead has a coefficient of linear expansion of $3 \times 10^{-5}/°C$ and its density at 20 °C is 11.0 g/cm^3. Find the density of lead at 200 °C.

Since $b = 3a$, the coefficient of volume expansion of lead is

$$b = 3 \times 3 \times 10^{-5}/°C = 9 \times 10^{-5}/°C$$

At 20 °C the volume of a mass m of lead is $V_0 = m/d_0$, and at 200 °C it is

$$V = \frac{m}{d} = V_0 + \Delta V = V_0 + b V_0\, \Delta T$$

Substituting $V_0 = m/d_0$ yields

$$\frac{m}{d} = \frac{m}{d_0} + \frac{bm\,\Delta T}{d_0} = \frac{m}{d_0}(1 + b\,\Delta T)$$

$$d = \frac{d_0}{1 + b\,\Delta T}$$

Here $\Delta T = 200\ °\text{C} - 20\ °\text{C} = 180\ °\text{C}$, and so

$$d = \frac{11\ \text{g/cm}^3}{1 + 9 \times 10^{-5}/°\text{C} \times 180\ °\text{C}} = 10.8\ \text{g/cm}^3$$

19.4. How much water overflows when a Pyrex vessel filled to the brim with 1 liter (1000 cm^3) of water at 20 °C is heated to 90 °C? The coefficients of volume expansion of Pyrex and water are $9 \times 10^{-6}/°\text{C}$ and $2.1 \times 10^{-4}/°\text{C}$ respectively.

A cavity in an object expands or contracts by the same amount as a solid body with the composition of the object and the same volume as the cavity. Hence the volume of water that overflows is

$$V_w - V_P = b_w V_0\,\Delta T - b_P V_0\,\Delta T = (b_w - b_P)V_0\,\Delta T$$

$$= (210 \times 10^{-6}/°\text{C} - 9 \times 10^{-6}/°\text{C}) \times 1000\ \text{cm}^3 \times 70\ °\text{C}$$

$$= 14.1\ \text{cm}^3$$

19.5. Find the force associated with the expansion of a steel beam whose cross-sectional area is 10 in^2 when its temperature increases from 50 °F to 95 °F. Young's modulus for steel is $Y = 2.9 \times 10^7$ lb/in^2 and $a = 6.7 \times 10^{-6}/°\text{F}$.

The change in the beam's length ΔL due to thermal expansion is given by

$$\Delta L = aL_0\,\Delta T$$

To obtain the same change in length by mechanical means would involve applying a tension force F such that

$$F = \frac{YA\,\Delta L}{L_0}$$

according to Chapter 13. Thus the amount of force associated with the expansion is

$$F = \frac{YA\,\Delta L}{L_0} = \frac{YA(aL_0\,\Delta T)}{L_0} = YaA\,\Delta T$$

The force depends on the cross section of the beam but not on its length. For the beam under consideration here

$$F = YaA\,\Delta T = 2.9 \times 10^7\ \frac{\text{lb}}{\text{in}^2} \times 6.7 \times 10^{-6}/°\text{F} \times 10\ \text{in}^2 \times 45\ °\text{F}$$

$$= 8.74 \times 10^4\ \text{lb}$$

which is nearly 44 tons.

19.6. A 1-liter sample of nitrogen at 0 °C and 1 atm pressure is compressed to 0.5 liter. If the temperature is unchanged, what happens to the pressure of the sample?

Since $p_2 = p_1 V_1/V_2$ and here $V_1/V_2 = 2$, the pressure doubles to 2 atm.

19.7. A steel cylinder contains 3 ft^3 of air at a gauge pressure of 200 lb/in^2. What volume would this amount of air occupy at sea-level atmospheric pressure of 15 lb/in^2?

The absolute pressure of the air in the tank is

$$p_1 = \text{gauge pressure} + \text{atmospheric pressure} = 200\ \text{lb/in}^2 + 15\ \text{lb/in}^2$$

$$= 215\ \text{lb/in}^2$$

and its volume is $V_1 = 3$ ft^3. When $p_2 = 15$ lb/in^2,

$$V_2 = \frac{p_1 V_1}{p_2} = \frac{215\ \text{lb/in}^2 \times 3\ \text{ft}^3}{15\ \text{lb/in}^2} = 43\ \text{ft}^3$$

19.8. What additional volume of air at atmospheric pressure must be pumped into the tank of Problem 19.7 in order to raise the gauge pressure to 300 lb/in²?

When the absolute pressure of the air in the tank is

$$p_1 = 300 \text{ lb/in}^2 + 15 \text{ lb/in}^2 = 315 \text{ lb/in}^2$$

the equivalent volume of air at $p_2 = 15$ lb/in² is

$$V_2 = \frac{p_1 V_1}{p_2} = \frac{315 \text{ lb/in}^2 \times 3 \text{ ft}^3}{15 \text{ lb/in}^2} = 63 \text{ ft}^3$$

Hence the additional volume of air is 63 ft³ − 43 ft³ = 20 ft³.

19.9. Nitrogen boils at − 196 °C. What is this temperature on the absolute scale?

$$T_K = T_C + 273 = -196 + 273 = 77 \text{ K}$$

19.10. The surface temperature of the sun is about 6000 K. What is the Celsius equivalent of this temperature?

$$T_C = T_K - 273 = 5727 \text{ °C}$$

19.11. Ethyl alcohol freezes at − 173 °F and boils at 172 °F. What are these temperatures on the Rankine scale?

$$T_R = T_F + 460° = -173° + 460° = 287 \text{ °R}$$
$$T_R = 172° + 460° = 632 \text{ °R}$$

19.12. To what temperature must a gas sample initially at 0 °C and atmospheric pressure be heated if its volume is to double while its pressure remains the same?

Since $T_1 = 0$ °C = 273 K and $V_2 = 2V_1$, from Charles's law

$$T_2 = \frac{T_1 V_2}{V_1} = \frac{273 \text{ K} \times 2V_1}{V_1} = 546 \text{ K} = 273 \text{ °C}$$

19.13. The tire of a car contains air at an absolute pressure of 35 lb/in² when its temperature is 50 °F. If the tire's volume does not change, what is the pressure when the temperature is 120 °F?

The first step is to convert the temperatures to their Rankine equivalents:
$$T_1 = 50° + 460° = 510 \text{ °R}, \qquad T_2 = 120° + 460° = 580 \text{ °R}$$

Since $p_1 = 35$ lb/in² and $V_1 = V_2$, from the ideal gas law $p_1 V_1 / T_1 = p_2 V_2 / T_2$ we have

$$p_2 = \frac{T_2 p_1}{T_1} = \frac{580 \text{ °R} \times 35 \text{ lb/in}^2}{510 \text{ °R}} = 40 \text{ lb/in}^2$$

19.14. A tank whose capacity is 0.1 m³ contains helium at an absolute pressure of 10 atm and a temperature of 20 °C. A rubber weather balloon is inflated with this helium. (*a*) The gas cools as it expands, and when the pressure of the helium in the balloon is 1 atm, its temperature is − 40 °C. Find the volume of the balloon. (*b*) Eventually the helium in the balloon absorbs heat from the air around it and returns to 20 °C. Find the volume of the balloon at this time.

(*a*) $T_1 = 20$ °C = 293 K, $T_2 = -40$ °C = 233 K, $V_1 = 0.1$ m³, $p_1 = 10$ atm, $p_2 = 1$ atm. From the ideal gas law,

$$V_2 = \frac{T_2 p_1 V_1}{T_1 p_2} = \frac{233 \text{ K} \times 10 \text{ atm} \times 0.1 \text{ m}^3}{293 \text{ K} \times 1 \text{ atm}} = 0.8 \text{ m}^3$$

The volume of the balloon is therefore 0.7 m³ at this time, since the tank retains 0.1 m³ of the helium after the expansion.

(b) Here $p_1 = 10$ atm, $p_3 = 1$ atm, $V_1 = 0.1$ m³, and, since $T_1 = T_3$, Boyle's law can be used. We have

$$V_3 = \frac{p_1 V_1}{p_3} = \frac{10 \text{ atm} \times 0.1 \text{ m}^3}{1 \text{ atm}} = 1 \text{ m}^3$$

The volume of the balloon is therefore 0.9 m³, assuming it is still attached to the tank.

19.15. A gas sample occupies 5 ft³ at 60 °F and atmospheric pressure.   (a) Find its volume at 200 °F and a gauge pressure of 50 lb/in².   (b) Find its gauge pressure when it has been compressed to 1 ft³ and the temperature has been reduced to 0 °F.

(a) Here $T_1 = 60° + 460° = 520$ °R, $V_1 = 5$ ft³, and $p_1 = 15$ lb/in².   When

$$T_2 = 200° + 460° = 660 \text{ °R} \qquad \text{and} \qquad p_2 = 50 \text{ lb/in}^2 + 15 \text{ lb/in}^2 = 65 \text{ lb/in}^2$$

the volume V_2 is

$$V_2 = \frac{T_2 p_1 V_1}{T_1 p_2} = \frac{660 \text{ °R} \times 15 \text{ lb/in}^2 \times 5 \text{ ft}^3}{520 \text{ °R} \times 65 \text{ lb/in}^2} = 1.46 \text{ ft}^3$$

(b) Now $T_2 = 0° + 460° = 460$ °R and $V_2 = 1$ ft³.   Hence the new absolute pressure is

$$p_2 = \frac{T_2 p_1 V_1}{T_1 V_2} = \frac{460 \text{ °R} \times 15 \text{ lb/in}^2 \times 5 \text{ ft}^3}{520 \text{ °R} \times 1 \text{ ft}^3} = 66 \text{ lb/in}^2$$

The new gauge pressure is $(66 - 15)$ lb/in² = 51 lb/in².

19.16. The density of air at 0 °C and 1 atm pressure is 1.293 kg/m³.   Find its density at 100 °C and 2 atm pressure.

At a pressure of $p_1 = 1$ atm and an absolute temperature of $T_1 = 0 + 273 = 273$ K the mass of a volume of air V_1 of density d_1 is $m = d_1 V_1$.   At $p_2 = 2$ atm and $T_2 = 100 + 273 = 373$ K the volume V_2 is

$$V_2 = \frac{m}{d_2} = \frac{T_2 p_1 V_1}{T_1 p_2}$$

Since $m = d_1 V_1$,

$$\frac{d_1 V_1}{d_2} = \frac{T_2 p_1 V_1}{T_1 p_2}$$

and so

$$d_2 = \frac{d_1 T_1 p_2}{T_2 p_1} = \frac{1.293 \text{ kg/m}^3 \times 273 \text{ K} \times 2 \text{ atm}}{373 \text{ K} \times 1 \text{ atm}} = 1.893 \text{ kg/m}^3$$

Supplementary Problems

19.17. A steel bridge is 500 m long at 0 °C.   By how much does it expand when the temperature becomes 35 °C?   The coefficient of linear expansion of steel is 1.2×10^{-5}/°C.

19.18. A brass rod 4 ft long expands by $\frac{1}{16}$ in. when heated from 70 °F to 200 °F.   Find the coefficient of linear expansion of brass.

19.19. How much mercury ($b = 1.8 \times 10^{-4}$/°C) overflows when a glass vessel ($b = 2 \times 10^{-5}$/°C) filled to the brim with 50 cm³ of mercury at 15 °C is heated to 75 °C?

19.20. The density of mercury is 830 lb/ft³ at 70 °F.   Find its density at 0 °F.   The coefficient of volume expansion of mercury is 1.0×10^{-4}/°F.

19.21. A load of 2000 kg is placed on a vertical steel beam 5 m high whose cross-sectional area is 30 cm². The temperature is 20 °C. (a) By how much is the beam compressed? (b) At what temperature will the beam return to its original height? For steel, $Y = 2 \times 10^{11}$ N/m² and $a = 1.2 \times 10^{-5}/°F$; 1 m² = 10⁴ cm².

19.22. Aluminum has a coefficient of linear expansion of $1.3 \times 10^{-5}/°F$ and a Young's modulus of 1.0×10^7 lb/in²; for steel the respective values are $6.7 \times 10^{-6}/°F$ and 2.9×10^7 lb/in². If identical bars of the two metals undergo the same temperature change, which bar has the greater force associated with its change in length?

19.23. A compressor pumps 50 liters of air at a pressure of 1 atm into an 8-liter tank. What is the absolute pressure (in atm) of the air in the tank?

19.24. How much air at a pressure of 1 atm can be stored in a 2-m³ tank which can safely withstand a pressure of 5×10^5 N/m²?

19.25. What is the Celsius equivalent of a temperature of 500 K?

19.26. What is the Kelvin equivalent of a temperature of 500 °C?

19.27. What is the Fahrenheit equivalent of a temperature of 500 °R?

19.28. What is the Rankine equivalent of a temperature of 500 °F?

19.29. A tire contains 1 ft³ of air at a gauge pressure of 28 lb/in². How much additional air at atmospheric pressure must be pumped into the tire to raise the pressure to 36 lb/in² at the same temperature?

19.30. A tire contains air at a gauge pressure of 28 lb/in² at 60 °F. If the tire's volume does not change, what will the gauge pressure be when its temperature is 100 °F?

19.31. The weight density of carbon dioxide is 0.1234 lb/ft³ at 32 °F and 1 atm pressure. Find its weight density at 80 °F and 10 atm pressure.

19.32. A gas sample occupies 4 m³ at an absolute pressure of 2×10^5 N/m² and a temperature of 320 K. Find its volume (a) at the same pressure and a temperature of 400 K, and (b) at the same temperature and a pressure of 4×10^4 N/m².

19.33. A gas sample occupies a volume of 1 m³ at a temperature of 27 °C and a pressure of 1 atm. Find its volume (a) at 127 °C and 0.5 atm; (b) at 127 °C and 2 atm; (c) at −73 °C and 0.5 atm; and (d) at −73 °C and 2 atm.

Answers to Supplementary Problems

19.17.	21 cm	**19.23.**	6.25 atm	**19.29.**	0.53 ft³
19.18.	$1.0 \times 10^{-5}/°F$	**19.24.**	9.87 m³	**19.30.**	31.3 lb/in²
19.19.	0.48 cm³	**19.25.**	227 °C	**19.31.**	1.124 lb/ft³
19.20.	836 lb/ft³	**19.26.**	773 K	**19.32.**	(a) 5 m³ (b) 20 m³
19.21.	(a) 0.163 mm (b) 22.7 °C	**19.27.**	40 °F	**19.33.**	(a) 2.67 m³ (b) 0.67 m³
19.22.	The steel bar.	**19.28.**	960 °R		(c) 1.33 m³ (d) 0.33 m³

Chapter 20

Kinetic Theory of Matter

KINETIC THEORY OF GASES

The *kinetic theory of gases* holds that a gas is composed of very small particles, called molecules, that are in constant random motion. The molecules are far apart relative to their dimensions and do not interact with one another except in collisions.

The pressure a gas exerts is due to the impacts of its molecules; there are so many molecules in even a small gas sample that the individual blows appear as a continuous force. Boyle's law is readily understood in terms of the kinetic theory of gases. Expanding a gas sample means that its molecules must travel farther between successive impacts on the container walls and that the impacts are spread over a larger area. Hence an increase in volume means a decrease in pressure, and vice versa.

MOLECULAR ENERGY

According to the kinetic theory of gases, the average kinetic energy of the molecules of a gas is proportional to the absolute temperature of the gas. This relationship is usually expressed in the form

$$KE_{av} = \frac{3}{2} kT$$

where k = Boltzmann's constant = 1.38×10^{-23} J/K. Actual molecular energies vary considerably on either side of KE_{av}.

At absolute zero, 0 K, gas molecules would be at rest, which is why this is such a significant temperature. At any temperature, all gases have the same average molecular energy. Therefore, in a gas whose molecules are heavy, the molecules move more slowly on the average than do those in a gas at the same temperature whose molecules are light.

Charles's law follows directly from the above interpretation of temperature. Compressing a gas causes its temperature to rise because molecules rebound from the inward-moving walls of the container with increased energy, just as a tennis ball rebounds with greater energy when struck by a racket. Similarly, expanding a gas causes its temperature to fall because molecules rebound from the outward-moving walls with decreased energy.

SOLIDS AND LIQUIDS

The molecules of a solid are close enough together to exert forces on one another that hold the entire assembly to a definite size and shape. As in the case of a gas, the molecules are in constant motion, but they vibrate about fixed locations instead of moving randomly. The molecules of a liquid continually move around past one another more or less freely, which enables the liquid to flow, but their spacing does not change and so the volume of a given liquid sample does not vary.

When a solid melts, the original ordered arrangement of its molecules changes to the random arrangement of molecules in a liquid. To accomplish the change, the molecules must be pulled apart against the forces holding them in place, which requires energy. The heat of fusion of a solid represents this energy. When a liquid boils, the heat of vaporization represents the energy needed to pull its molecules entirely free of one another so that a gas is formed.

ATOMS AND MOLECULES

Elements are the fundamental substances of which all matter in bulk is composed.   There are 105 known elements, of which a number are not found in nature but have been prepared in the laboratory. Elements cannot be transformed into one another by ordinary chemical or physical means, but two or more elements can combine to form a *compound*, which is a substance whose properties are different from those of its constituent elements.

The ultimate particles of an element are called *atoms*, and those of a compound which exists in the gaseous state are called *molecules*.   The molecules of a compound consist of the atoms of the elements that compose it joined together in a specific arrangement; each molecule of water, for instance, contains two hydrogen atoms and one oxygen atom, as its symbol H_2O indicates.   Many compounds in the solid and liquid states do not consist of individual molecules, as discussed later.   Elemental gases may consist of atoms (helium, He; argon, Ar) or of molecules (hydrogen, H_2; oxygen, O_2).

The masses of atoms and molecules are expressed in *atomic mass units* (u), where

$$1 \text{ atomic mass unit} = 1 \text{ u} = 1.660 \times 10^{-27} \text{ kg}$$

The mass of a molecule is the sum of the masses of the atoms of which it is composed; thus $m(H_2O) = 2m(H) + m(O)$.

THE MOLE

The samples of matter used in both industry and the laboratory involve so many atoms or molecules that counting them is out of the question.   Instead the mass of a particular sample is used as a measure of its quantity, and it is necessary to relate this mass to the number of atoms or molecules present in the sample.

When the masses of two samples of different substances stand in the same proportion as their molecular masses, they contain the same number of molecules.   To make use of this fact, the *mole* is defined as follows: A mole of any substance is that amount of it whose mass is equal to its molecular mass expressed in grams instead of atomic mass units.   Thus a mole of water has a mass of 18 g since a water molecule has a mass of 18 u.   A mole of an elemental substance that consists of individual atoms rather than molecules is that amount of it whose mass is equal to its atomic mass expressed in grams.   In the SI system the amount of a substance corresponding to a mole is taken as a basic unit and written as 1 mol.

The number of molecules in a mole of any substance is *Avogadro's number N*, whose value is

$$N = 6.023 \times 10^{23} \text{ molecules/mole}$$

The number of molecules in a sample of a substance is the number of moles it contains multiplied by N.

MOLAR VOLUME

Equal volumes of all gases, under the same conditions of temperature and pressure, contain the same number of molecules and therefore the same number of moles.   This observation is most useful stated in reverse: under given conditions of temperature and pressure, the volume of a gas is proportional to the number of moles present.

For convenience, a temperature of 0 °C (273 K) and a pressure of 1 atm ($1.013 \times 10^5 \text{ N/m}^2 = 14.7$ lb/in^2) are taken as the standard temperature and pressure (STP); Charles's and Boyle's laws permit measurements made at other temperatures and pressures to be reduced to their equivalents at STP. Experimentally it is found that one mole of any gas at STP occupies a volume of 22.4 liters.   Thus the *molar volume* of a gas is 22.4 liters at STP. This observation makes it possible to deal with gas volumes in chemical reactions. If a certain reaction is known to produce 2.5 moles of a gas, for instance, we know that at STP the volume of the gas will be

$$V = \text{moles of gas} \times \text{molar volume} = 2.5 \text{ moles} \times 22.4 \text{ liters/mole} = 56 \text{ liters}$$

UNIVERSAL GAS CONSTANT

According to the ideal gas law (Chapter 19), the pressure, volume, and temperature of a gas sample obey the relationship $pV/T = $ constant. We can find the value of the constant in terms of the number of moles n of gas in the sample by making use of the fact that the molar volume at STP is 22.4 liters. At STP we have $T = 0$ °C $= 273$ K, $p = 1$ atm, and $V = n \times 22.4$ liters/mole, so that

$$\frac{pV}{T} = \frac{n \times 1 \text{ atm} \times 22.4 \text{ liters/mole}}{273 \text{ K}} = nR$$

where R, the *universal gas constant*, has the value

$$R = 0.0821 \text{ atm-liter/mole-K}$$

In SI units, in which the unit of p is the N/m^2 and the unit of V is the m^3,

$$R = 8.31 \text{ J/mol-K}$$

The complete ideal gas law is usually written in the form

$$pV = nRT$$

Solved Problems

20.1. Gas molecules have velocities comparable with those of rifle bullets, yet we all know that a gas with a strong odor, such as ammonia, takes several seconds to diffuse through a room. Why?

Gas molecules collide frequently with one another, which means that a particular molecule follows a long, very complicated path in going from one place to another.

20.2. Explain the evaporation of a liquid at a temperature below its boiling point on the basis of the kinetic theory of matter.

At any moment in a liquid, some molecules are moving faster and others are moving slower than the average. The fastest ones are able to escape from the liquid surface despite the attractive forces exerted by the other molecules; this constitutes evaporation. The warmer the liquid, the greater the number of very fast molecules, and the more rapidly evaporation takes place. Since the molecules that remain behind are the slower ones, the liquid has a lower temperature than before (unless heat has been added to it from an outside source during the process).

20.3. Find the mass of (*a*) the water molecule, H_2O, and (*b*) the ethyl alcohol molecule, C_2H_6O. The atomic masses of H, C, and O are respectively 1.008 u, 12.01 u, and 16.00 u.

(*a*)
$$2H = 2 \times 1.008 \text{ u} = 2.02 \text{ u}$$
$$1O = 1 \times 16.00 \text{ u} = \underline{16.00 \text{ u}}$$
$$18.02 \text{ u}$$

$$m(H_2O) = 18.02 \text{ u} \times 1.66 \times 10^{-27} \text{ kg/u} = 2.99 \times 10^{-26} \text{ kg}$$

(*b*)
$$2C = 2 \times 12.01 \text{ u} = 24.02 \text{ u}$$
$$6H = 6 \times 1.008 \text{ u} = 6.05 \text{ u}$$
$$1O = 1 \times 16.00 \text{ u} = \underline{16.00 \text{ u}}$$
$$46.07 \text{ u}$$

$$m(C_2H_6O) = 46.07 \text{ u} \times 1.66 \times 10^{-27} \text{ kg/u} = 7.65 \times 10^{-26} \text{ kg}$$

20.4. How many H_2O molecules are present in 1 kg of water?

$$\text{Molecules of } H_2O = \frac{\text{mass of } H_2O}{\text{mass of } H_2O \text{ molecule}}$$

$$= \frac{1 \text{ kg}}{2.99 \times 10^{-26} \text{ kg}} = 3.34 \times 10^{25} \text{ molecules}$$

20.5. What is the average kinetic energy of the molecules of any gas at 100 °C?

The absolute temperature corresponding to 100 °C is

$$T_K = T_C + 273 = 373 \text{ K}$$

The average kinetic energy at this temperature is

$$KE_{av} = \frac{3}{2} kT = \frac{3}{2} \times 1.38 \times 10^{-23} \text{ J/K} \times 373 \text{ K} = 7.72 \times 10^{-21} \text{ J}$$

20.6. What is the average velocity of the molecules in a sample of oxygen at 100 °C? The mass of an oxygen molecule is 5.3×10^{-26} kg.

Since $KE_{av} = \frac{1}{2} mv_{av}^2 = \frac{3}{2} kT$,

$$v_{av} = \sqrt{\frac{3kT}{m}}$$

Here $T = 100$ °C $= 373$ K, and so

$$v_{av} = \sqrt{\frac{3kT}{m}} = \sqrt{\frac{3 \times 1.38 \times 10^{-23} \text{ J/K} \times 373 \text{ K}}{5.3 \times 10^{-26} \text{ kg}}} = \sqrt{29.1 \times 10^4} \text{ m/s}$$

$$= 5.4 \times 10^2 \text{ m/s} = 540 \text{ m/s}$$

20.7. A certain tank holds 1 g of hydrogen at 0 °C and another identical tank holds 1 g of oxygen at 0 °C. The mass of an oxygen molecule is 16 times greater than that of a hydrogen molecule. (*a*) Which tank contains more molecules? How many more? (*b*) Which gas exerts the greater pressure? How much greater? (*c*) In which gas do the molecules have greater average energies? How much greater? (*d*) In which gas do the molecules have greater average velocities? How much greater?

(*a*) There are 16 times more hydrogen molecules.

(*b*) The hydrogen pressure is 16 times greater because there are 16 times more molecules to exert force on the container walls.

(*c*) The average molecular energies are the same in both gases because their temperatures are the same.

(*d*) The average velocity of the hydrogen molecules is $\sqrt{16} = 4$ times more than that of the oxygen molecules because the hydrogen molecules are 16 times lighter and $v_{av} = \sqrt{3kT/m}$.

20.8. (*a*) What volume does 1 g of ammonia, NH_3, occupy at STP? (*b*) What volume does it occupy at 100 °C and a pressure of 1.2 atm?

(*a*) The molecular mass of NH_3 is

$$1N = 1 \times 14.01 \text{ u} = 14.01 \text{ u}$$
$$3H = 3 \times 1.008 \text{ u} = 3.02 \text{ u}$$
$$\overline{17.03 \text{ u}} = 17.03 \text{ g/mole}$$

so the number of moles in 1 g of NH_3 is

$$\text{Moles of } NH_3 = \frac{\text{mass of } NH_3}{\text{molecular mass of } NH_3} = \frac{1 \text{ g}}{17.03 \text{ g/mole}} = 0.0587 \text{ mole}$$

The volume at STP is therefore

$$\text{Volume of NH}_3 = \text{moles of NH}_3 \times \text{molar volume}$$
$$= 0.0587 \text{ mole} \times 22.4 \text{ liters/mole} = 1.32 \text{ liters}$$

(b) From the ideal gas law,

$$\frac{p_1 V_1}{T_1} = \frac{p_2 V_2}{T_2} \quad \text{or} \quad V_2 = \frac{p_1 V_1 T_2}{p_2 T_1}$$

Here $p_1 = 1$ atm, $V_1 = 1.32$ liters, $T_1 = 0$ °C $= 273$ K and $p_2 = 1.2$ atm, $V_2 = ?$, $T_2 = 100$ °C $= 373$ K. Hence

$$V_2 = \frac{1 \text{ atm} \times 1.32 \text{ liters} \times 373 \text{ K}}{1.2 \text{ atm} \times 273 \text{ K}} = 1.50 \text{ liters}$$

20.9. What is the mass of 40 liters of uranium hexafluoride, UF_6, at 500 °C and 4 atm pressure?

The most direct way to solve this problem is to use the ideal gas law to find the number of moles of UF_6 present in the sample. Since $pV = nRT$ and $T = 500$ °C $= 773$ K we have

$$n = \frac{pV}{RT} = \frac{4 \text{ atm} \times 40 \text{ liters}}{0.0821 \text{ atm-liter/mole-K} \times 773 \text{ K}} = 2.52 \text{ moles}$$

The molecular mass of UF_6 is

$$1U = 1 \times 238.03 \text{ u} = 238.03 \text{ u}$$
$$6F = 6 \times 19.00 \text{ u} = 114.00 \text{ u}$$
$$\overline{352.03 \text{ u}} = 352.03 \text{ g/mole}$$

so the mass of UF_6 is

$$\text{Mass of } UF_6 = \text{moles of } UF_6 \times \text{molecular mass of } UF_6$$
$$= 2.52 \text{ moles} \times 352.03 \text{ g/mole} = 887 \text{ g}$$

20.10. Find the density in g/liter of ethylene, C_2H_4, at STP.

At STP one mole of any gas occupies 22.4 liters. The molecular mass of C_2H_4 is

$$2C = 2 \times 12.01 \text{ u} = 24.02 \text{ u}$$
$$4H = 4 \times 1.008 \text{ u} = 4.03 \text{ u}$$
$$\overline{28.05 \text{ u}} = 28.05 \text{ g/mole}$$

One mole of C_2H_4 therefore has a density at STP of

$$d = \frac{m}{V} = \frac{28.05 \text{ g}}{22.4 \text{ liters}} = 1.25 \text{ g/liter}$$

20.11. What is the density of oxygen at 20 °C and 5 atm pressure?

It is simplest here to use the ideal gas law to find the mass of 1 liter of O_2 under the specified conditions. The number of moles in 1 liter of O_2 at $T = 20$ °C $= 293$ K and $p = 5$ atm is, from $pV = nRT$,

$$n = \frac{pV}{RT} = \frac{5 \text{ atm} \times 1 \text{ liter}}{0.0821 \text{ atm-liter/mole-K} \times 293 \text{ K}} = 0.208 \text{ mole}$$

Since the molecular mass of O_2 is 2×16.00 u $= 32.00$ u $= 32.00$ g/mole, the mass here is

$$\text{Mass of } O_2 = \text{moles of } O_2 \times \text{molecular mass of } O_2$$
$$= 0.208 \text{ mole} \times 32.00 \text{ g/mole} = 6.66 \text{ g}$$

Hence the density of the gas is

$$g = \frac{m}{V} = \frac{6.66 \text{ g}}{1 \text{ liter}} = 6.66 \text{ g/liter}$$

There are 10^3 g per kg and 10^3 liters per m^3, which means that the density in SI units is 6.66 kg/m^3.

20.12. A sample of an unknown gas has a mass of 28.1 g and occupies 4.8 liters at STP. What is its molecular mass?

Since 1 mole of any gas occupies 22.4 liters at STP, this sample must consist of

$$4.8 \text{ liters}/(22.4 \text{ liters}/\text{mole}) = 0.214 \text{ mole}$$

Hence

$$\text{Molecular mass} = \frac{\text{mass of sample}}{\text{moles of sample}} = \frac{28.1 \text{ g}}{0.214 \text{ mole}} = 131 \text{ g}/\text{mole} = 131 \text{ u}$$

Supplementary Problems

20.13. The volume of a gas sample is enlarged. Why does the pressure the gas exerts decrease?

20.14. The temperature of a gas sample is raised. Why does the pressure the gas exerts increase?

20.15. Find the mass of the propane molecule, C_3H_8, and that of the glucose molecule, $C_6H_{12}O_6$.

20.16. The atomic mass of copper is 63.54 u. Find the number of atoms present in 100 g of copper.

20.17. A gas sample at 0 °C is heated until the average energy of its molecules doubles. What is its new temperature?

20.18. A gas sample at 0 °C is heated until the average velocity of its molecules doubles. What is its new temperature?

20.19. At room temperature oxygen molecules have an average velocity of about 1000 mi/hr. (a) What is the average velocity of hydrogen molecules, whose mass is 1/16 that of oxygen molecules, at this temperature? (b) What is the average velocity of sulfur dioxide molecules, whose mass is twice that of oxygen molecules, at this temperature?

20.20. Mercury is a gas at 500 °C. (a) What is the average energy of mercury atoms at this temperature? (b) What is the average velocity of mercury atoms at this temperature? The mass of a mercury atom is 3.3×10^{-25} kg.

20.21. (a) What volume does 8.2 moles of fluorine, F_2, occupy at STP? (b) What volume does it occupy at 40 °C and a pressure of 2.5 atm?

20.22. (a) What volume does 5 g of methane, CH_4, occupy at STP? (b) What volume does it occupy at 0 °C and a pressure of 0.5 atm? (c) What volume does it occupy at 80 °C and a pressure of 2 atm?

20.23. (a) What volume does 20 g of CO_2 occupy at STP? (b) What volume does it occupy at -20 °C and a pressure of 4 atm?

20.24. What is the mass of 4 liters of ammonia, NH_3, at STP?

20.25. (a) Find the mass of 12 liters of chlorine, Cl_2, at 40 °C and 0.8 atm pressure. (b) What is its density under those conditions?

20.26. A balloon is filled with 50 m³ (5×10^4 liters) of hydrogen at STP. What is the mass of the hydrogen?

20.27. Find the density of sulfur dioxide, SO_2, at STP.

20.28. Find the density of nitrogen, N_2, at 120 °C and a pressure of 66,600 N/m^2.

20.29. One liter of an unknown gas has a mass of 2.9 g at STP.   Find its molecular mass.

Answers to Supplementary Problems

20.13. The pressure decreases partly because the gas molecules now must travel farther between impacts on the container walls and partly because these impacts are now distributed over a larger area.

20.14. The pressure increases partly because the gas molecules move faster than before and therefore strike the walls more often and partly because each impact yields a greater force than before.

20.15. 7.32×10^{-26} kg; 2.99×10^{-25} kg

20.16. 9.48×10^{23} atoms

20.17. 273 °C

20.18. 819 °C

20.19. (a) 4000 mi/hr   (b) 707 mi/hr

20.20. (a) 1.6×10^{-20} J   (b) 311 m/s

20.21. (a) 184 liters   (b) 84.4 liters

20.22. (a) 6.98 liters   (b) 13.96 liters   (c) 4.51 liters

20.23. (a) 10.2 liters   (b) 2.36 liters

20.24. 3.04 g

20.25. (a) 26.5 g   (b) 2.21 g/liter = 2.21 kg/m^3

20.26. 4.5 kg

20.27. 2.86 g/liter = 2.86 kg/m^3

20.28. 0.571 kg/m^3

20.29. 65 u

Chapter 21

Thermodynamics

MECHANICAL EQUIVALENT OF HEAT

It is possible to transform other types of energy into internal energy and vice versa. The ratio between an energy unit and a heat unit is called the *mechanical equivalent of heat* and has the values in the metric and British systems of

$$\mathcal{J} = 4185 \ \frac{\text{J}}{\text{kcal}}$$

$$\mathcal{J} = 778 \ \frac{\text{ft-lb}}{\text{Btu}}$$

HEAT ENGINES

To convert internal energy into mechanical energy is much more difficult than the reverse, and perfect efficiency is impossible. A *heat engine* is a device or system that can perform this conversion; the human body and the earth's atmosphere are heat engines, as well as gasoline and diesel motors, aircraft jet engines, and steam turbines.

All heat engines operate by absorbing heat from a reservoir of some kind at a high temperature, performing work, and then giving off heat to a reservoir of some kind at a lower temperature. According to the principle of conservation of energy, the work done in a complete cycle that returns the engine to its original state is equal to the difference between the heat absorbed and the heat given off; this statement constitutes the *first law of thermodynamics*.

SECOND LAW OF THERMODYNAMICS

Internal energy resides in the kinetic energies of randomly moving atoms and molecules, whereas the output of a heat engine appears in the ordered motions of a piston or a wheel. Since all physical systems in the universe tend to go in the opposite direction, from order to disorder, no heat engine can completely convert heat into mechanical energy or, in general, into work. This fundamental principle leads to the *second law of thermodynamics*: It is impossible to construct a continuously operating engine that takes heat from a source and performs an exactly equivalent amount of work.

Because some of the heat input to a heat engine must be wasted, and because heat flows from a hot reservoir to a cold one, every heat engine must have a low-temperature reservoir for exhaust heat to go to as well as a high-temperature reservoir from which the input heat is to come.

ENGINE EFFICIENCY

The efficiency of an ideal heat engine (often called a *Carnot engine*) in which there are no losses due to such practical difficulties as friction depends only upon the temperatures at which heat is absorbed and exhausted. If heat is absorbed at the absolute temperature T_1 and is given off at the absolute temperature T_2, the efficiency of such an engine is

$$\text{Efficiency} = \frac{\text{work output}}{\text{heat input}} = \frac{W}{Q_1} = 1 - \frac{T_2}{T_1}$$

The smaller the ratio between T_2 and T_1, the more efficient the engine. Because no reservoir can exist at a temperature of 0 K or 0 °R, which is absolute zero, no heat engine can be 100% efficient.

REFRIGERATION

A *refrigerator* is a heat engine that operates backwards to extract heat from a low-temperature reservoir and transfer it to a high-temperature reservoir. Because the natural tendency of heat is to flow from a hot region to a cold one, energy must be provided to a refrigerator to reverse the flow, and this energy adds to the heat exhausted by the refrigerator.

The unit of refrigeration capacity is the *ton*, which is that rate of heat extraction that can freeze 1 ton of water at 32 °F to ice at 32 °F per day. Since the heat of fusion of water is 144 Btu/lb,

$$1 \text{ refrigeration ton} = 12,000 \text{ Btu/hr}$$

Solved Problems

21.1. In an effort to cool a kitchen during the summer, the refrigerator door is left open and the kitchen's door and windows are closed. What will happen?

Since no refrigerator can be completely efficient, more heat is exhausted by the refrigerator into the kitchen than is extracted from the kitchen. The net effect, then, is to increase the kitchen's temperature.

21.2. What are isobaric, isothermal, and adiabatic processes?

An *isobaric* process is one that takes place at constant pressure. An *isothermal* process is one that takes place at constant temperature. An *adiabatic* process is one that takes place in a system so insulated from its surroundings that heat neither enters nor leaves the system during the process.

21.3. A sample of a gas expands from V_1 to V_2. Is the amount of work done by the gas greatest when the expansion is (*a*) isobaric, (*b*) isothermal, or (*c*) adiabatic? How does the temperature vary during each expansion?

(*a*) At the constant pressure p the work done is $p(V_2 - V_1)$ and is the greatest of the three expansions. The temperature must increase during the expansion in order to maintain the pressure constant despite the increase in volume.

(*b*) Since $pV/T = $ constant, during an expansion at constant temperature the pressure must drop as V increases and the work done is accordingly less than in (*a*).

(*c*) In an adiabatic expansion the temperature must drop since all the work done is at the expense of the internal energy of the gas. The final pressure is therefore lower than in (*a*) or (*b*) and the least amount of work is done.

21.4. When a pound of coal is burned, 14,000 Btu of heat is liberated. How many ft-lb of energy is this equivalent to?

$$E = \mathcal{J}Q = 778 \, \frac{\text{ft-lb}}{\text{Btu}} \times 14,000 \text{ Btu} = 1.09 \times 10^7 \text{ ft-lb}$$

21.5. An ice cube at 0 °C is dropped on the ground and melts to water at 0 °C. If all the kinetic energy of the ice went into melting it, from what height did it fall?

$$\mathcal{J} \times \text{mass of ice} \times \text{heat of fusion} = \text{initial potential energy of ice}$$
$$\mathcal{J}mL_f = mgh$$
$$h = \frac{\mathcal{J}L_f}{g} = \frac{4185 \text{ J/kcal} \times 80 \text{ kcal/kg}}{9.8 \text{ m/s}^2} = 3.4 \times 10^4 \text{ m}$$

21.6. A 1-MW (10^6 W) generating plant has an overall efficiency of 40%. How much fuel oil whose heat of combustion is 11,000 kcal/kg does the plant burn each day?

In one day the plant produces

$$W = Pt = 10^6 \text{ W} \times 3600 \text{ s/hr} \times 24 \text{ hr/day} = 8.64 \times 10^{10} \text{ J}$$

of electric energy.   Since its efficiency is 0.4 and Eff = work output/heat input, we have

$$\text{Heat input} = \frac{\text{work output}}{\text{Eff}} = \frac{8.64 \times 10^{10} \text{ J}}{0.4} = 2.16 \times 10^{11} \text{ J}$$

To convert the heat input to its equivalent in kcal we divide by 4185 J/kcal to get

$$\text{Heat input} = \frac{2.16 \times 10^{11} \text{ J}}{4.185 \times 10^3 \text{ J/kcal}} = 5.16 \times 10^7 \text{ kcal}$$

The fuel required to supply this amount of heat is

$$m = \frac{5.16 \times 10^7 \text{ kcal}}{1.1 \times 10^4 \text{ kcal/kg}} = 4.7 \times 10^3 \text{ kg}$$

21.7. A total of 0.8 kg of water at 20 °C is placed in a 1-kW electric kettle.   How long a time is needed to raise the temperature of the water to 100 °C?

The required heat is

$$Q = mc \, \Delta T = 0.8 \text{ kg} \times 1 \text{ kcal/kg-°C} \times (100 \text{ °C} - 20 \text{ °C}) = 64 \text{ kcal}$$

The energy equivalent of this amount of heat is

$$E = \mathcal{J}Q = 4185 \text{ J/kcal} \times 64 \text{ kcal} = 2.68 \times 10^5 \text{ J}$$

Since $P = E/t$ and $P = 1 \text{ kW} = 10^3 \text{ J/s}$ here, the time needed is

$$t = \frac{E}{P} = \frac{2.68 \times 10^5 \text{ J}}{10^3 \text{ J/s}} = 2.68 \times 10^2 \text{ s} = 268 \text{ s} = 4.5 \text{ min}$$

21.8. Steam enters a turbine at a velocity of 2500 ft/s and emerges at 400 ft/s.   If 20 tons of steam pass through the turbine per hour, find its power output if the mechanical efficiency is 100%.

The work done by the turbine in a certain period of time equals the difference between the kinetic energy of the steam when it enters and the kinetic energy when it leaves.   Since $m = w/g$ and $w = 40{,}000$ lb, the work done per hour is

$$W = \tfrac{1}{2}mv_1^2 - \tfrac{1}{2}mv_2^2 = \frac{w}{2g}\left(v_1^2 - v_2^2\right)$$

$$= \frac{40{,}000 \text{ lb}}{2 \times 32 \text{ ft/s}^2}\left[(2500 \text{ ft/s})^2 - (400 \text{ ft/s})^2\right] = 3.8 \times 10^9 \text{ ft-lb}$$

There are 3600 s in an hour and 1 hp = 550 ft-lb/s, so

$$P = \frac{W}{t} = \frac{3.8 \times 10^9 \text{ ft-lb}}{3600 \text{ s} \times 550 \text{ (ft-lb/s)/hp}} = 1922 \text{ hp}$$

21.9. Show that the power output of each cylinder of a reciprocating engine of any kind (steam, gasoline, diesel) is given by the formula

$$P \text{ (hp)} = \frac{pLAn}{33{,}000}$$

where

$\quad p$ = mean effective pressure on piston during power stroke in lb/in^2

$\quad L$ = length of piston travel in ft

$\quad A$ = area of piston in in^2

$\quad n$ = number of power strokes per minute

In general, $P = Fs/t$. Here F is the force exerted on the piston during each power stroke by the mean effective pressure p, so since $p = F/A$, $F = pA$. The distance traveled by the piston per power stroke is L, and the distance it covers per minute is therefore $s/t = Ln$. Hence

$$P \text{ (ft-lb/min)} = \frac{Fs}{t} = pLAn$$

To convert from ft-lb/min to hp, we note that

$$1 \text{ hp} = 550 \text{ ft-lb/s} \times 60 \text{ s/min} = 33,000 \text{ ft-lb/min}$$

The result is

$$P \text{ (hp)} = \frac{pLAn}{33,000}$$

21.10. The six-cylinder, four-cycle gasoline engine of a car has pistons 3.4 in. in diameter whose stroke (length of travel) is 4 in. If the mean effective pressure on the pistons during the power stroke is 70 lb/in^2, find the number of horsepower developed by the engine when it operates at 2000 rpm.

The area of each piston is

$$A = \frac{\pi d^2}{4} = \frac{\pi \times (3.4 \text{ in.})^2}{4} = 9.1 \text{ in}^2$$

and its stroke is $L = 4$ in. $= 0.33$ ft. In a four-cycle engine, a power stroke occurs in each cylinder once every other revolution, so here $n = 1000$ strokes/min. Since there are six cylinders, the power output is

$$P \text{ (hp)} = 6 \times \frac{pLAn}{33,000} = \frac{6 \times 70 \text{ lb/in}^2 \times 0.33 \text{ ft} \times 9.1 \text{ in}^2 \times 1000/\text{min}}{33,000}$$

$$= 38.2 \text{ hp}$$

21.11. (a) Find the maximum possible efficiency of an engine that absorbs heat at a temperature of 327 °C and exhausts heat at a temperature of 127 °C. (b) What is the maximum amount of work (in joules) the engine can perform per kcal of heat input?

(a) Here $T_1 = 327$ °C $+ 273 = 600$ K and $T_2 = 127$ °C $+ 273 = 400$ K. Hence

$$\text{Eff} = 1 - \frac{T_2}{T_1} = 1 - \frac{400 \text{ K}}{600 \text{ K}} = \frac{1}{3} = 33\%$$

(b) Since 1 kcal $= 4185$ J,

$$\text{Work} = \text{Eff} \times \text{heat input} = \frac{1}{3} \times 4185 \text{ J} = 1395 \text{ J}$$

21.12. Three designs are proposed for an engine which is to operate between 500 K and 300 K. Design A is claimed to produce 3000 J of work per kcal of heat input, B is claimed to produce 2000 J, and C is claimed to produce 1000 J. Which design would you choose?

The efficiency of an ideal engine operating between $T_1 = 500$ K and $T_2 = 300$ K is

$$\text{Eff} = 1 - \frac{T_2}{T_1} = 1 - \frac{300 \text{ K}}{500 \text{ K}} = 0.40 = 40\%$$

Since 1 kcal $= 4185$ J, the claimed efficiencies of the proposed engines are

$$\text{Eff (A)} = \frac{\text{work output}}{\text{heat input}} = \frac{3000 \text{ J}}{4185 \text{ J}} = 0.72 = 72\%$$

$$\text{Eff (B)} = \frac{2000 \text{ J}}{4185 \text{ J}} = 0.48 = 48\%$$

$$\text{Eff (C)} = \frac{1000 \text{ J}}{4185 \text{ J}} = 0.24 = 24\%$$

Both A and B claim efficiencies greater than that of an ideal engine and hence could not possibly work as stated. Design C is therefore the only possible choice.

21.13. A steam engine is being planned which is to use steam at 400 °F and whose efficiency is to be 20%. Find the maximum temperature at which the spent steam can emerge.

The intake temperature is $T_1 = 400 \text{ °F} + 460° = 860 \text{ °R}$. We proceed as follows:

$$\text{Eff} = 1 - \frac{T_2}{T_1} \qquad \frac{T_2}{T_1} = 1 - \text{Eff}$$

$$T_2 = T_1(1 - \text{Eff}) = (860 \text{ °R})(1 - 0.20) = 688 \text{ °R}$$

The maximum exhaust temperature is therefore $T_2 = 688 \text{ °R} - 460° = 228 \text{ °F}$.

21.14. A refrigerator which is half as efficient as an ideal refrigerator extracts heat from a storage chamber at 0 °F and exhausts it at 100 °F. How many ft-lb of work per Btu extracted does this refrigerator require?

We begin by finding Q_1/W for an ideal refrigerator. Since

$$T_1 = 0 \text{ °F} + 460° = 460 \text{ °R} \qquad T_2 = 100 \text{ °F} + 460° = 560 \text{ °R}$$

$$\frac{Q_1}{W} = \frac{T_2}{T_2 - T_1} = \frac{560 \text{ °R}}{560 \text{ °R} - 460 \text{ °R}} = 5.6$$

Here $Q_1 = 1$ Btu and, since 1 Btu = 778 ft-lb,

$$W = \frac{Q_1}{5.6} = \frac{778 \text{ ft-lb}}{5.6} = 139 \text{ ft-lb}$$

This refrigerator is half as efficient as an ideal refrigerator, so the required work per Btu is twice as great, or 278 ft-lb.

21.15. The refrigerator of Problem 21.14 has a capacity of 2 tons. How much power is required to operate its compressor?

Since the refrigerator requires 278 ft-lb of work per Btu of heat extracted,

$$P = \frac{W}{t} = 2 \text{ tons} \times 12{,}000 \, \frac{\text{Btu/hr}}{\text{ton}} \times 278 \, \frac{\text{ft-lb}}{\text{Btu}} \times \frac{1}{3600 \text{ s/hr}} = 1853 \text{ ft-lb/s}$$

In terms of hp,

$$P = \frac{1853 \text{ ft-lb/s}}{550 \text{ (ft-lb/s)/hp}} = 3.37 \text{ hp}$$

21.16. An ice-making plant has an output of 1 ton of ice at 15 °F per hour. If the water reaching the plant is at 60 °F, find the refrigeration capacity of the plant assuming that no heat losses occur.

$$Q_1 = \text{heat extraction per hour to cool water from 60 °F to 32 °F}$$

$$= mc_w \, \Delta T = 2000 \text{ lb} \times 1 \text{ Btu/lb-°F} \times 28 \text{ °F} = 5.6 \times 10^4 \text{ Btu}$$

$$Q_2 = \text{heat extracted per hour to freeze water at 32 °F}$$

$$= mL_f = 2000 \text{ lb} \times 144 \text{ Btu/lb} = 28.8 \times 10^4 \text{ Btu}$$

$$Q_3 = \text{heat extracted per hour to cool ice from 32 °F to 15 °F}$$

$$= mc_{\text{ice}}T = 2000 \text{ lb} \times 0.5 \text{ Btu/lb-°F} \times 17 \text{ °F} = 1.7 \times 10^4 \text{ Btu}$$

The total amount of heat to be extracted per hour is therefore

$$Q = Q_1 + Q_2 + Q_3 = 36.1 \times 10^4 \text{ Btu}$$

Since 1 refrigeration ton = 12,000 Btu/hr, the capacity required is

$$\frac{36.1 \times 10^4 \text{ Btu/hr}}{1.2 \times 10^4 \text{ (Btu/hr)/ton}} = 30.1 \text{ tons}$$

Supplementary Problems

21.17. Why is it impossible for a ship to use the internal energy of seawater to operate its engine?

21.18. (a) How many kcal per hour are given off by a 100-W light bulb?   (b) How many Btu per hour?

21.19. How many hp is a refrigeration ton equivalent to?   The heat of fusion of water is 144 Btu/lb and 1 hp = 550 ft-lb/s.

21.20. A typical gum drop contains 35 kcal of energy.   If this energy were used to raise an 80-kg man above the ground, how high would he go?

21.21. A lead bullet traveling at 200 m/s strikes a tree and comes to a stop.   If half the heat produced is retained by the bullet, by how much does its temperature increase?   ($c_{\text{lead}} = 0.03$ kcal/kg-°C)

21.22. One and a half kg of water at 10 °C in a 300-g aluminum kettle is placed on a 2-kW electric hot plate. What is the temperature of the water after 3 min?   ($c_{\text{aluminum}} = 0.22$ kcal/kg-°C)

21.23. A 2400-hp diesel locomotive burns 160 gal of fuel per hour.   If the heat of combustion of the diesel oil used is 1.2×10^5 Btu/gal, find the efficiency of the engine.

21.24. Steam enters a turbine engine at 550 °C and emerges at 90 °C.   The engine has an actual overall efficiency of 35%.   What percentage of its ideal efficiency is this?

21.25. An engine absorbs 500 kcal of heat at 600 K and exhausts 300 kcal of heat at 300 K.   (a) What is its efficiency?   (b) If it were an ideal engine, what would its efficiency be and how much heat would it exhaust?

21.26. At what temperature must an ideal engine absorb heat if its efficiency is 33% and it exhausts heat at 250 °F?

21.27. A three-cylinder two-stroke diesel engine has pistons 4.25 in. in diameter whose stroke is 5 in.   If the engine develops 85 hp at 1800 rpm, find the mean effective pressure on the piston during the power stroke.

21.28. Steam enters a 100,000-hp turbine at 2600 ft/s and emerges at 300 ft/s.   Assuming perfect mechanical efficiency, how much steam passes through the turbine per minute?

21.29. An ideal refrigerator extracts heat from a freezer at −20 °C and exhausts it at 50 °C.   How many kcal of heat are extracted per joule of work input?

21.30. The specific heat capacity of ice cream is 0.78 Btu/lb-°F when liquid and 0.45 Btu/lb-°F when frozen. The ice cream freezes at 28 °F and has a heat of fusion of 126 Btu/lb.   Find the refrigeration capacity needed to produce 1 ton of ice cream at 10 °F per day from a mix initially at 65 °F.

Answers to Supplementary Problems

21.17. There would be no suitable low-temperature reservoir to absorb the waste heat from the engine.

21.18. (*a*) 86 kcal   (*b*) 342 Btu

21.19. 4.7 hp

21.20. 187 m

21.21. 80 °C

21.22. 65 °C

21.23. 32%

21.24. 63%

21.25. (*a*) 40%   (*b*) 50%; 250 kcal

21.26. 600 °F

21.27. 88 lb/in^2

21.28. 31,664 lb

21.29. 0.0011 kcal/J

21.30. 1.13 tons

Chapter 22

Heat Transfer

CONDUCTION

The three mechanisms by which heat can be transferred from one place to another are conduction, convection, and radiation.

In *conduction*, heat is carried by means of collisions between rapidly moving molecules at the hot end of a body of matter and the slower molecules at the cold end. Some of the kinetic energy of the fast molecules passes to the slow molecules, and the result of successive collisions is a flow of heat through the body of matter. Solids, liquids, and gases all conduct heat. Conduction is poorest in gases because their molecules are relatively far apart and so interact less frequently than in the case of solids and liquids. Metals are the best conductors of heat because some of their electrons are able to move about relatively freely and can travel past many atoms between collisions.

The rate at which heat is conducted through a slab of a particular material is proportional to the area A of the slab and to the temperature difference ΔT between its sides, and inversely proportional to the slab's thickness d (Fig. 22-1). The amount of heat Q that flows through the slab in the time t is given by

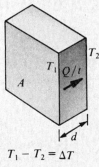

$$\text{Rate of heat conduction} = \frac{Q}{t} = \frac{kA\,\Delta T}{d}$$

where k, the *thermal conductivity* of the material, is a measure of its ability to conduct heat. In the SI system, the correct unit of k is the W/m-K, but the kcal/m-s-°C is more often used. In the British system, the usual unit of k is the Btu/(ft²-hr-°F/in) since A is customarily expressed in ft² and d in inches.

$T_1 - T_2 = \Delta T$

Fig. 22-1

CONVECTION

In *convection*, a volume of hot fluid (gas or liquid) moves from one region to another carrying internal energy with it. When a pan of water is heated on a stove, for instance, the hot water at the bottom expands slightly so that its density decreases, and the buoyancy of this water causes it to rise to the surface while colder, denser water descends to take its place at the bottom.

RADIATION

In *radiation*, energy is carried by the *electromagnetic waves* emitted by every object. Electromagnetic waves, of which light, radio waves, and X-rays are examples, travel at the velocity of light (3×10^8 m/s = 186,000 mi/s) and require no material medium for their passage. The better an object absorbs radiation, the better it emits radiation. A perfect absorber of radiation is called a *blackbody*, and it is accordingly the best radiator.

The rate at which an object whose surface area is A and whose absolute temperature is T emits radiation is given by the *Stefan-Boltzmann law*:

$$R = \frac{P}{A} = e\sigma T^4$$

The constant σ (Greek letter *sigma*) has the value 5.67×10^{-8} W/m²-K⁴. The *emissivity* e has a value between 0 (for a perfect reflector, hence a nonradiator) and 1 (for a blackbody), depending on the nature of the radiating surface.

With increasing temperature, the predominant wavelength of the radiation emitted by a body decreases. Thus a hot body that glows red is cooler than one that glows bluish-white since red light has a longer wavelength than blue light (see Chapter 30). A body at room temperature emits radiation that is chiefly in the infrared part of the spectrum, to which the eye is not sensitive.

Solved Problems

22.1. If all objects radiate electromagnetic energy, why do not the objects around us in everyday life grow colder and colder?

Every object also absorbs electromagnetic energy from its surroundings, and if both object and surroundings are at the same temperature, energy is emitted and absorbed at the same rate. When an object is at a higher temperature than its surroundings and heat is not supplied to it, it radiates more energy than it absorbs and cools down to the temperature of its surroundings.

22.2. The thermal conductivities of brick and pine wood are respectively 5.0 and 0.8 Btu/(ft²-hr-°F/in). What thickness of brick has the same insulating ability as 2 in. of pine?

When the ratio k/d is the same for the two materials, their insulating abilities will also be the same. Hence

$$d_{\text{brick}} = k_{\text{brick}} \times \frac{d_{\text{pine}}}{k_{\text{pine}}} = \frac{5.0 \times 2 \text{ in.}}{0.8} = 12.5 \text{ in.}$$

22.3. The thermal conductivity of ice is 5.2×10^{-4} kcal/m-s-°C. At what rate is heat lost by the water in a 6 m × 10 m outdoor swimming pool covered by a layer of ice 1 cm thick if the water is at a temperature of 0 °C and the surrounding air is at a temperature of -10 °C?

Here $A = 6$ m × 10 m = 60 m², $d = 0.01$ m, and $\Delta T = 10$ °C. Hence

$$\frac{Q}{t} = \frac{kA\Delta T}{d} = 5.2 \times 10^{-4} \frac{\text{kcal}}{\text{m-s-°C}} \times \frac{60 \text{ m}^2 \times 10 \text{ °C}}{0.01 \text{ m}} = 31.2 \text{ kcal/s}$$

22.4. The handle of a freezer door 5 in. thick is attached by two brass bolts $\frac{1}{4}$ in. in diameter that pass through the entire door and are secured on the inside by nuts. The interior of the freezer is maintained at 0 °F and the room temperature is 65 °F; the thermal conductivity of brass is 730 Btu/(ft²-hr-°F/in). Find the heat lost per hour through the bolts.

The total cross-sectional area of the two bolts is

$$A = 2 \times \pi r^2 = 2 \times \pi \times \left(\frac{0.125 \text{ in.}}{12 \text{ in/ft}} \right)^2 = 6.82 \times 10^{-4} \text{ ft}^2$$

and their length is $d = 5$ in. Since $\Delta T = 65$ °F,

$$\frac{Q}{t} = \frac{kA \Delta T}{d} = 730 \frac{\text{Btu}}{\text{ft}^2\text{-hr-°F/in}} \times \frac{6.82 \times 10^{-4} \text{ ft}^2 \times 65 \text{ °F}}{5 \text{ in.}} = 6.47 \text{ Btu/hr}$$

22.5. According to *Newton's law of cooling*, the rate at which a hot object loses heat to the surrounding air is approximately proportional to the temperature difference between the object and the air. If a cup of coffee cools from 150 °F to 140 °F in 1 min in a room at 70 °F, how long will it take to cool from 100 °F to 90 °F?

The average temperature difference between the coffee and the air in the first period of cooling is 145 °F − 70 °F = 75 °F and in the second period it is 95 °F − 70 °F = 25 °F. Since the temperature difference in the second period is 1/3 that of the first, the rate of cooling will be 1/3 as great, and the coffee will take 3 min to cool the 10 °F from 100 °F to 90 °F.

22.6. A copper ball 2 cm in radius is heated in a furnace to 400 °C. If its emissivity is 0.3, at what rate does it radiate energy?

The surface area of the ball is

$$A = 4\pi r^2 = 4\pi \times (0.02 \text{ m})^2 = 0.005 \text{ m}^2$$

and its absolute temperature is $T = 400 \text{ °C} + 273 = 673 \text{ K}$. Hence

$$P = e\sigma A T^4 = 0.3 \times 5.67 \times 10^{-8} \text{ W/m}^2\text{-K}^4 \times 0.005 \text{ m}^2 \times (673 \text{ K})^4$$
$$= 17.4 \text{ W}$$

22.7. The sun radiates energy at the rate of $6.5 \times 10^7 \text{ W/m}^2$ from its surface. Assuming that the sun radiates like a blackbody (which is approximately true), find its surface temperature.

The emissivity of a blackbody is $e = 1$. From $R = \sigma T^4$ we have

$$T = \sqrt[4]{\frac{R}{\sigma}} = \sqrt[4]{\frac{6.5 \times 10^7 \text{ W/m}^2}{5.67 \times 10^{-8} \text{ W/m}^2\text{-K}^4}} = 5800 \text{ K}$$

Supplementary Problems

22.8. Outdoors in the winter, why does a piece of metal feel colder than a piece of wood?

22.9. Why is it desirable to paint hot-water pipes with aluminum paint?

22.10. An icebox whose walls consist of a 4-in. thickness of pine wood is to be replaced by a more modern one using glass wool insulation between pine inner and outer walls $\frac{1}{2}$ in. thick. What thickness of glass wool will give the same degree of insulation as before? The thermal conductivities of pine and glass wool are respectively 0.8 and 0.27 Btu/(ft²-hr-°F/in).

22.11. A yacht has an aluminum hull 5 mm thick whose underwater area is 100 m². How much heat is conducted from the yacht's interior to the water per hour if the interior is at a temperature of 20 °C and the water is at a temperature of 12 °C? The thermal conductivity of aluminum is 0.057 kcal/m-s-°C.

22.12. How many Btu per day are conducted through a glass window 5 ft × 8 ft × 0.3 in. whose inner and outer faces are at the respective temperatures of 65 °F and 40 °F? The thermal conductivity of glass is 5.5 Btu/(ft²-hr-°F/in).

22.13. A 3-kg beef roast requires 13 min to be heated from 40 °C to 45 °C in an oven maintained at 200 °C. How long will the same roast take to be heated from 60 °C to 65 °C in the same oven?

22.14. A small hole leading into a cavity behaves like a blackbody because any radiation that falls on it is trapped inside by multiple reflections until it is absorbed. How many watts are radiated from a hole 1 cm in diameter in the wall of a furnace whose interior temperature is 650 °C?

22.15. A blackbody is at a temperature of 500 °C. What should its temperature be in order that it radiate twice as much energy per second?

Answers to Supplementary Problems

22.8.   Metals are much better conductors of heat than wood and therefore conduct heat away from the hand more rapidly.

22.9.   Such paint gives a finish that reflects most of the light that falls on it.   Since a poor absorber of radiation is also a poor emitter of radiation, a pipe painted in this way radiates heat at the minimum rate.

22.10. 1.0 in.   **22.12.** 4.4×10^5 Btu   **22.14.** 3.23 W

22.11. 3.28×10^7 kcal   **22.13.** 15 min   **22.15.** 646 °C

Chapter 23

Electricity

ELECTRIC CHARGE

Electric charge, like mass, is one of the basic properties of certain of the elementary particles of which all matter is composed. There are two kinds of charge, *positive charge* and *negative charge*. The positive charge in ordinary matter is carried by *protons*, the negative charge by *electrons*. Charges of the same sign repel each other, charges of opposite sign attract each other.

The unit of charge is the *coulomb* (C). The charge of the proton is $+1.6 \times 10^{-19}$ C and the charge of the electron is -1.6×10^{-19} C. All charges in nature occur in multiples of $\pm e = \pm 1.6 \times 10^{-19}$ C.

According to the principle of *conservation of charge*, the net electric charge in an isolated system always remains constant. (Net charge means the total positive charge minus the total negative charge.) When matter is created from energy, equal amounts of positive and negative charge always come into being, and when matter is converted to energy, equal amounts of positve and negative charge disappear.

COULOMB'S LAW

The force one charge exerts on another is given by *Coulomb's law*:

$$\text{Electric force} = F = k \frac{q_1 q_2}{r^2}$$

where q_1 and q_2 are the magnitudes of the charges, r is the distance between them, and k is a constant whose value in free space is

$$k = 9.0 \times 10^9 \frac{\text{N-m}^2}{\text{C}^2}$$

The value of k in air is slightly greater. The constant k is sometimes replaced by

$$k = \frac{1}{4\pi\varepsilon_0}$$

where ε_0, the *permittivity of free space*, has the value

$$\varepsilon_0 = 8.85 \times 10^{-12} \frac{\text{C}^2}{\text{N-m}^2}$$

(ε is the Greek letter *epsilon*.)

ATOMIC STRUCTURE

An atom of any element consists of a small, positively charged *nucleus* with a number of electrons some distance away. The nucleus is composed of protons (charge $+e$, mass $= 1.673 \times 10^{-27}$ kg) and neutrons (uncharged, mass $= 1.675 \times 10^{-27}$ kg); the number of protons in the nucleus is normally equal to the number of electrons around it, so that the atom as a whole is electrically neutral. The forces between atoms that hold them together as solids and liquids are electrical in origin. The mass of the electron is 9.1×10^{-31} kg.

IONS

Under certain circumstances an atom may lose one or more electrons and become a *positive ion*, or it may gain one or more electrons and become a *negative ion*. Many solids consist of positive and negative ions rather than of atoms or molecules. An example is ordinary table salt, which is made up of positive sodium ions (Na^+) and negative chlorine ions (Cl^-). Solutions of such solids in water also contain ions. Sparks, flames, and X-rays are among the influences that can ionize gases. Ions of opposite sign in a gas come together soon after being formed and the excess electrons on the negative ions pass to the positive ones to form neutral molecules. A gas can be maintained in an ionized state by passing an electric current through it (as in a neon sign) or by bombarding it with X-rays or ultraviolet light (as in the upper atmosphere of the earth, where the radiation comes from the sun).

ELECTRIC FIELD

An *electric field* is a region of space in which a charge would be acted upon by an electric force. An electric field may be produced by one or more charges, and it may be uniform or it may vary in magnitude and/or direction from place to place.

If a charge q at a certain point is acted upon by the force $\mathbf{F}$, the electric field $\mathbf{E}$ at that point is defined as the ratio between $\mathbf{F}$ and q:

$$\mathbf{E} = \frac{\mathbf{F}}{q}$$
$$\text{Electric field} = \frac{\text{force}}{\text{charge}}$$

Electric field is a vector quantity whose direction is that of the force on a positive charge. The unit of $\mathbf{E}$ is the newton/coulomb (N/C) or, more commonly, the equivalent volt/meter (V/m).

The advantage of knowing the electric field at some point is that we can at once establish the force on *any* charge q placed there, which is

$$\mathbf{F} = q\mathbf{E}$$
$$\text{Force} = \text{charge} \times \text{electric field}$$

ELECTRIC LINES OF FORCE

Lines of force are a means of describing a force field, such as an electric field, by using imaginary lines to indicate the direction and magnitude of the field. The direction of an electric line of force at any point is the direction in which a positive charge would move if placed there, and lines of force are drawn close together where the field is strong and far apart where the field is weak (Fig. 23-1).

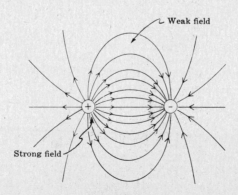

Fig. 23-1

POTENTIAL DIFFERENCE

The *potential difference* V between two points in an electric field is the amount of work needed to take a charge of 1 C from one of the points to the other.   Thus

$$V = \frac{W}{q}$$

$$\text{Potential difference} = \frac{\text{work}}{\text{charge}}$$

The unit of potential difference is the *volt* (V):

$$1 \text{ volt} = 1 \frac{\text{joule}}{\text{coulomb}}$$

The potential difference between two points in a uniform electric field **E** is equal to the product of E and the distance s between the points in a direction parallel to **E**:

$$V = Es$$

Since an electric field is usually produced by applying a potential difference between two metal plates s apart, this equation is most useful in the form

$$E = \frac{V}{s}$$

$$\text{Electric field} = \frac{\text{potential difference}}{\text{distance}}$$

A battery uses chemical reactions to produce a potential difference between its terminals; a generator uses electromagnetic induction (Chapter 28) for this purpose.

Solved Problems

23.1. An iron atom has 26 protons in its nucleus.   (*a*) How many electrons does this atom contain? (*b*) How many electrons does the Fe^{3+} ion contain?

   (*a*)   Since a normal atom is electrically neutral, the number of negatively charged electrons it contains equals the number of positively charged protons in its nucleus.   The iron atom therefore contains 26 electrons.

   (*b*)   The symbol Fe^{3+} represents an iron atom that has a net charge of $+3e$, which means that it has lost three of its usual complement of electrons.   The Fe^{3+} ion therefore contains 23 electrons.

23.2. A pith ball has a charge of $+10^{-12}$ C.   (*a*) Does it contain an excess or a deficiency of electrons compared with its normal state of electrical neutrality?   (*b*) How many electrons?

   (*a*)   Since the pith ball is positively charged, it has fewer electrons than are needed to balance the positive charge of its nuclear protons.

   (*b*)   The charge on an electron is $e = 1.6 \times 10^{-19}$ C.   Hence

$$\text{Number of electrons} = \frac{q}{e} = \frac{10^{-12} \text{ C}}{1.6 \times 10^{-19} \text{ C/electron}} = 6.25 \times 10^6 \text{ electrons}$$

23.3. What is the magnitude and direction of the force on a charge of $+4 \times 10^{-9}$ C that is 5 cm from a charge of $+5 \times 10^{-8}$ C?

   Since 5 cm $= 5 \times 10^{-2}$ m, we have from Coulomb's law

$$F = k \frac{q_1 q_2}{r^2} = \frac{9 \times 10^9 \text{ N-m}^2/\text{C}^2 \times 4 \times 10^{-9} \text{ C} \times 5 \times 10^{-8} \text{ C}}{(5 \times 10^{-2} \text{ m})^2} = 7.2 \times 10^{-4} \text{ N}$$

The force is directed away from the $+5 \times 10^{-8}$ C charge since both charges are positive.

23.4. A hydrogen atom consists of a proton (charge $+e$) and an electron (charge $-e$) that are an average of 5.3×10^{-11} m apart. Find the attractive force between them.

Since $e = 1.6 \times 10^{-19}$ C, we have from Coulomb's law

$$F = k\frac{q_1 q_2}{r^2} = \frac{9 \times 10^9 \text{ N-m}^2/\text{C}^2 \times (1.6 \times 10^{-19} \text{ C})^2}{(5.3 \times 10^{-11} \text{ m})^2} = 8.2 \times 10^{-8} \text{ N}$$

23.5. Two charges, one of $+5 \times 10^{-7}$ C and the other of -2×10^{-7} C, attract each other with a force of 100 N. How far apart are they?

From Coulomb's law we have

$$r = \sqrt{\frac{kq_1 q_2}{F}} = \sqrt{\frac{9 \times 10^9 \text{ N-m}^2/\text{C}^2 \times 5 \times 10^{-7} \text{ C} \times 2 \times 10^{-7} \text{ C}}{10^2 \text{ N}}}$$

$$= \sqrt{90 \times 10^{-7}} \text{ m} = \sqrt{9 \times 10^{-6}} \text{ m} = 3 \times 10^{-3} \text{ m} = 3 \text{ mm}$$

23.6. Two charges repel each other with a force of 10^{-5} N when they are 20 cm apart. (a) What is the force on each of them when they are 5 cm apart? (b) When they are 100 cm apart?

(a) Since F is proportional to $1/r^2$, the force increases when the charges are brought closer together to $(20/5)^2 = 16$ times what it was before, namely to 1.6×10^{-4} N.

(b) The force decreases when the charges are moved apart to $(20/100)^2 = 0.04$ times what it was before, namely to 4×10^{-7} N.

23.7. Under what circumstances, if any, is the gravitational attraction between two protons equal to their electrical repulsion?

Since the proton mass is 1.67×10^{-27} kg, the gravitational force between two protons that are a distance r apart is

$$F_{\text{grav}} = \frac{Gm_1 m_2}{r^2} = \frac{6.67 \times 10^{-11} \text{ N-m}^2/\text{kg}^2 \times (1.67 \times 10^{-27} \text{ kg})^2}{r^2} = \frac{1.86 \times 10^{-64}}{r^2} \text{ N}$$

The electric force between the protons is

$$F_{\text{elec}} = \frac{kq_1 q_2}{r^2} = \frac{9 \times 10^9 \text{ N-m}^2/\text{C}^2 \times (1.6 \times 10^{-19} \text{ C})^2}{r^2} = \frac{2.3 \times 10^{-28}}{r^2} \text{ N}$$

At every separation r, the electric force between the protons is greater than the gravitational force between them by a factor of more than 10^{36}; the forces are never equal.

23.8. A test charge of $+1 \times 10^{-6}$ C is placed halfway between a charge of $+5 \times 10^{-6}$ C and a charge of $+3 \times 10^{-6}$ C that are 20 cm apart (Fig. 23-2). Find the magnitude and direction of the force on the test charge.

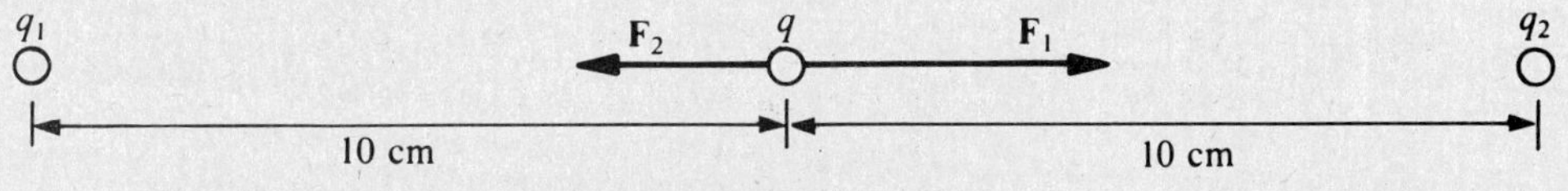

Fig. 23-2

The force exerted on the test charge q by the charge q_1 is

$$F_1 = \frac{kqq_1}{r_1^2} = \frac{9 \times 10^9 \text{ N-m}^2/\text{C}^2 \times 1 \times 10^{-6} \text{ C} \times 5 \times 10^{-6} \text{ C}}{(0.1 \text{ m})^2} = +4.5 \text{ N}$$

This force is taken to be positive because it acts to the right. The force exerted by the charge q_2 on q is

$$F_2 = \frac{kqq_2}{r_2^2} = \frac{9 \times 10^9 \text{ N-m}^2/\text{C}^2 \times 1 \times 10^{-6} \text{ C} \times 3 \times 10^{-6} \text{ C}}{(0.1 \text{ m})^2} = -2.7 \text{ N}$$

This force is taken to be negative because it acts to the left. The net force on the test charge q is

$$F = F_1 + F_2 = +4.5\ \text{N} - 2.7\ \text{N} = +1.8\ \text{N}$$

and it acts to the right, that is, toward the $+3 \times 10^{-6}$ C charge.

23.9. (a) What is the electric field a distance r from a charge q? (b) What is the electric field that acts upon the electron in a hydrogen atom, which is 5.3×10^{-11} m from the proton that is the atom's nucleus?

(a) The force F that the charge q exerts on a test charge q_0 when it is r away is $F = kqq_0/r^2$. From the definition of electric field, then,

$$E = \frac{F}{q_0} = k\frac{q}{r^2}$$

(b) Here $q = e = 1.6 \times 10^{-19}$ C, and so

$$E = k\frac{q}{r^2} = \frac{9 \times 10^9\ \text{N-m}^2/\text{C}^2 \times 1.6 \times 10^{-19}\ \text{C}}{(5.3 \times 10^{-11}\ \text{m})^2} = 5.1 \times 10^{11}\ \text{V/m}$$

23.10. The electric field in a certain neon sign is 5000 V/m. (a) What is the force this field exerts on a neon ion of mass 3.3×10^{-26} kg and charge $+e$? (b) What is the acceleration of the ion?

(a) The force on the neon ion is

$$F = qE = eE = 1.6 \times 10^{-19}\ \text{C} \times 5 \times 10^3\ \text{V/m} = 8 \times 10^{-16}\ \text{N}$$

(b) According to the second law of motion $F = ma$, and so here

$$a = \frac{F}{m} = \frac{8 \times 10^{-16}\ \text{N}}{3.3 \times 10^{-26}\ \text{kg}} = 2.4 \times 10^{10}\ \text{m/s}^2$$

23.11. How strong an electric field is required to exert a force on a proton equal to its weight at sea level?

The electric force on the proton is $F = eE$ and its weight is mg. Hence $eE = mg$ and

$$E = \frac{mg}{e} = \frac{1.67 \times 10^{-27}\ \text{kg} \times 9.8\ \text{m/s}^2}{1.6 \times 10^{-19}\ \text{C}} = 1.02 \times 10^{-7}\ \text{V/m}$$

23.12. The potential difference between a certain thundercloud and the ground is 7×10^6 V. Find the energy dissipated when a charge of 50 C is transferred from the cloud to the ground in a lightning stroke.

$$W = qV = 50\ \text{C} \times 7 \times 10^6\ \text{V} = 3.5 \times 10^8\ \text{J}$$

23.13. A potential difference of 20 V is applied across two parallel metal plates and an electric field of 500 V/m is produced. How far apart are the plates?

Since $E = V/s$, here

$$s = \frac{V}{E} = \frac{20\ \text{V}}{500\ \text{V/m}} = 0.04\ \text{m} = 4\ \text{cm}$$

23.14. (a) What potential difference must be applied across two metal plates 15 cm apart if the electric field between them is to be 600 V/m? (b) What is the force on a charge of 10^{-10} C in this field? (c) How much kinetic energy will the charge have when it has moved through 5 cm in the field starting from rest?

(a) $$V = Es = 600\ \text{V/m} \times 0.15\ \text{m} = 90\ \text{V}$$

(b) $$F = qE = 10^{-10}\ \text{C} \times 600\ \text{V/m} = 6 \times 10^{-8}\ \text{N}$$

(c) Since the KE of the charge is equal to the work done on it by the electric field when it travels 0.05 m,

$$\text{KE} = W = Fs = 6 \times 10^{-8}\ \text{N} \times 0.05\ \text{m} = 3 \times 10^{-9}\ \text{J}$$

23.15. What potential difference must be applied to produce an electric field that can accelerate an electron to a velocity of 10^7 m/s?

The kinetic energy of such an electron is

$$\text{KE} = \tfrac{1}{2}mv^2 = \tfrac{1}{2} \times 9.1 \times 10^{-31}\,\text{kg} \times \left(10^7\,\text{m/s}\right)^2 = 4.6 \times 10^{-17}\,\text{J}$$

This KE is equal to the work W that must be done on the electron by the electric field and so, since $W = qV$ in general, we have here

$$V = \frac{W}{q} = \frac{\text{KE}}{e} = \frac{4.6 \times 10^{-17}\,\text{J}}{1.6 \times 10^{-19}\,\text{C}} = 2.9 \times 10^2\,\text{V} = 290\,\text{V}$$

23.16. A 12-V storage battery is being charged at the rate of 15 C/s. (a) How much power is being used to charge the battery? (b) How much energy is stored in the battery if it is charged at this rate for 1 hr?

(a) The work done to transfer the charge q from one set of the battery's electrodes to the other set against the potential difference V is $W = Vq$. Since power is the rate at which work is being done, here

$$P = \frac{W}{t} = \frac{Vq}{t} = 12\,\text{V} \times 15\,\text{C/s} = 180\,\text{W}$$

(b) The work done in $t = 1$ hr $= 3600$ s is

$$W = Pt = 180\,\text{W} \times 3600\,\text{s} = 6.48 \times 10^5\,\text{J}$$

If the charging process is perfectly efficient, this amount of energy will be stored in the battery as a result.

Supplementary Problems

23.17. An oxygen atom has 8 protons in its nucleus. (a) How many electrons does this atom contain? (b) How many electrons does the O^{--} ion contain?

23.18. What information is provided by a sketch of the lines of force of an electric field?

23.19. Why is it impossible for the lines of force of an electric field to cross one another?

23.20. A rod with a charge of $+q$ at one end and $-q$ at the other is placed in a uniform electric field whose direction is parallel to the rod. How does the rod behave?

23.21. The rod of Problem 23.20 is placed in a uniform electric field whose direction is perpendicular to the rod. How does the rod behave?

23.22. A billion (10^9) electrons are added to a neutral pith ball. What is its charge?

23.23. What is the force between two $+1$-C charges located 1 m apart?

23.24. What is the magnitude and direction of the force on a charge of $+2 \times 10^{-7}$ C that is 0.3 m from a charge of -5×10^{-7} C?

23.25. Two electrons repel each other with a force of 10^{-8} N. How far apart are they?

23.26. Two charges attract each other with a force of 10^{-6} N when they are 1 cm apart. (*a*) How far apart should they be for the force between them to be 10^{-4} N? (*b*) 10^{-8} N?

23.27. A test charge of $+2 \times 10^{-7}$ C is located 5 cm to the right of a charge of $+1 \times 10^{-6}$ C and 10 cm to the left of a charge of -1×10^{-6} C. The three charges lie on a straight line. Find the force on the test charge.

23.28. A charge of $+1 \times 10^{-7}$ C and a charge of $+3 \times 10^{-7}$ C are 40 cm apart. (*a*) Where should a charge of $+q$ be placed on the line between these charges so that no net force acts on it? (*b*) Where should a charge of $-q$ be so placed?

23.29. How much force is exerted on a charge of 10^{-6} C by an electric field of 50 V/m?

23.30. An electron is present in an electric field of 10^4 V/m. (*a*) Find the force on the electron. (*b*) Find the electron's acceleration.

23.31. Two charges of $+10^{-6}$ C are located 1 cm apart. (*a*) What is the force on a charge of $+10^{-8}$ C halfway between them? (*b*) What is the force on a charge of -10^{-8} C at the same place? (*c*) What must be true of the strength of the electric field halfway between the two $+10^{-6}$-C charges?

23.32. A potential difference of 100 V is applied by a battery across a pair of metal plates 5 cm apart. (*a*) What is the electric field between the plates? (*b*) How much force does a charge of $+10^{-8}$ C experience in this field? (*c*) How much kinetic energy does this charge acquire when it goes from the positive plate to the negative plate?

23.33. A charge of -2×10^{-9} C in an electric field between two parallel metal plates 4 cm apart is acted upon by a force of 10^{-4} N. (*a*) What is the strength of the field? (*b*) What is the potential difference between the plates?

23.34. A proton is accelerated by a potential difference of 15,000 V. What is its kinetic energy?

23.35. A particle of charge 10^{-12} C starts to move from rest in an electric field of 500 V/m. (*a*) What is the force on the particle? (*b*) How much kinetic energy will it have when it has moved 1 cm in the field?

23.36. When a certain 12-V storage battery is charged, a total of 10^5 C is transferred from one set of its electrodes to the other set. Find the energy stored in the battery.

Answers to Supplementary Problems

23.17. (*a*) 8 electrons   (*b*) 6 electrons

23.18. Such a sketch shows how the magnitude and direction of the field vary in space. At a given point, the direction of the field is given by the direction of the nearest lines of force, and the relative magnitude of the field is indicated by how close together the lines of force are in the vicinity of the point.

23.19. By definition, a line of force represents the path a positively charged particle would follow in an electric field, and such a particle can travel in only one direction at any point.

23.20. The force exerted by the field on the $-q$ charge is equal and opposite to the force exerted on the $+q$ charge, so the rod does not move since the two forces have the same line of action.

23.21. The equal and opposite forces exerted on the charges now cause the rod to rotate until it is parallel to the electric field.

23.22. -1.6×10^{-10} C

23.23. 9×10^9 N; repulsive

23.24. 10^{-2} N directed toward the other charge

23.25. 1.5×10^{-10} m

23.26. (a) 0.1 cm   (b) 10 cm

23.27. 0.9 N directed to the right

23.28. (a) 14.6 cm from the charge of $+1 \times 10^{-7}$ C   (b) the same location

23.29. 5×10^{-5} N

23.30. (a) 1.6×10^{-15} N   (b) 1.8×10^{15} m/s^2

23.31. (a) 0   (b) 0   (c) $E = 0$

23.32. (a) 2000 V   (b) 2×10^{-5} N   (c) 10^{-6} J

23.33. (a) 5×10^4 V/m   (b) 2000 V

23.34. 2.4×10^{-15} J

23.35. (a) 5×10^{-10} N   (b) 5×10^{-12} J

23.36. 1.2×10^6 J

Chapter 24

Electric Current

ELECTRIC CURRENT

A flow of charge from one place to another constitutes an *electric current*. The direction of a current is conventionally considered to be that in which positive charge would have to move to produce the same effects as the actual current. Thus a current is always supposed to go from the positive terminal of a battery or generator to its negative terminal.

Electric currents in metal wires always consist of flows of electrons; such currents are assumed to occur in the direction opposite to that in which the electrons move. Since a positive charge going one way is for most purposes equivalent to a negative charge going the other way, this assumption makes no practical difference. Both positive and negative charges move when a current is present in a liquid or gaseous conductor.

If an amount of charge q passes a given point in a conductor in the time interval t, the current in the conductor is

$$I = \frac{q}{t}$$

$$\text{Electric current} = \frac{\text{charge}}{\text{time interval}}$$

The unit of electric current is the *ampere* (A), where

$$1 \text{ ampere} = 1 \frac{\text{coulomb}}{\text{second}}$$

OHM'S LAW

In order for a current to exist in a conductor, there must be a potential difference between its ends, just as a difference in height between source and outlet is necessary for a river current to exist. In the case of a metallic conductor, the current is proportional to the applied potential difference: doubling V causes I to double, tripling V causes I to triple, and so forth. This relationship is known as *Ohm's law* and is expressed in the form

$$I = \frac{V}{R}$$

$$\text{Electric current} = \frac{\text{potential difference}}{\text{resistance}}$$

The quantity R is a constant for a given conductor and is called its *resistance*. The unit of resistance is the *ohm* (Ω), where

$$1 \text{ ohm} = 1 \frac{\text{volt}}{\text{ampere}}$$

The greater the resistance of a conductor, the less the current when a certain potential difference is applied.

Ohm's law is not a physical principle but is an experimental relationship that most metals obey over a wide range of values of V and I.

RESISTIVITY

The resistance of a conductor that obeys Ohm's law is given by

$$R = \rho \frac{L}{A}$$

where L is the length of the conductor, A is its cross-sectional area, and ρ (Greek letter *rho*) is the *resistivity* of the material of the conductor. In the SI system, the unit of resistivity is the Ω-m.

In engineering practice using British units, it is customary to use as the unit of area the *circular mil* (CM). A circular mil is the area of a circle whose diameter is 1 mil, where 1 mil = 1/1000 in. = 0.001 in. = 10^{-3} in. Thus the area in CM of a circle is equal to the square of its diameter in mils:

$$A\ (\text{CM}) = \left[\, d\ (\text{mils})\right]^{2}$$

The use of the CM simplifies calculations of resistance since most conductors have circular cross sections and a factor of π is thereby eliminated. When the length of a conductor is given in ft and its area in CM, the unit of resistivity is the Ω-CM/ft.

The resistivities of most materials vary with temperature. If R is the resistance of a conductor at a particular temperature, then the change in its resistance ΔR when the temperature changes by ΔT is approximately proportional to both R and ΔT, so that

$$\Delta R = \alpha R\ \Delta T$$

The quantity α is the *temperature coefficient of resistance* of the material.

ELECTRIC POWER

The rate at which work is done to maintain an electric current is given by the product of the current I and the potential difference V:

$$P = IV$$

Power = current $\times$ potential difference

When I is in amperes and V is in volts, P will be in watts.

If the conductor or device through which a current passes obeys Ohm's law, the power consumed may be expressed in the alternative forms

$$P = IV = I^{2}R = \frac{V^{2}}{R}$$

Solved Problems

24.1. Since electric current is a flow of charge, why are two wires rather than a single one used to carry current?

If a single wire were used, charge of one sign or the other (depending upon the situation) would be permanently transferred from the source of current to the appliance at the far end of the wire. In a short time so much charge would have been transferred that the source would be unable to shift further charge against the repulsive force of the charge piled up at the appliance. Thus a single wire cannot carry a current continuously. The use of two wires, on the other hand, enables charge to be circulated from source to appliance and back, so that a continuous one-way flow of energy can take place.

24.2. Which solids are good electrical conductors and which are good insulators? How well do these substances conduct heat?

All metals are good electrical conductors. All nonmetallic solids are good insulators, for instance glass, wood, plastics, rubber. In general, solids that are good conductors of electricity are also good conductors of heat, and solids that are good electrical insulators are poor conductors of heat. Metals are good conductors of heat and electricity because both are transferred through a metal by the freely moving electrons that are a characteristic feature of its structure.

24.3. A wire carries a current of 1 A. How many electrons pass any point in the wire each second?

The electron charge is of magnitude $e = 1.6 \times 10^{-19}$ C and so a current of 1 A = 1 C/s corresponds to a flow of

$$\frac{1 \text{ C/s}}{1.6 \times 10^{-19} \text{ C/electron}} = 6.3 \times 10^{18} \text{ electrons/s}$$

24.4. A 120-V toaster has a resistance of 12 Ω. What must be the minimum rating of the fuse in the electrical circuit to which the toaster is connected?

The current in the toaster is

$$I = \frac{V}{R} = \frac{120 \text{ V}}{12 \text{ }\Omega} = 10 \text{ A}$$

so this must be the rating of the fuse.

24.5. A 120-V electric heater draws a current of 25 A. What is its resistance?

$$R = \frac{V}{I} = \frac{120 \text{ V}}{25 \text{ A}} = 4.8 \text{ }\Omega$$

24.6. What is the resistance of a copper wire 0.5 mm in diameter and 20 m long? The resistivity of copper is 1.7×10^{-8} Ω-m.

The wire's cross-sectional area is πr^2, where $r = 0.25$ mm $= 2.5 \times 10^{-4}$ m. Hence,

$$R = \rho \frac{L}{A} = \frac{1.7 \times 10^{-8} \text{ }\Omega\text{-m} \times 20 \text{ m}}{\pi (2.5 \times 10^{-4} \text{ m})^2} = 1.73 \text{ }\Omega$$

24.7. A platinum wire 80 cm long is to have a resistance of 0.1 Ω. What should its diameter be? The resistivity of platinum is 1.1×10^{-7} Ω-m.

Since $R = \rho L/A = \rho L/\pi r^2$,

$$r = \sqrt{\frac{\rho L}{\pi R}} = \sqrt{\frac{1.1 \times 10^{-7} \text{ }\Omega\text{-m} \times 0.8 \text{ m}}{\pi \times 0.1 \text{ }\Omega}} = 5.3 \times 10^{-4} \text{ m} = 0.53 \text{ mm}$$

The wire's diameter should therefore be $2r = 1.06$ mm.

24.8. What is the resistance of 200 ft of No. 30 copper wire? The resistivity of copper is 9.6 Ω-CM/ft and the diameter of No. 30 wire in the American Wire Gage scale is 0.010 in.

Since 0.001 in. = 1 mil, No. 30 wire has a diameter of 10 mils and a cross-sectional area in CM of

$$A \text{ (CM)} = [d \text{ (mils)}]^2 = 100 \text{ CM}$$

The resistance of the wire is therefore

$$R = \rho \frac{L}{A} = 9.6 \frac{\Omega\text{-CM}}{\text{ft}} \times \frac{200 \text{ ft}}{100 \text{ CM}} = 19.2 \text{ }\Omega$$

24.9. A copper wire has a resistance of 10.0 Ω at 20 °C. (a) What will its resistance be at 80 °C? (b) At 0 °C? The temperature coefficient of resistance of copper is 0.004/°C.

(a)   Here $R = 10.0\ \Omega$ and $T = 60\ °C$.   Hence the wire's change in resistance is

$$\Delta R = \alpha R\ \Delta T = 0.004/°C \times 10.0\ \Omega \times 60\ °C = 2.4\ \Omega$$

and the resistance at 80 °C will be $R + \Delta R = 12.4\ \Omega$.

(b)   Here $\Delta T = -20\ °C$, and so

$$\Delta R = \alpha R\ \Delta T = 0.004/°C \times 10.0\ \Omega \times (-20\ °C) = -0.8\ \Omega$$

The resistance at 0 °C will be $R + \Delta R = 9.2\ \Omega$.

24.10. A *resistance thermometer* makes use of the variation of the resistance of a conductor with temperature.   If the resistance of such a thermometer with a platinum element is 5 Ω at 20 °C and 16 Ω when inserted in a furnace, find the temperature of the furnace.   The value of α for platinum is 0.0036/°C.

Here $R = 5\ \Omega$ and $\Delta R = 16\ \Omega - 5\ \Omega = 11\ \Omega$.   Since $\Delta R = \alpha R\ \Delta T$,

$$\Delta T = \frac{\Delta R}{\alpha R} = \frac{11\ \Omega}{0.0036/°C \times 5\ \Omega} = 611\ °C$$

The temperature of the furnace is $T + \Delta T = 20\ °C + 611\ °C = 631\ °C$.

24.11. The current through a 50-Ω resistance is 2 A.   How much power is dissipated as heat?

$$P = I^2R = (2\ A)^2 \times 50\ \Omega = 200\ W$$

24.12. A 2-kW water heater is to be connected to a 240-V power line whose circuit breaker is rated at 10 A.   Will the breaker open when the heater is switched on?

The heater draws a current of

$$I = \frac{P}{V} = \frac{2000\ W}{240\ V} = 8\tfrac{1}{3}\ A$$

Since this current is less than 10 A, the breaker will not open.

24.13. The starting motor of a certain car develops 1 hp when it turns over the engine.   How much current does it draw from a 12-V battery?

Here $P = 1\ hp = 746\ W = IV$, and so

$$I = \frac{P}{V} = \frac{746\ W}{12\ V} = 62\ A$$

24.14. A 12-V storage battery is charged by a current of 20 A for 1 hr.   (a) How much power is required to charge the battery at this rate?   (b) How much energy has been provided during the process?

(a) $$P = IV = 20\ A \times 12\ V = 240\ W$$

(b) $$W = Pt = 240\ W \times 3600\ s = 8.64 \times 10^5\ J$$

24.15. The 12-V battery of a certain car has a capacity of 80 A-hr, which means that it can furnish a current of 80 A for 1 hr, a current of 40 A for 2 hr, and so forth.   (a) How much energy is stored in the battery?   (b) If the car's lights require 60 W of power, how long can the battery keep them lit when the engine (and hence its generator) is not running?

(a)   The 80 A-hr capacity of the battery is a way to express the amount of charge it can transfer from one of its terminals to the other.   Here the amount of charge is

$$q = 80\ \text{A-hr} \times 3600\ s/hr = 2.88 \times 10^5\ \text{A-s} = 2.88 \times 10^5\ C$$

and so the energy the battery can provide is

$$W = qV = 2.88 \times 10^5\ C \times 12\ V = 3.46 \times 10^6\ J$$

(b)   Since

$$P = \frac{W}{t}, \qquad t = \frac{W}{P} = \frac{3.46 \times 10^6 \text{ J}}{60 \text{ W}} = 5.8 \times 10^4 \text{ s} = 16 \text{ hr}$$

Supplementary Problems

24.16. Bends in a pipe slow down the flow of water through it. Do bends in a wire increase its electrical resistance?

24.17. How many electrons pass through the filament of a 75-W, 120-V light bulb per second?

24.18. Find the current in a 200-Ω resistor when the potential difference across it is 40 V.

24.19. An electric water heater draws 10 A of current from a 240-V power line. What is its resistence?

24.20. The resistance of a 10-ft length of 20-mil iron wire is found to be 1.8 Ω. What is the resistivity of iron?

24.21. Find the resistance of 8 m of aluminum wire 0.1 mm in diameter. The resistivity of aluminum is 2.6×10^{-8} Ω-m.

24.22. How long should a copper wire 0.4 mm in diameter be for it to have a resistance of 10 Ω? The resistivity of copper is 1.7×10^{-8} Ω-m.

24.23. The temperature coefficient of resistance of carbon is $-0.0005/°C$. If the resistance of a carbon resistor is 1000 Ω at 0 °C, find its resistance at 120 °C.

24.24. An iron wire has a resistance of 0.20 Ω at 20 °C and a resistance of 0.30 Ω at 110 °C. Find the temperature coefficient of resistance of the iron used in the wire.

24.25. What is the resistance of a 750-W, 120-V electric iron?

24.26. How much power is developed by an electric motor which draws a current of 4 A when operated at 240 V? How many horsepower is this?

24.27. What is the current in a 100-W light bulb when it is operated at 120 V?

24.28. A 32-V storage battery has a capacity of 10^6 J. How long can it supply a current of 5 A?

24.29. The 12-V battery of a car is required to be able to operate the 1.5-kW starting motor for a total of at least 10 min. (a) What should the minimum capacity of the battery be (in A-hr)? (b) How much energy is stored in such a battery?

24.30. A light bulb whose power is 100 W when operated at 240 V is instead connected to a 120-V source. (a) What is the current in the bulb? (b) How much power does it dissipate?

24.31. Currents of 5 A pass through two resistors, one of which has a potential difference of 100 V across it and the other of which has a potential difference of 300 V across it. (a) Compare the rates at which charge passes through each resistor. (b) Compare the rates at which energy is dissipated by each resistor.

Answers to Supplementary Problems

24.16. Bends in a wire have no effect on its electrical resistance because the electrons whose motion constitutes an electric current are extremely small with very little mass and therefore can change direction readily.

24.17. 3.9×10^{18} electrons

24.18. 0.2 A

24.19. 24 Ω

24.20. 72 Ω-CM/ft

24.21. 26.5 Ω

24.22. 74 m

24.23. 940 Ω

24.24. 0.0056/°C

24.25. 19.2 Ω

24.26. 960 W; 1.3 hp

24.27. 0.83 A

24.28. 6250 s = 1 hr 44 min 10 s

24.29. (a) 21 A-hr   (b) 9.0×10^{5} J

24.30. (a) 0.21 A   (b) 25 W

24.31. (a) Since the currents are the same, charge flows at the same rate through each resistor.   (b) Since $P = IV$, energy is dissipated by the second resistor three times as fast as by the first resistor.

Chapter 25

Direct Current Circuits

RESISTORS IN SERIES

The equivalent resistance of a set of resistors depends upon the way in which they are connected as well as upon their values. If the resistors are joined in *series*, that is, consecutively (Fig. 25-1), the equivalent resistance R of the combination is the sum of the individual resistances:

$$R = R_1 + R_2 + R_3 + \cdots \qquad \text{(series resistors)}$$

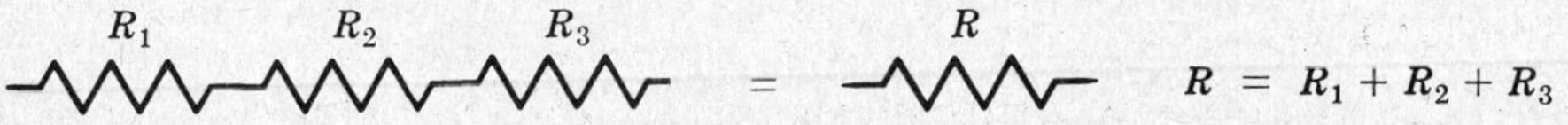

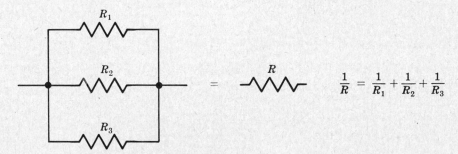

Fig. 25-1

RESISTORS IN PARALLEL

In a *parallel* set of resistors, the corresponding terminals of the resistors are connected together (Fig. 25-2). The reciprocal $1/R$ of the equivalent resistance of the combination is the sum of the reciprocals of the individual resistances:

$$\frac{1}{R} = \frac{1}{R_1} + \frac{1}{R_2} + \frac{1}{R_3} + \cdots \qquad \text{(parallel resistors)}$$

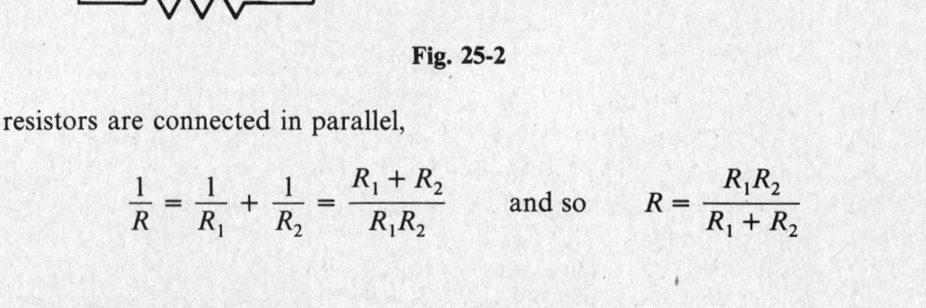

Fig. 25-2

If only two resistors are connected in parallel,

$$\frac{1}{R} = \frac{1}{R_1} + \frac{1}{R_2} = \frac{R_1 + R_2}{R_1 R_2} \qquad \text{and so} \qquad R = \frac{R_1 R_2}{R_1 + R_2}$$

EMF AND INTERNAL RESISTANCE

The work done per coulomb on the charge passing through a battery, generator, or other source of electrical energy is called the *electromotive force*, or *emf*, of the source. The emf ls equal to the potential difference across the terminals of the source when no current flows. When a current I flows, this potential difference is less than the emf because of the *internal resistance* of the source. If the internal resistance is r, then a potential drop of Ir occurs within the source. The terminal voltage

V across a source of emf $\mathcal{E}$ whose internal resistance is r when it provides a current of I is therefore

$$V = \mathcal{E} - Ir$$
Terminal voltage = emf − potential drop due to internal resistance

When a battery or generator of emf $\mathcal{E}$ is connected to an external resistance R, the total resistance in the circuit is $R + r$, and the current that flows is

$$I = \frac{\mathcal{E}}{R + r}$$

$$\text{Current} = \frac{\text{emf}}{\text{external resistance} + \text{internal resistance}}$$

KIRCHHOFF'S RULES

The current that flows in each branch of a complex circuit can be found by applying *Kirchhoff's rules* to the circuit. The first rule applies to *junctions* of three or more wires (Fig. 25-3) and is a consequence of conservation of charge. The second rule applies to *loops*, which are closed conducting paths in the circuit, and is a consequence of conservation of energy. The rules are:

1. The sum of the currents that flow into a junction is equal to the sum of the currents that flow out of the junction.

2. The sum of the emf's around a loop is equal to the sum of the IR potential drops around the loop.

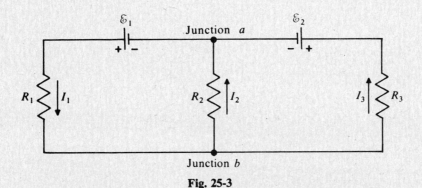

Fig. 25-3

The procedure for applying Kirchhoff's rules is as follows. First, a direction is arbitrarily chosen for the current in each resistance, as in Fig. 25-3. If the choice is correct, the current will be found to be positive; if not, the current will be found to be negative, which means that the actual current is in the opposite direction. Second, in going around a loop (which can be done either clockwise or counterclockwise) an emf is considered positive if the − terminal of its source is met first, negative if the + terminal is met first. Third, an IR drop is considered positive if the current in the resistance R is in the same direction as the path being followed, negative if the direction is opposite to the path.

In the case of the circuit shown in Fig. 25-3, Kirchhoff's first rule when applied to either junction a or junction b yields

$$I_1 = I_2 + I_3$$
The second rule applied to loop 1, shown in Fig. 25-4(a), and proceeding counterclockwise, yields
$$\mathcal{E}_1 = I_1 R_1 + I_2 R_2$$

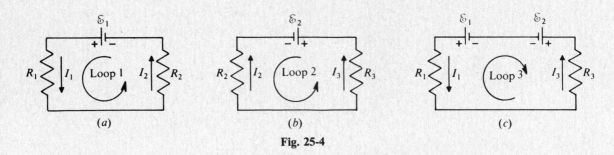

(a)   (b)   (c)

Fig. 25-4

This rule applied to loop 2, shown in Fig. 25-4(b), and again proceeding counterclockwise, yields

$$-\mathcal{E}_2 = -I_2R_2 + I_3R_3$$

There is also a third loop, namely the outside one shown in Fig. 25-4(c), which must similarly obey Kirchhoff's second rule. For the sake of variety we now proceed clockwise and obtain

$$-\mathcal{E}_1 + \mathcal{E}_2 = -I_3R_3 - I_1R_1$$

Note that this last equation is just the negative sum of the two preceding equations. Thus, we may use the junction equation and *any two* of the loop equations to solve for the unknown currents I_1, I_2, and I_3.

AMMETERS AND VOLTMETERS

An *ammeter* is an instrument that measures current. The lower the resistance of an ammeter, the better, since this resistance affects the circuit whose current is being measured. In practice, the meter itself (usually a *galvanometer*, in which magnetic forces produced by a current rotate a pointer) is used in parallel with a low-resistance *shunt* which carries nearly all the current, leaving a small fraction to pass through the higher-resistance meter.

A *voltmeter* is an instrument that measures potential difference. The higher the resistance of a voltmeter, the better, since its presence across a circuit element reduces the current through that element and thus changes the potential difference being measured. In practice, the meter itself (again it is usually a galvanometer) is used in series with a high resistance.

Solved Problems

25.1. To reduce the brightness of a light bulb, should an auxiliary resistance be connected in series with it or in parallel?

    In series, because in this way the current in the bulb is reduced; if the resistor were connected in parallel with the bulb, the current in the bulb and hence its brightness would not be affected.

25.2. It is desired to limit the current in a 50-Ω resistor to 10 A when it is connected to a 600-V power source. How should an auxiliary resistor be connected in the circuit and what should its resistance be?

    For a current of 10 A, the total resistance in the circuit should be

$$R = \frac{V}{I} = \frac{600 \text{ V}}{10 \text{ A}} = 60 \text{ }\Omega$$

Hence a 10-Ω resistor should be connected in series with the 50-Ω resistor to give a total of 60 Ω.

25.3. (a) What is the equivalent resistance of three 5-Ω resistors connected in series? (b) If a potential difference of 60 V is applied across the combination, what is the current in each resistor?

(a) $$R = R_1 + R_2 + R_3 = 5\,\Omega + 5\,\Omega + 5\,\Omega = 15\,\Omega$$

(b) The current in the entire circuit is

$$I = \frac{V}{R} = \frac{60\text{ V}}{15\,\Omega} = 4\text{ A}$$

Since the resistors are in series, this current flows through each of them.

25.4. (a) What is the equivalent resistance of three 5-Ω resistors connected in parallel? (b) If a potential difference of 60 V is applied across the combination, what is the current in each resistor?

(a) $$\frac{1}{R} = \frac{1}{R_1} + \frac{1}{R_2} + \frac{1}{R_3} = \frac{1}{5\,\Omega} + \frac{1}{5\,\Omega} + \frac{1}{5\,\Omega} = \frac{3}{5\,\Omega}$$

$$R = \frac{5}{3}\,\Omega = 1.67\,\Omega$$

(b) Since each resistor has a potential difference of 60 V across it, the current in each one is

$$I = \frac{V}{R} = \frac{60\text{ V}}{5\,\Omega} = 12\text{ A}$$

25.5. Two 240-Ω light bulbs are connected in parallel to a 120-V power source. (a) What is the current in each bulb? (b) How much power does each bulb dissipate?

(a) The potential difference across each bulb is 120 V. Hence the current in each bulb is

$$I = \frac{V}{R} = \frac{120\text{ V}}{240\,\Omega} = 0.5\text{ A}$$

(b) $$P = I^2R = (0.5\text{ A})^2 \times 240\,\Omega = 60\text{ W}$$

25.6. Two 240-Ω light bulbs are connected in series with a 120-V power source. (a) What is the current in each bulb? (b) How much power does each bulb dissipate?

(a) The equivalent resistance of the two bulbs is

$$R = R_1 + R_2 = 240\,\Omega + 240\,\Omega = 480\,\Omega$$

The current in the circuit is therefore

$$I = \frac{V}{R} = \frac{120\text{ V}}{480\,\Omega} = 0.25\text{ A}$$

and, since the bulbs are in series, the current passes through each of them.

(b) $$P = I^2R = (0.25\text{ A})^2 \times 240\,\Omega = 15\text{ W}$$

25.7. A circuit has a resistance of 50 Ω. How can it be reduced to 20 Ω?

To obtain an equivalent resistance of $R = 20\,\Omega$, a resistor R_2 must be connected in parallel with the circuit of $R_1 = 50\,\Omega$. To find R_2 we proceed as follows:

$$\frac{1}{R} = \frac{1}{R_1} + \frac{1}{R_2} \qquad \frac{1}{R_2} = \frac{1}{R} - \frac{1}{R_1} = \frac{R_1 - R}{R_1 R}$$

$$R_2 = \frac{R_1 R}{R_1 - R} = \frac{50\,\Omega \times 20\,\Omega}{50\,\Omega - 20\,\Omega} = 33.3\,\Omega$$

25.8. Find the equivalent resistance of the circuit shown in Fig. 25-5(a).

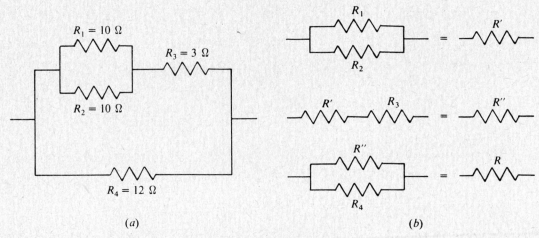

(a)   (b)

Fig. 25-5

Figure 25-5(b) shows how the original circuit is decomposed into its series and parallel parts, each of which is treated in turn.   The equivalent resistance of R_1 and R_2 is

$$R' = \frac{R_1 R_2}{R_1 + R_2} = \frac{10\ \Omega \times 10\ \Omega}{10\ \Omega + 10\ \Omega} = 5\ \Omega$$

This equivalent resistance is in series with R_3, and so

$$R'' = R' + R_3 = 5\ \Omega + 3\ \Omega = 8\ \Omega$$

Finally R'' is in parallel with R_4, hence the equivalent resistance of the entire circuit is

$$R = \frac{R'' R_4}{R'' + R_4} = \frac{8\ \Omega \times 12\ \Omega}{8\ \Omega + 12\ \Omega} = 4.8\ \Omega$$

25.9. A potential difference of 20 V is applied to the circuit of Fig. 25-5.   Find the current through each resistor and the current through the entire circuit.

Because resistor R_4 has the full 20-V potential difference across it,

$$I_4 = \frac{V}{R_4} = \frac{20\ \text{V}}{12\ \Omega} = 1.67\ \text{A}$$

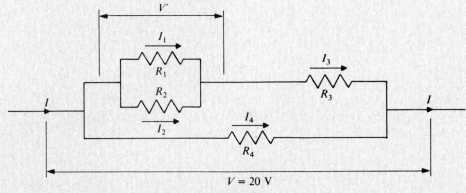

$V = 20$ V

Fig. 25-6

From Fig. 25-6 we see that the current I_3 also flows through the entire upper branch of the circuit, whose equivalent resistance is $R'' = 8\ \Omega$.   Hence

$$I_3 = \frac{V}{R''} = \frac{20\ \text{V}}{8\ \Omega} = 2.5\ \text{A}$$

The potential difference V' across R_1 and R_2 is

$$V' = V - I_3 R_3 = 20 \text{ V} - 2.5 \text{ A} \times 3 \ \Omega = 12.5 \text{ V}$$

Hence the current I_1 is

$$I_1 = \frac{V'}{R_1} = \frac{12.5 \text{ V}}{10 \ \Omega} = 1.25 \text{ A}$$

and the current I_2 is

$$I_2 = \frac{V'}{R_2} = \frac{12.5 \text{ V}}{10 \ \Omega} = 1.25 \text{ A}$$

The current through the entire circuit is

$$I = \frac{V}{R} = \frac{20 \text{ V}}{4.8 \ \Omega} = 4.17 \text{ A}$$

We note that $I = I_3 + I_4$ and that $I_3 = I_1 + I_2$, as they should.

25.10. A dry cell of emf 1.5 V and internal resistance 0.05 Ω is connected to a flashlight bulb whose resistance is 0.4 Ω. Find the current in the circuit.

$$I = \frac{\mathcal{E}}{R + r} = \frac{1.5 \text{ V}}{0.4 \ \Omega + 0.05 \ \Omega} = 3.33 \text{ A}$$

25.11. A battery whose emf is 45 V is connected to a 20-Ω resistance and a current of 2.1 A flows. (*a*) Find the internal resistance of the battery. (*b*) Find the terminal voltage of the battery.

(*a*) From $I = \mathcal{E} /(R + r)$ we obtain

$$r = \frac{\mathcal{E}}{I} - R = \frac{45 \text{ V}}{2.1 \text{ A}} - 20 \ \Omega = 21.4 \ \Omega - 20 \ \Omega = 1.4 \ \Omega$$

(*b*) $$V = \mathcal{E} - Ir = 45 \text{ V} - 2.1 \text{ A} \times 1.4 = 42 \text{ V}$$

25.12. A generator has an emf of 120 V and an internal resistance of 0.2 Ω. (*a*) How much current does the generator supply when the terminal voltage is 115 V? (*b*) How much power does it supply? (*c*) How much power is dissipated in the generator itself?

(*a*) From $V = \mathcal{E} - Ir$ we obtain

$$I = \frac{\mathcal{E} - V}{r} = \frac{120 \text{ V} - 115 \text{ V}}{0.2 \ \Omega} = 25 \text{ A}$$

(*b*) $$P = IV = 25 \text{ A} \times 115 \text{ V} = 2875 \text{ W}$$

(*c*) $$P = I^2 r = (25 \text{ A})^2 \times 0.2 \ \Omega = 125 \text{ W}$$

25.13. When a source of emf whose internal resistance is r is connected to an external load of resistance R, the power delivered to R will be a maximum when $R = r$. Verify this statement by calculating the power delivered by a battery of emf 10 V and internal resistance 0.5 Ω when it is connected to (*a*) 0.25-Ω, (*b*) 0.5-Ω, and (*c*) 1-Ω resistors.

(*a*)
$$I_1 = \frac{\mathcal{E}}{R_1 + r} = \frac{10 \text{ V}}{0.25 \ \Omega + 0.5 \ \Omega} = 13.3 \text{ A}$$
$$P_1 = I_1^2 R_1 = (13.3 \text{ A})^2 \times 0.25 \ \Omega = 44 \text{ W}$$

(*b*)
$$I_2 = \frac{\mathcal{E}}{R_2 + r} = \frac{10 \text{ V}}{0.5 \ \Omega + 0.5 \ \Omega} = 10 \text{ A}$$
$$P_2 = I_2^2 R_2 = (10 \text{ A})^2 \times 0.5 \ \Omega = 50 \text{ W}$$

(c)

$$I_3 = \frac{\mathcal{E}}{R_3 + r} = \frac{10 \text{ V}}{1 \, \Omega + 0.5 \, \Omega} = 6.7 \text{ A}$$

$$P_3 = I_3^2 R_3 = (6.7 \text{ A})^2 \times 1 \, \Omega = 44 \text{ W}$$

25.14. A source of what potential difference is required in order to charge a battery of $\mathcal{E} = 6$ V and $r = 0.1 \, \Omega$ at a rate of 10 A?

The required potential difference must equal the emf of the battery *plus* the Ir drop in its internal resistance. Hence

$$V_{\text{applied}} = \mathcal{E} + Ir = 6 \text{ V} + 10 \text{ A} \times 0.1 \, \Omega = 7 \text{ V}$$

25.15. Two batteries in parallel, one of emf 6 V and internal resistance 0.5 Ω and the other of emf 8 V and internal resistance 0.6 Ω, are connected to an external 10-Ω resistor, as in Fig. 25-7(*a*). Find the current in the external resistor.

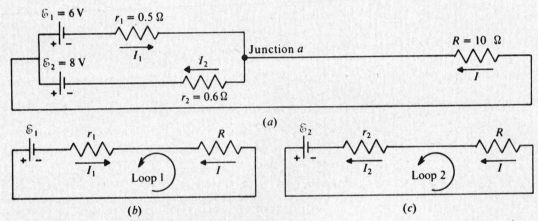

Fig. 25-7

Directions for the currents I, I_1, and I_2 are assumed as in Fig. 25-7(*a*). At junction *a* Kirchhoff's first law gives

$$I_2 = I + I_1$$

For the loop of Fig. 25-7(*b*), which we go around counterclockwise,

$$\mathcal{E}_1 = IR - I_1 r_1$$

and for the loop of Fig. 25-7(*c*), which we also go around counterclockwise,

$$\mathcal{E}_2 = IR + I_2 r_2$$

Since $I_2 = I + I_1$,

$$\mathcal{E}_2 = IR + Ir_2 + I_1 r_2$$

We now solve this equation and the first loop equation for I_1, set the two expressions for I_1 equal, and then solve for I:

$$I_1 = \frac{IR - \mathcal{E}_1}{r_1} \quad \text{and} \quad I_1 = \frac{\mathcal{E}_2 - IR - Ir_2}{r_2}$$

$$\frac{IR - \mathcal{E}_1}{r_1} = \frac{\mathcal{E}_2 - IR - Ir_2}{r_2}$$

$$I(Rr_2 + Rr_1 + r_2 r_1) = \mathcal{E}_2 r_1 + \mathcal{E}_1 r_2$$

$$I = \frac{\mathcal{E}_2 r_1 + \mathcal{E}_1 r_2}{Rr_2 + Rr_1 + r_2 r_1} = \frac{8 \text{ V} \times 0.5 \, \Omega + 6 \text{ V} \times 0.6 \, \Omega}{10 \, \Omega \times 0.6 \, \Omega + 10 \, \Omega \times 0.5 \, \Omega + 0.6 \, \Omega \times 0.5 \, \Omega}$$

$$= 0.673 \text{ A}$$

25.16. Find the currents in the three resistors of the circuit shown in Fig. 25-8(a). The internal resistances of the emf sources are included in R_1 and R_3.

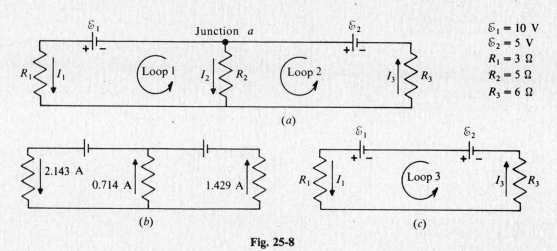

Fig. 25-8

We assume the current directions shown in the figure. Applying Kirchhoff's first rule to junction a yields

$$I_3 = I_1 + I_2$$

Next we analyze the two inner loops with the help of Kirchhoff's second rule. Proceeding counterclockwise in loop 1 yields

$$\mathcal{E}_1 = I_1R_1 - I_2R_2$$

and proceeding counterclockwise in loop 2 yields

$$\mathcal{E}_2 = I_2R_2 + I_3R_3$$

We now have three equations that relate the unknown quantities I_1, I_2, and I_3. One way to proceed (there are others, equally suitable) is to substitute $I_3 = I_1 + I_2$ in the second loop equation, which gives

$$\mathcal{E}_2 = I_2R_2 + I_1R_3 + I_2R_3$$

$$I_1 = \frac{\mathcal{E}_2 - I_2R_2 - I_2R_3}{R_3}$$

From the first loop equation,

$$I_1 = \frac{\mathcal{E}_1 + I_2R_2}{R_1}$$

Since these two expressions must be equal,

$$\frac{\mathcal{E}_1 + I_2R_2}{R_1} = \frac{\mathcal{E}_2 - I_2R_2 - I_2R_3}{R_3}$$

At this point we substitute the values of the various emf's and resistances and solve for I_2. This substitution can also be done earlier or later in a calculation of this kind, whatever seems most convenient. The calculation proceeds as follows:

$$\frac{10\text{ V} + (5\ \Omega)I_2}{3\ \Omega} = \frac{5\text{ V} - (5\ \Omega)I_2 - (6\ \Omega)I_2}{6\ \Omega}$$

$$\frac{10\text{ V}}{3\ \Omega} + \left(\frac{5\ \Omega}{3\ \Omega}\right)I_2 = \frac{5\text{ V}}{6\ \Omega} - \left(\frac{5\ \Omega}{6\ \Omega}\right)I_2 - \left(\frac{6\ \Omega}{6\ \Omega}\right)I_2$$

$$\left(\frac{5}{3} + \frac{5}{6} + \frac{6}{6}\right)I_2 = \left(\frac{5}{6} - \frac{10}{3}\right)\text{ A}$$

$$I_2 = -0.714\text{ A}$$

The minus sign means that the current I_2 is in the opposite direction to the one shown in the figure. From the first loop equation,

$$I_1 = \frac{\mathcal{E}_1 + I_2 R_2}{R_1} = \frac{10\text{ V} - 0.714\text{ A} \times 5\ \Omega}{3\ \Omega} = 2.143\text{ A}$$

Finally we find I_3 from the junction equation:

$$I_3 = I_1 + I_2 = 2.143\text{ A} - 0.714\text{ A} = 1.429\text{ A}$$

The actual currents are shown in Fig. 25-8(b).

As a check on the calculation we can apply Kirchhoff's second rule to the outside loop of the circuit, shown in Fig. 25-8(c). Proceeding counterclockwise,

$$\mathcal{E}_2 + \mathcal{E}_1 = I_1 R_1 + I_3 R_3$$
$$5\text{ V} + 10\text{ V} = 2.143\text{ A} \times 3\ \Omega + 1.429\text{ A} \times 6\ \Omega$$
$$15\text{ V} = 15\text{ V}$$

25.17. Find the currents in the three resistors of the circuit shown in Fig. 25-9(a). The internal resistances of the emf sources are included in the resistances shown.

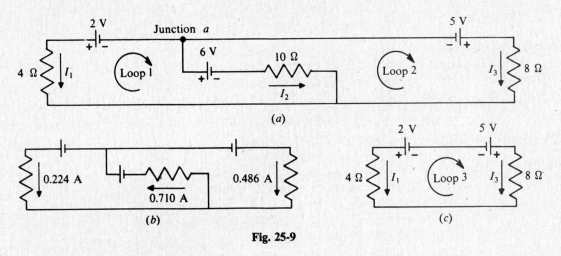

Fig. 25-9

Let us now work directly from the numerical values of the emf's and resistances in the figure. Applying Kirchhoff's first rule to junction a, assuming the current directions shown, yields

$$I_1 + I_2 + I_3 = 0$$

Clearly one or two current directions are incorrect, but this makes no difference since the result will be a negative current in those cases. Next we apply Kirchhoff's second rule to loop 1 and proceed clockwise:

$$-2\text{ V} - 6\text{ V} = (10\ \Omega)I_2 - (4\ \Omega)I_1$$

The emf's are considered negative because we encountered their $+$ terminals first. Solving for I_2 gives

$$I_2 = \frac{-8\text{ V}}{10\ \Omega} + \left(\frac{4\ \Omega}{10\ \Omega} \right) I_1 = -0.8\text{ A} + 0.4 I_1$$

Now we proceed clockwise in loop 2 and obtain

$$6\text{ V} + 5\text{ V} = (8\ \Omega)I_3 - (10\ \Omega)I_2$$

Substituting $I_3 = -I_1 - I_2$ and solving for I_2 yields

$$11\text{ V} = -(8\ \Omega)I_1 - (8\ \Omega)I_2 - (10\ \Omega)I_2$$
$$I_2 = \frac{-11\text{ V}}{18\ \Omega} - \left(\frac{8\ \Omega}{18\ \Omega} \right) I_1 = -0.611\text{ A} - 0.444\ I_1$$

Setting equal the two expressions for I_2 and solving for I_1,

$$-0.8 \text{ A} + 0.4 \, I_1 = -0.611 \text{ A} - 0.444 \, I_1$$
$$0.844 \, I_1 = 0.189 \text{ A}$$
$$I_1 = 0.224 \text{ A}$$

From the first loop equation

$$I_2 = -0.8 \text{ A} + 0.4 \, I_1 = -0.710 \text{ A}$$

and so

$$I_3 = -I_1 - I_2 = -0.224 \text{ A} + 0.710 \text{ A}$$
$$= 0.486 \text{ A}$$

The actual currents are shown in Fig. 25-9(b).

Again we check the results by using the outside loop of the circuit as shown in Fig. 25-9(c). Proceeding clockwise,

$$-2 \text{ V} + 5 \text{ V} = 8 \, \Omega \times 0.486 \text{ A} - 4 \, \Omega \times 0.224 \text{ A}$$
$$3 \text{ V} = 3 \text{ V}$$

25.18. A galvanometer which measures currents from 0 to 1 mA (1 mA = 1 milliampere = 0.001 A) has a resistance of 40 Ω. How can this galvanometer be used to measure currents from 0 to 1 A?

What is needed here is a shunt resistor that will carry 0.999 A when the total current is 1.000 A (Fig. 25-10). To find the value of the shunt resistance, we note that the potential difference V across both R_{meter} and R_{shunt} is the same, so that

$$V = I_{\text{meter}} R_{\text{meter}} = I_{\text{shunt}} R_{\text{shunt}}$$

Since the meter current is to be 0.001 A when $I_{\text{shunt}} = 0.999$ A,

$$R_{\text{shunt}} = \frac{I_{\text{meter}}}{I_{\text{shunt}}} \times R_{\text{meter}}$$

$$= \frac{0.001 \text{ A}}{0.999 \text{ A}} \times 40 \, \Omega = 0.04 \, \Omega$$

A 0.04-Ω resistor in parallel with the meter will permit it to measure currents from 0 to 1 A.

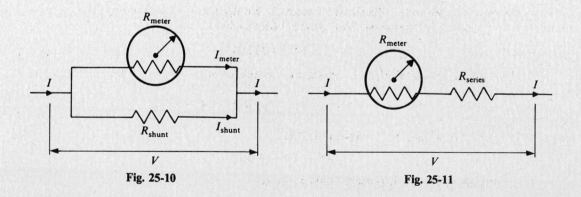

Fig. 25-10   Fig. 25-11

25.19. The galvanometer of Problem 25.18 is to be used to measure potential differences from 0 to 1 V. How can this be done?

What is needed now is a resistor in series with the meter that will limit the current to 0.001 A when the applied potential difference is 1 V (Fig. 25-11). The equivalent resistance of meter and resistor must therefore be

$$R = R_{meter} + R_{series} = \frac{V}{I}$$

and so

$$R_{series} = \frac{V}{I} - R_{meter} = \frac{1\ V}{0.001\ A} - 40\ \Omega = 960\ \Omega$$

A 960-Ω resistor in series with the meter will permit it to measure potential differences from 0 to 1 V.

25.20. A voltmeter whose resistance is 1000 Ω is connected across a resistor and the combination is connected in series with an ammeter (Fig. 25-12). When a potential difference is applied, the voltmeter reads 40 V and the ammeter reads 0.05 A. What is the resistance of the resistor?

At first glance it would seem that the resistance is simply

$$R' = \frac{V}{I} = \frac{40\ V}{0.05\ A} = 800\ \Omega$$

However, the voltmeter's own resistance is significant here since some of the current in the circuit is diverted through it, and $R' = 800\ \Omega$ is actually the equivalent resistance of R and R_{meter} in parallel. Hence

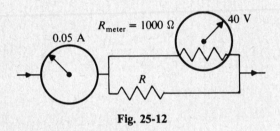

Fig. 25-12

$$\frac{1}{R'} = \frac{1}{R} + \frac{1}{R_{meter}}, \qquad \frac{1}{R} = \frac{1}{R'} - \frac{1}{R_{meter}} = \frac{R_{meter} - R'}{R_{meter}R'}$$

$$R = \frac{R_{meter}R'}{R_{meter} - R'} = \frac{1000\ \Omega \times 800\ \Omega}{1000\ \Omega - 800\ \Omega} = 4000\ \Omega$$

Supplementary Problems

25.21. A number of light bulbs are to be connected to a single power outlet. Will they provide more illumination if connected in series or in parallel? Why?

25.22. It is desired to have a current of 20 A in a 5-Ω resistor when it is connected to an 80-V battery. Is there any way in which an auxiliary resistor can be connected in the circuit to increase the current in the 5-Ω resistor to this value? If so, what should its resistance be?

25.23. (a) Find the equivalent resistance of four 60-Ω resistors connected in series. (b) If a potential difference of 12 V is applied across the combination, what is the current in each resistor?

25.24. (a) Find the equivalent resistance of four 60-Ω resistors connected in parallel. (b) If a potential difference of 12 V is applied across the combination, what is the current in each resistor?

25.25. You have three 2-Ω resistors. List the various resistances you can provide with them.

25.26. A 100-Ω resistor and a 200-Ω resistor are connected in series with a 40-V power source. (a) What is the current in each resistor? (b) How much power does each one dissipate?

25.27. A 100-Ω resistor and a 200-Ω resistor are connected in parallel with a 40-V power source.   (*a*) What is the current in each resistor?   (*b*) How much power does each one dissipate?

25.28. What resistance should be connected in parallel with a 1000-Ω resistor to produce an equivalent resistance of 200 Ω?

25.29. A 5-Ω resistor is connected in parallel with a 15-Ω resistor.   When a potential difference is applied to the combination, which resistor will carry the greater current?   What will the ratio of the currents be?

25.30. (*a*) Find the equivalent resistance of the circuit shown in Fig. 25-13.   (*b*) If a potential difference of 20 V is applied to the circuit, find the current in each resistor.

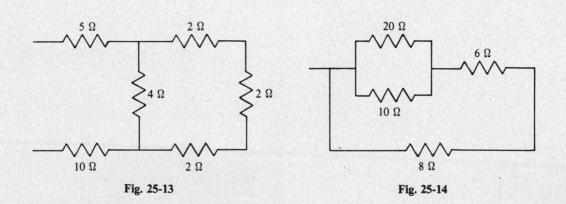

Fig. 25-13   Fig. 25-14

25.31. (*a*) Find the equivalent resistance of the circuit shown in Fig. 25-14.   (*b*) If a potential difference of 100 V is applied to the circuit, find the current in each resistor.

25.32. (*a*) Find the equivalent resistance of the circuit shown in Fig. 25-15.   (*b*) A 6-V battery whose internal resistance is 1 Ω is connected to the circuit. Find the current in each resistor.

25.33. Two batteries in parallel, each of emf 10 V and internal resistance 0.5 Ω, are connected to an external 20-Ω resistor.   Find the current in the external resistor.

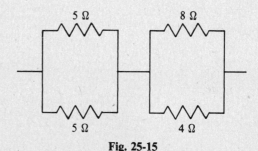

Fig. 25-15

25.34. A dry cell has an emf of 1.5 V and an internal resistance of 0.08 Ω.   (*a*) Find the current when the cell's terminals are connected together.   (*b*) Find the current when the cell is connected to a 5-Ω resistance.

25.35. A certain "12-V" storage battery actually has an emf of 13.2 V and an internal resistance of 0.01 Ω.   What is the terminal voltage of the battery when it delivers 80 A to the starter motor of a car engine?

25.36. A generator whose emf is 240 V has a terminal voltage of 220 V when it delivers a current of 50 A.   (*a*) Find the internal resistance of the generator.   (*b*) Find the power supplied by the generator.   (*c*) Find the power dissipated within the generator.

25.37. A storage battery of emf 34 V and internal resistance 0.1 Ω is to be charged at a rate of 20 A from a 110-V source.   What series resistance is needed in the circuit?

25.38. Find the currents in the resistors of the circuit shown in Fig. 25-16. The internal resistances of the emf sources are included in the external resistances.

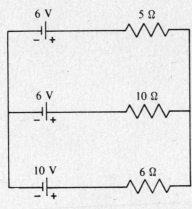

Fig. 25-16

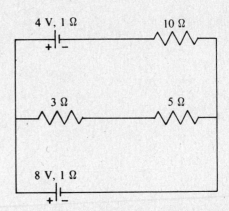

Fig. 25-17

25.39. Find the current in the resistors of the circuit shown in Fig. 25-17. The internal resistances of the emf sources must be taken into account.

25.40. A galvanometer has a resistance of 20 Ω and a range of 0–5 mA. (a) What shunt resistance is needed to convert the meter to a 0–100 mA ammeter? (b) What series resistance is needed to convert the meter to a 0–10 V voltmeter?

25.41. A 0–0.5 A ammeter has an equivalent resistance of 0.1 Ω. (a) If this meter is used as a voltmeter, what range of potential differences can it measure? (b) What series resistance is needed for the meter to have a range of 0–1000 V?

25.42. You have a 0–10 mA galvanometer whose resistance is 20 Ω and a separate 20-Ω resistor. What are the maximum ranges of the ammeter and voltmeter that can be assembled from the meter and resistor?

25.43. A 2000-Ω voltmeter reads 10 V when it is in parallel with a resistor of unknown resistance. At the same time an ammeter in series with the combination reads 0.1 A. Find the unknown resistance.

Answers to Supplementary Problems

25.21. In parallel, because this way each bulb has the maximum voltage across it and hence the maximum power dissipation.

25.22. There is no way in which an auxiliary resistor can be connected to increase the current.

25.23. (a) 240 Ω   (b) 0.05 A

25.24. (a) 15 Ω   (b) 0.2 A

25.25. 0.67 Ω; 1 Ω; 2 Ω; 3 Ω; 4 Ω; 6 Ω

25.26. (a) 0.133 A; 0.133 A   (b) 1.78 W; 3.55 W

25.27. (a) 0.4 A; 0.2 A   (b) 16 W; 8 W

25.28. 250 Ω

25.29. The 5-Ω resistor will carry a current three times greater than that carried by the 15-Ω resistor.

25.30. (*a*) The equivalent resistance is 17.4 Ω.   (*b*) The current in the 5-Ω and 10-Ω resistors is 1.15 A; in the three 0.2-Ω resistors, 0.46 A; and in the 4-Ω resistor, 0.69 A.

25.31. (*a*) The equivalent resistance is 4.90 Ω.   (*b*) The current in the 20-Ω resistor is 2.63 A; in the 10-Ω resistor, 5.26 A; in the 6-Ω resistor, 7.89 A; and in the 8-Ω resistor, 12.50 A.

25.32. (*a*) The equivalent resistance is 5.167 Ω.   (*b*) The current in each 5-Ω resistor is 0.581 A; in the 8-Ω resistor, 0.387 A; and in the 4-Ω resistor, 0.774 A.

25.33. 0.494 A

25.34. (*a*) 19 A   (*b*) 0.3 A

25.35. 12.4 V

25.36. (*a*) 0.4 Ω   (*b*) 11 kW   (*c*) 1 kW

25.37. 3.7 Ω

25.38. The current in the 5-Ω resistor is 0.286 A to the left, the current in the 10-Ω resistor is 0.143 A to the left, and the current in the 6-Ω resistor is 0.429 A to the right.

25.39. The current in the 10-Ω resistor is 0.935 A to the left and the currents in the 3-Ω and 5-Ω resistors are both 0.785 A to the left.

25.40. (*a*) 1.05 Ω   (*b*) 1980 Ω

25.41. (*a*) 0–0.05 V   (*b*) 1999.9 Ω

25.42. 0–20 mA; 0–0.4 V

25.43. 105 Ω

Chapter 26

Capacitance

CAPACITANCE

A *capacitor* is a system that stores energy in the form of an electric field. In its simplest form a capacitor consists of a pair of parallel metal plates separated by air or other insulating material.

The potential difference V between the plates of a capacitor is directly proportional to the charge Q on either of them, so the ratio Q/V is always the same for a particular capacitor. This ratio is called the *capacitance C* of the capacitor:

$$C = \frac{Q}{V}$$

$$\text{Capacitance} = \frac{\text{charge on either plate}}{\text{potential difference between plates}}$$

The unit of capacitance is the *farad* (F), where 1 farad = 1 coulomb/volt. Since the farad is too large for practical purposes, the *microfarad* and *picofarad* are commonly used, where

$$1 \text{ microfarad} = 1\mu F = 10^{-6} \text{ F}$$

$$1 \text{ picofarad} = 1pF = 10^{-12} \text{ F}$$

A charge of 10^{-6} C on each plate of a 1-μF capacitor will produce a potential difference of $V = Q/C = 1$ V between the plates.

PARALLEL-PLATE CAPACITOR

A capacitor that consists of parallel plates each of area A separated by the distance d has a capacitance of

$$C = K\varepsilon_0 \frac{A}{d}$$

The constant ε_0 is the permittivity of free space mentioned in Chapter 23; its value is

$$\varepsilon_0 = 8.85 \times 10^{-12} \text{ C}^2/\text{N-m}^2 = 8.85 \times 10^{-12} \text{ F/m}$$

The quantity K is the *dielectric constant* of the material between the capacitor plates; the greater K is, the more effective the material is in diminishing an electric field. For free space, $K = 1$; for air, $K = 1.0006$; a typical value for glass is $K = 6$; and for water, $K = 80$.

CAPACITORS IN COMBINATION

The *equivalent capacitance* of a set of capacitors connected together is the capacitance of the single capacitor that can replace the set without changing the properties of any circuit it is part of. The equivalent capacitance of a set of capacitors joined in series (Fig. 26-1) is

$$\frac{1}{C} = \frac{1}{C_1} + \frac{1}{C_2} + \frac{1}{C_3} + \cdots \qquad \text{(capacitors in series)}$$

If there are only two capacitors in series,

$$\frac{1}{C} = \frac{1}{C_1} + \frac{1}{C_2} = \frac{C_1 + C_2}{C_1 C_2} \qquad \text{and so} \qquad C = \frac{C_1 C_2}{C_1 + C_2}$$

In a parallel set of capacitors (Fig. 26-2) the equivalent capacitance is the sum of the individual capacitances:

$$C = C_1 + C_2 + C_3 + \cdots \qquad \text{(capacitors in parallel)}$$

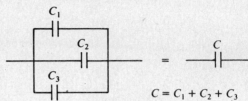

$$\frac{1}{C} = \frac{1}{C_1} + \frac{1}{C_2} + \frac{1}{C_3}$$

$$C = C_1 + C_2 + C_3$$

Fig. 26-1   **Fig. 26-2**

ENERGY OF A CHARGED CAPACITOR

In order to produce the electric field in a charged capacitor, work must be done to separate the positive and negative charges. This work is stored as electric potential energy in the capacitor. The potential energy W of a capacitor of capacitance C whose charge is Q and whose potential difference is V is given by

$$W = \tfrac{1}{2}QV = \tfrac{1}{2}CV^2 = \frac{1}{2}\frac{Q^2}{C}$$

TIME CONSTANT

When a capacitor is being charged, at any moment the potential difference across it due to the charge already on its plates is in the opposite direction to the applied emf and thus tends to oppose the flow of additional charge. For this reason a capacitor does not acquire its final charge the instant it is

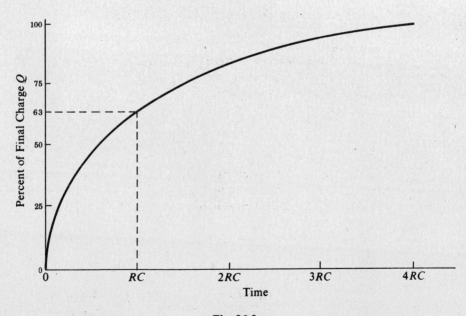

Fig. 26-3

connected to a battery or other source of emf. The product RC of the resistance R in the circuit and the capacitance C governs the rate at which the capacitor reaches its ultimate charge of $Q = CV$. After a time equal to RC has elapsed, the charge on the capacitor will be 63% of its final value (Fig. 26-3). If the capacitor is discharged through a resistance R, its charge will fall to 37% of its original value after $t = RC$. The quantity RC is called the *time constant* of the circuit; the smaller the time constant, the more rapidly a capacitor in a circuit can be charged or discharged.

Solved Problems

26.1. A 200-pF capacitor is connected to a 100-V battery. Find the charge on the capacitor's plates.

$$Q = CV = 200 \times 10^{-12} \text{ F} \times 100 \text{ V} = 2 \times 10^{-8} \text{ C}$$

26.2. A capacitor has a charge of 5×10^{-4} C when the potential difference across its plates is 300 V. Find its capacitance.

$$C = \frac{Q}{V} = \frac{5 \times 10^{-4} \text{ C}}{300 \text{ V}} = 1.67 \times 10^{-6} \text{ F} = 1.67 \ \mu\text{F}$$

26.3. Three capacitors whose capacitances are 1 μF, 2 μF, and 3 μF are connected in series. Find the equivalent capacitance of the combination.

$$\frac{1}{C} = \frac{1}{C_1} + \frac{1}{C_2} + \frac{1}{C_3} = \frac{1}{1 \ \mu\text{F}} + \frac{1}{2 \ \mu\text{F}} + \frac{1}{3 \ \mu\text{F}} = \frac{11}{6 \ \mu\text{F}}$$

Hence

$$C = \tfrac{6}{11} \ \mu\text{F} = 0.545 \ \mu\text{F}$$

26.4. The three capacitors of Problem 26.3 are connected in parallel. Find the equivalent capacitance of the combination.

$$C = C_1 + C_2 + C_3 = 1 \ \mu\text{F} + 2 \ \mu\text{F} + 3 \ \mu\text{F} = 6 \ \mu\text{F}$$

26.5. A 2-μF and a 3-μF capacitor are connected in series. (*a*) What is their equivalent capacitance? (*b*) A potential difference of 500 V is applied to the combination. Find the charge on each capacitor and the potential difference across it.

(*a*) $$C = \frac{C_1 C_2}{C_1 + C_2} = \frac{2 \ \mu\text{F} \times 3 \ \mu\text{F}}{2 \ \mu\text{F} + 3 \ \mu\text{F}} = 1.2 \ \mu\text{F}$$

(*b*) The charge on the combination is

$$Q = CV = 1.2 \times 10^{-6} \text{ F} \times 500 \text{ V} = 6 \times 10^{-4} \text{ C}$$

The same charge is present on each capacitor (Fig. 26-4). Hence the potential difference across the 2-μF capacitor is

$$V_1 = \frac{Q}{C_1} = \frac{6 \times 10^{-4} \text{ C}}{2 \times 10^{-6} \text{ F}} = 300 \text{ V}$$

and that across the 3-μF capacitor is

$$V_2 = \frac{Q}{C_2} = \frac{6 \times 10^{-4} \text{ C}}{3 \times 10^{-6} \text{ F}} = 200 \text{ V}$$

As a check we note that $V_1 + V_2 = 500$ V.

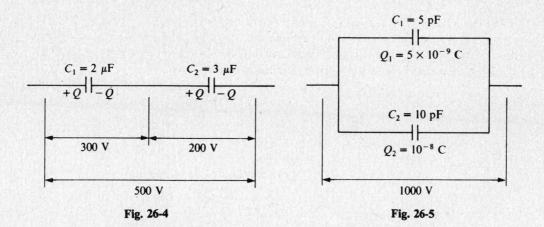

Fig. 26-4   Fig. 26-5

26.6. A 5-pF and a 10-pF capacitor are connected in parallel. (*a*) What is their equivalent capacitance? (*b*) A potential difference of 1000 V is applied to the combination. Find the charge on each capacitor and the potential difference across it.

(*a*) $$C = C_1 + C_2 = 5 \text{ pF} + 10 \text{ pF} = 15 \text{ pF}$$

(*b*) The same potential difference $V = 1000$ V is across each capacitor (Fig. 26-5). The charge on the 5-pF capacitor is

$$Q_1 = C_1 V = 5 \times 10^{-12} \text{ F} \times 10^3 \text{ V} = 5 \times 10^{-9} \text{ C}$$

and that on the 10-pF capacitor is

$$Q_2 = C_2 V = 10 \times 10^{-12} \text{ F} \times 10^3 \text{ V} = 10^{-8} \text{ C}$$

26.7. How much energy is stored in a 50-pF capacitor when it is charged to a potential difference of 200 V?

$$W = \tfrac{1}{2} C V^2 = \tfrac{1}{2} \times 50 \times 10^{-12} \text{ F} \times (200 \text{ V})^2 = 10^{-6} \text{ J}$$

26.8. A 100-μF capacitor is to have an energy content of 50 J in order to operate a flashlamp. (*a*) What voltage is required to charge the capacitor? (*b*) How much charge passes through the flashlamp?

(*a*) Since $W = \tfrac{1}{2} C V^2$,

$$V = \sqrt{\frac{2W}{C}} = \sqrt{\frac{2 \times 50 \text{ J}}{10^{-4} \text{ F}}} = 1000 \text{ V}$$

(*b*) $$Q = CV = 10^{-4} \text{ F} \times 10^3 \text{ V} = 0.1 \text{ C}$$

26.9. A parallel-plate capacitor has plates 5 cm square and 0.1 mm apart. Find its capacitance (*a*) in air and (*b*) with mica of $K = 6$ between the plates.

(*a*) In air K is very nearly 1, and so

$$C = K\epsilon_0 \frac{A}{d} = 1 \times 8.85 \times 10^{-12} \frac{\text{F}}{\text{m}} \times \frac{(0.05 \text{ m})^2}{10^{-4} \text{ m}} = 2.21 \times 10^{-10} \text{ F} = 221 \text{ pF}$$

(*b*) With mica between the plates the capacitance will be $K = 6$ times greater, or
$$C = 6 \times 221 \text{ pF} = 1326 \text{ pF}$$

26.10. A parallel-plate capacitor has a capacitance of 2 μF in air and 4.6 μF when immersed in benzene. What is the dielectric constant of benzene?

Since C is proportional to K, in general

$$\frac{C_1}{K_1} = \frac{C_2}{K_2}$$

for the same capacitor. Here, with $K_1 = K_{air} = 1$, the dielectric constant K_2 of benzene is

$$K_2 = K_1 \frac{C_2}{C_1} = 1 \times \frac{4.6\ \mu F}{2\ \mu F} = 2.3$$

26.11. A 10-μF capacitor with air between its plates is connected to a 50-V source and then disconnected. (*a*) What are the charge on the capacitor and the potential difference across it? (*b*) The space between the plates of the charged capacitor is filled with Teflon ($K = 2.1$). What are the charge on the capacitor and the potential difference across it now?

(*a*) The capacitor's charge is

$$Q = CV = 10 \times 10^{-6}\ F \times 50\ V = 5 \times 10^{-4}\ C$$

The potential difference across it remains 50 V after it is disconnected.

(*b*) The presence of another dielectric does not change the charge on the capacitor. Since its capacitance is now

$$C_2 = \frac{K_2}{K_1} C_1$$

and $V = Q/C$, the new potential difference is

$$V_2 = \frac{Q}{C_2} = \frac{K_1}{K_2} \frac{Q}{C_1} = \frac{K_1}{K_2} V_1 = \frac{1}{2.1} \times 50\ V = 23.8\ V$$

26.12. A 20-μF capacitor is connected to a 45-V battery through a circuit whose resistance is 2000 Ω. (*a*) What is the final charge on the capacitor? (*b*) How long does it take for the charge to reach 63% of its final value?

(*a*) $$Q = CV = 20 \times 10^{-6}\ F \times 45\ V = 9 \times 10^{-4}\ C$$

(*b*) $$t = RC = 2000\ \Omega \times 20 \times 10^{-6}\ F = 0.04\ s$$

Supplementary Problems

26.13. Verify that RC has the dimensions of time.

26.14. A 10-μF capacitor has a potential difference of 250 V across it. What is the charge on the capacitor?

26.15. A capacitor has a charge of 0.002 C when it is connected across a 100-V battery. Find its capacitance.

26.16. Three capacitors whose capacitances are 5 μF, 10 μF, and 20 μF are connected in series. Find the equivalent capacitance of the combination.

26.17. The three capacitors of Problem 26.16 are connected in parallel. Find the equivalent capacitance of the combination.

26.18. A 20-pF and a 25-pF capacitor are connected in parallel and a potential difference of 100 V is applied to the combination.   Find the charge on each capacitor and the potential difference across it.

26.19. A 50-pF and a 75-pF capacitor are connected in series and a potential difference of 250 V is applied to the combination.   Find the charge on each capacitor and the potential difference across it.

26.20. A 5-μF capacitor has a potential difference of 1000 V.   What is its potential energy?

26.21. The plates of a parallel-plate capacitor have areas of 40 cm^2 and are separated by 0.2 mm of waxed paper ($K = 2.2$).   Find the capacitance.

26.22. The waxed paper is removed from between the plates of the capacitor of Problem 26.21.   Find the new capacitance.

26.23. The plates of a parallel-plate capacitor of capacitance C are moved closer together until they are half their original separation.   What is the new capacitance?

26.24. The dielectric between the plates of a certain 80-μF capacitor has a resistance of 10^9 Ω.   If the capacitor is charged and then disconnected, how long will it take for the charge on the capacitor to fall to 37% of its original value?

Answers to Supplementary Problems

26.13. From their definitions, $R = V/I = Vt/Q$ and $C = Q/V$.   Hence $RC = Vt/Q \times Q/V = t$.

26.14. 0.0025 C

26.15. 20 μF

26.16. 2.86 μF

26.17. 35 μF

26.18. $Q_1 = 2 \times 10^{-9}$ C, $Q_2 = 2.5 \times 10^{-9}$ C, $V_1 = V_2 = 100$ V

26.19. $Q_1 = Q_2 = 7.5 \times 10^{-9}$ C, $V_1 = 150$ V, $V_2 = 100$ V

26.20. 2.5 J

26.21. 3.9 pF

26.22. 5.9 pF

26.23. $2C$

26.24. 8×10^4 s = 22.2 hr

Chapter 27

Magnetism

THE NATURE OF MAGNETISM

Two electric charges at rest exert forces on each other according to Coulomb's law. When the charges are in motion, the forces are different, and it is customary to attribute the differences to *magnetic forces* that occur between moving charges in addition to the electric forces between them. In this interpretation, the total force on a charge Q at a certain time and place can be divided into two parts: an electric force that depends only upon the value of Q, and a magnetic force that depends upon the velocity v of the charge as well as upon Q.

In reality there is only a single interaction between charges, the *electromagnetic interaction*. The theory of relativity provides the link between electric and magnetic forces: just as the mass of an object moving with respect to an observer is greater than when it is at rest, so the electric force between two charges appears altered to an observer when the charges are moving with respect to him. Magnetism is not distinct from electricity in the way that, for example, gravitation is. Despite the unity of the electromagnetic interaction, however, it is convenient for many purposes to treat electric and magnetic effects separately.

MAGNETIC FIELD

A *magnetic field* **B** is present wherever a magnetic force acts on a moving charge. The direction of **B** at a certain place is that along which a charge can move without experiencing a magnetic force; along any other direction the charge would be acted upon by such a force. The magnitude of **B** is equal numerically to the force on a charge of 1 C moving at 1 m/s perpendicular to **B**.

The unit of magnetic field is the *tesla* (T), where

$$1 \text{ tesla} = 1 \frac{\text{newton}}{\text{ampere-meter}}$$

The tesla is sometimes called the *weber*/m^2. The *gauss*, equal to 10^{-4} T, is another unit of magnetic field sometimes used.

MAGNETIC FIELD OF A STRAIGHT CURRENT

The magnetic field a distance s from a long, straight current I has the magnitude

$$B = \frac{\mu}{2\pi} \frac{I}{s} \qquad \text{(straight current)}$$

where μ is the *permeability* of the medium in which the magnetic field exists. The permeability of free space μ_0 has the value

$$\mu_0 = 4\pi \times 10^{-7} \text{ T-m/A} = 1.257 \times 10^{-6} \text{ T-m/A}$$

The value of μ in air is very nearly the same as μ_0 and is usually considered as equal to μ_0.

The lines of force of the magnetic field around a straight current are in the form of concentric circles around the current. To find the direction of **B**, place the thumb of the right hand in the direction of the current; the curled fingers of that hand then point in the direction of **B** (Fig. 27-1).

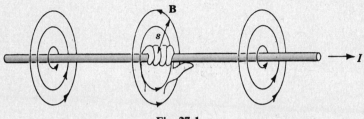

Fig. 27-1

MAGNETIC FIELD OF A CURRENT LOOP

The magnetic field at the center of a current loop whose radius is r has the magnitude

$$B = \frac{\mu}{2} \frac{I}{r} \quad \text{(current loop)}$$

The lines of force of **B** are perpendicular to the plane of the loop, as shown in Fig. 27-2(a). To find the direction of **B**, grasp the loop so the curled fingers of the right hand point in the direction of the current; the thumb of that hand then points in the direction of **B** [Fig. 27-2(b)].

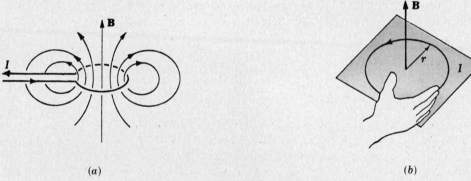

(a)

(b)

Fig. 27-2

A *solenoid* is a coil consisting of many loops of wire. If the turns are close together and the solenoid is long compared with its diameter, the magnetic field inside it is uniform and parallel to the axis with the magnitude

$$B = \mu \frac{N}{L} I \quad \text{(solenoid)}$$

In this formula N is the number of turns, L is the length of the solenoid, and I is the current. The direction of **B** is as shown in Fig. 27-3.

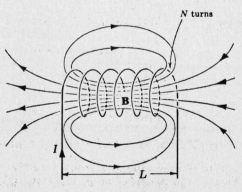

Fig. 27-3

THE EARTH'S MAGNETIC FIELD

The earth has a magnetic field which arises from electric currents in its liquid iron core. The field is like that which would be produced by a current loop centered a few hundred miles from the earth's center whose plane is tilted by 11° from the plane of the equator (Fig. 27-4). The *geomagnetic poles* are the points where the magnetic axis passes through the earth's surface. The magnitude of the earth's magnetic field varies from place to place; a typical sea-level value is 3×10^{-5} T.

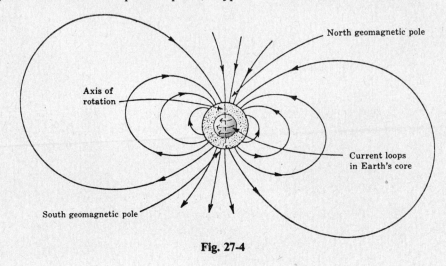

Fig. 27-4

MAGNETIC FORCE ON A MOVING CHARGE

The magnetic force on a moving charge Q in a magnetic field varies with the relative directions of **v** and **B**. When **v** is parallel to **B**, $F = 0$; when **v** is perpendicular to **B**, F has its maximum value of

$$F = QvB \quad (\mathbf{v} \perp \mathbf{B})$$

The direction of **F** in the case of a positive charge is given by the right-hand rule shown in Fig. 27-5; **F** is in the opposite direction when the charge is negative.

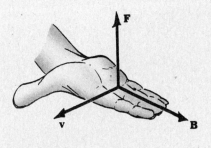 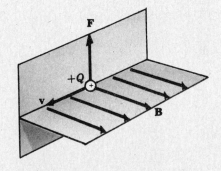

Fig. 27-5

MAGNETIC FORCE ON A CURRENT

Since a current consists of moving charges, it follows that a current-carrying wire will experience no force when parallel to a magnetic field **B** and maximum force when perpendicular to **B**. In the latter case, F has the value

$$F = ILB \quad (\mathbf{I} \perp \mathbf{B})$$

where I is the current and L is the length of wire in the magnetic field. The direction of the force is as shown in Fig. 27-6.

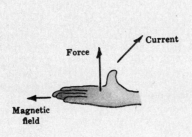

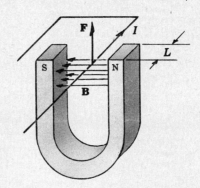

Fig. 27-6

Owing to the different forces exerted on each of its sides, a current loop in a magnetic field always tends to rotate so that its plane is perpendicular to **B**. This is the effect that underlies the operation of all electric motors.

FORCE BETWEEN TWO CURRENTS

Two parallel electric currents exert magnetic forces on each other (Fig. 27-7). If the currents are in the same direction, the forces are attractive; if the currents are in opposite directions, the forces are repulsive. The force per unit length F/L on each current depends upon the currents I_1 and I_2 and their separation s:

$$\frac{F}{L} = \frac{\mu_0}{2\pi}\frac{I_1 I_2}{s} \qquad \text{(parallel currents)}$$

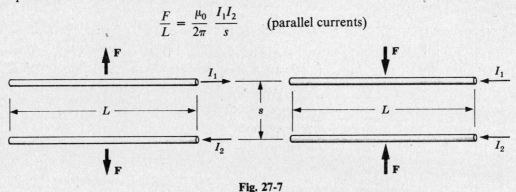

Fig. 27-7

FERROMAGNETISM

The magnetic field produced by a current is altered by the presence of a substance of any kind. Usually the change, which may be an increase or a decrease in **B**, is very small, but in certain cases there is an increase in **B** by hundreds or thousands of times. Substances that have the latter effect are called *ferromagnetic*; iron and iron alloys are familiar examples. An *electromagnet* is a solenoid with a ferromagnetic core to increase its magnetic field.

Ferromagnetism is a consequence of the magnetic properties of the electrons all atoms contain. An electron behaves in some respects as though it is a spinning charged sphere, and is therefore magnetically equivalent to a tiny current loop. In most substances the magnetic fields of the atomic electrons cancel each other out, but in ferromagnetic substances the cancellation is not complete and each atom has a certain magnetic field of its own. The atomic magnetic fields align themselves in groups called *domains* with an external magnetic field to produce a much stronger total **B**. When the external field is removed, the atomic magnetic fields may remain aligned to produce a *permanent magnet*. The field of a bar magnet has the same form as that of a solenoid because both fields are due to parallel current loops (Fig. 27-8).

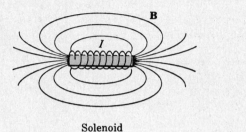

Solenoid   Bar magnet

Fig. 27-8

MAGNETIC INTENSITY

A substance which decreases the magnetic field of a current is called *diamagnetic*; it has a permeability μ that is less than μ_0. Copper and water are examples. A substance which increases the magnetic field of a current by a small amount is called *paramagnetic*; it has a permeability μ that is greater than μ_0. Aluminum is an example. Ferromagnetic substances have permeabilities hundreds or thousands of times greater than μ_0. Diamagnetic substances are repelled by magnets; paramagnetic and ferromagnetic ones are attracted by magnets.

Because different substances have different magnetic properties, it is useful to define a quantity called *magnetic intensity* **H** which is independent of the medium in which a magnetic field is located. The magnetic intensity in a place where the magnetic field is **B** and the permeability is μ is given by

$$H = \frac{B}{\mu}$$

$$\text{Magnetic intensity} = \frac{\text{magnetic field}}{\text{permeability of medium}}$$

The unit of **H** is the A/m. Magnetic intensity is sometimes called *magnetizing force* or *magnetizing field*.

The permeability of a ferromagnetic material at a given value of H varies both with H and with the previous degree of magnetization of the material. The latter effect is known as *hysteresis*.

Solved Problems

27.1. In what ways are electric and magnetic fields similar? In what ways are they different?

Similarities. Both fields originate in electric charges, and both fields can exert forces on electric charges.

Differences. All electric charges give rise to electric fields, but only a charge in motion relative to an observer gives rise to a magnetic field. Electric fields exert forces on all charges, but magnetic fields only exert forces on moving charges.

27.2. A positive charge is moving vertically upward when it enters a magnetic field directed to the north. In what direction is the force on the charge?

To apply the right-hand rule here, the fingers of the right hand are pointed north and the thumb of that hand is pointed upward. The palm of the hand faces west, which is therefore the direction of the force on the charge.

27.3. A stream of protons is moving parallel to a stream of electrons. Do the streams tend to come together or to move apart?

The electric force between the streams is attractive but the magnetic force is repulsive. Which of the forces is stronger depends upon how fast the particles are moving.

27.4. The ends of a bar magnet are traditionally called its "poles," with the end that tends to point north called the "north pole" and the end that tends to point south called the "south pole." It is observed that like poles of nearby magnets repel each other and that unlike poles attract. Explain this behavior in terms of the interaction of current loops.

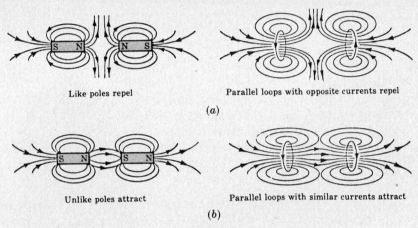

Like poles repel　　　　　　　　　　　　Parallel loops with opposite currents repel

(a)

Unlike poles attract　　　　　　　　　　Parallel loops with similar currents attract

(b)

Fig. 27-9

Bar magnets with like poles facing each other are equivalent to parallel current loops whose currents are in opposite directions [Fig. 27-9(a)]. Such loops repel. Bar magnets with opposite poles facing each other are equivalent to parallel current loops whose current loops are in the same direction [Fig. 27-9(b)]. Such loops attract.

27.5. How does a permanent magnet attract an unmagnetized iron object?

The presence of the magnet induces the atomic magnets in the object to line up with its field (Fig. 27-10), and the attraction of opposite poles then produces a net force on the object.

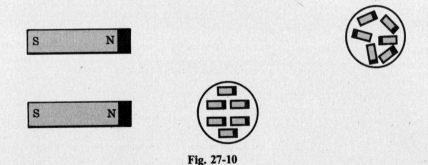

Fig. 27-10

27.6. An unmagnetized iron rod is placed inside a solenoid. The current in the solenoid is then increased from 0 to a maximum in one direction, decreased back to 0, increased to a maximum in the other direction, brought back to 0, and so on. Plot B versus H for the iron rod and discuss the shape of the resulting curve.

The required curve is shown in Fig. 27-11. At a, H and B are both 0. As H increases, B increases slowly at first, then rapidly, and finally levels off at a maximum value at b. The rod is now *saturated* and a further increase in H will not change B. Saturation occurs when all the magnetic domains in the rod are aligned with H. The curve from a to b is called the *magnetization curve* of the material.

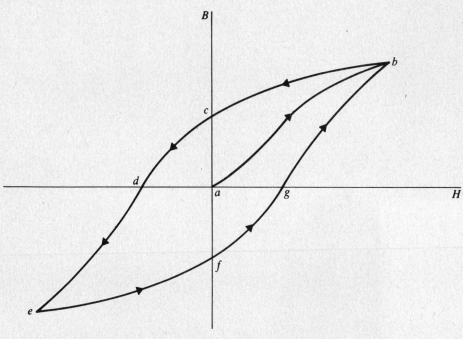

Fig. 27-11

When H is brought back to 0 from H_b, B lags behind, so that when $H = 0$, $B = B_c$. This is an example of hysteresis. The value of B_c is called the *retentivity* of the material. In order to demagnetize the rod completely, H must be reversed in direction and increased to H_d, the *coercive force*. The greater the retentivity, the stronger the residual magnetization of the rod; the greater the coercive force, the better able the rod will be to keep its magnetization despite the presence of strong magnetic fields. Thus a good material for a permanent magnet should have both a high retentivity and a high coercive force.

Increasing $-H$ beyond d produces an increasing $-B$ until the rod saturates again at e, where $B_e = -B_b$. When H is returned to 0, B again lags behind, and this time has the value B_f at $H = 0$, where $B_f = -B_c$. Increasing H then returns B to B_b to complete the *hysteresis loop bcdefgb*.

The area enclosed by a hysteresis loop is proportional to the energy dissipated as heat during each magnetization cycle. A broad hysteresis loop with high values of retentivity and coercive force is characteristic of a suitable material for a permanent magnet, since a great deal of work must be done to change its magnetization. On the other hand, a material with a narrow hysteresis loop is better for such applications as transformer cores which must undergo frequent reversals in magnetization; the smaller the area of the loop, the greater the efficiency of the transformer.

27.7. How can a magnetized piece of iron be demagnetized?

One method is to heat the iron, since all ferromagnetic materials lose their ability to retain magnetization beyond a certain temperature, which is about 760 °C in the case of iron. Another method is to bring the iron through a succession of hysteresis loops of smaller size, as in Fig. 27-12. To do this, the iron can be placed in a solenoid connected to a source of alternating current, and the current then gradually decreased to zero.

27.8. A proton is moving in a uniform magnetic field. Describe the path of the proton if its initial direction is (*a*) parallel to the field, (*b*) perpendicular to the field, and (*c*) at an intermediate angle to the field.

(*a*) There is no magnetic force on the proton so it continues to move in a straight line [Fig. 27-13(*a*)].

(*b*) The force on the proton is perpendicular to its velocity **v** and also perpendicular to **B**, hence it moves in a circle as in Fig. 27-13(*b*).

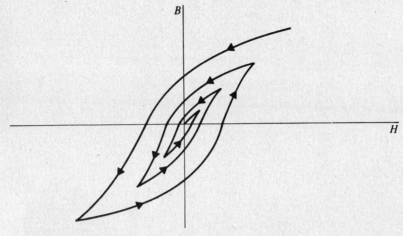

Fig. 27-12

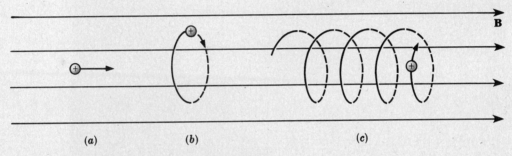

Fig. 27-13

(c)   The proton moves in the helical path of Fig. 27-13(c) because the component of **v** parallel to **B** is not changed while the component of **v** perpendicular to **B** leads to an inward force as in (b).

27.9.   A charged particle moving perpendicular to a uniform magnetic field follows a circular path. Find the radius of the circle.

The magnetic force QvB on the particle provides the centripetal force mv^2/r that keeps it moving in a circle of radius r. Hence

$$F_{\text{magnetic}} = F_{\text{centripetal}}$$

$$QvB = \frac{mv^2}{r}$$

$$r = \frac{mv}{QB}$$

The radius is directly proportional to the particle's momentum mv and inversely proportional to its charge Q and the magnetic field B.

27.10.   A cable 5 m above the ground carries a current of 100 A from east to west. Find the direction and magnitude of the magnetic field on the ground directly beneath the cable. (Neglect the earth's magnetic field.)

From the right-hand rule, the direction of the field is south. The magnitude of the field is

$$B = \frac{\mu_0}{2\pi}\frac{I}{s} = \frac{4\pi \times 10^{-7}\ \text{T-m/A} \times 100\ \text{A}}{2\pi \times 5\ \text{m}} = 4 \times 10^{-6}\ \text{T}$$

27.11. Two parallel wires 10 cm apart carry currents in the same direction of 8 A. What is the magnetic field halfway between them?

The magnetic field halfway between the wires is 0 because the fields of the currents are opposite in direction and have the same magnitude there.

27.12. Two parallel wires 10 cm apart carry currents in opposite directions of 8 A. What is the magnetic field halfway between them?

Here the magnetic fields of the two currents are in the same direction and hence add together. Since $s = 5$ cm $= 5 \times 10^{-2}$ m halfway between the wires, the field of each current there is

$$B = \frac{\mu_0}{2\pi} \frac{I}{s} = \frac{4\pi \times 10^{-7} \text{ T-m/A} \times 8 \text{ A}}{2\pi \times 5 \times 10^{-2} \text{ m}} = 3.2 \times 10^{-5} \text{ T}$$

and the total field is $2B = 6.4 \times 10^{-5}$ T.

27.13. A 100-turn flat circular coil has a radius of 5 cm. Find the magnetic field at the center of the coil when the current is 4 A.

Each turn of the coil acts as a separate loop in contributing to the total magnetic field. If there are N turns, the result is a field N times stronger than each turn produces by itself. Hence

$$B = \frac{\mu_0}{2\pi} \frac{NI}{r} = \frac{4\pi \times 10^{-7} \text{ T-m/A} \times 100 \times 4 \text{ A}}{2\pi \times 5 \times 10^{-2} \text{ m}} = 5.03 \times 10^{-3} \text{ T}$$

27.14. A solenoid 0.2 m long has 1000 turns of wire and is oriented with its axis parallel to the earth's magnetic field at a place where the latter is 2.5×10^{-5} T. What should the current in the solenoid be in order that its field exactly cancel the earth's field inside the solenoid?

The magnetic field inside an air-core solenoid is

$$B = \mu_0 \frac{N}{L} I$$

Here $N = 10^3$, $L = 0.2$ m, and $B = 2.5 \times 10^{-5}$ T, so

$$I = \frac{BL}{\mu_0 N} = \frac{2.5 \times 10^{-5} \text{ T} \times 0.2 \text{ m}}{4\pi \times 10^{-7} \text{ T-m/A} \times 10^3} = 0.004 \text{ A}$$

27.15. In a certain electric motor wires that carry a current of 5 A are perpendicular to a magnetic field of 0.8 T. What is the force on each cm of these wires?

$$F = ILB = 5 \text{ A} \times 0.01 \text{ m} \times 0.8 \text{ T} = 0.04 \text{ N}$$

27.16. The wires that supply current to a 120-V, 2-kW electric heater are 2 mm apart. What is the force per meter between the wires?

Since $P = IV$ the current in the wires is

$$I = I_1 = I_2 = \frac{P}{V} = \frac{2000 \text{ W}}{120 \text{ V}} = 16.7 \text{ A}$$

Since $s = 2$ mm $= 2 \times 10^{-3}$ m, the force between the wires is

$$\frac{F}{L} = \frac{\mu_0}{2\pi} \frac{I_1 I_2}{s} = \frac{4\pi \times 10^{-7} \text{ T-m/A} \times (16.7 \text{ A})^2}{2\pi \times 2 \times 10^{-3} \text{ m}} = 0.028 \text{ N/m}$$

The currents are in opposite directions, so the force is repulsive.

27.17. A sample of cast iron exhibits a magnetic field of $B = 0.50$ T when the magnetic intensity is $H = 10$ A/m. (a) Find the permeability of cast iron at this value of H. (b) What would the field be in air at this value of H?

(a) $$\mu = \frac{B}{H} = \frac{0.50 \text{ T}}{10 \text{ A/m}} = 0.05 \text{ T-m/A}$$

(b) $$B = \mu_0 H = 4\pi \times 10^{-7} \frac{\text{T-m}}{\text{A}} \times 10 \frac{\text{A}}{\text{m}} = 1.257 \times 10^{-5} \text{ T}$$

27.18. A solenoid 20 cm long is wound with 300 turns of wire and carries a current of 1.5 A. (a) What is the magnetic intensity H inside the solenoid? (b) What should the permeability of the core at this value of H be in order that the magnetic field inside be 0.6 T? How many times greater than μ_0 is this?

(a) $$H = \frac{B}{\mu} = \frac{N}{L} I = \frac{300}{0.2 \text{ m}} \times 1.5 \text{ A} = 2250 \text{ A/m}$$

(b) $$\frac{B}{H} = \frac{0.6 \text{ T}}{2250 \text{ A/m}} = 2.67 \times 10^{-4} \text{ T-m/A}$$

$$\frac{\mu}{\mu_0} = \frac{2.67 \times 10^{-4} \text{ T-m/A}}{4\pi \times 10^{-7} \text{ T-m/A}} = 212$$

Supplementary Problems

27.19. An observer is able to measure electric, magnetic, and gravitational fields. Which of these does he detect when (a) a proton moves past him, and (b) he moves past a proton?

27.20. A beam of electrons that are moving slowly at first are accelerated to higher and higher velocities. What happens to the diameter of the beam as this happens?

27.21. An electric current is flowing south along a power line. What is the direction of the magnetic field above it? Below it?

27.22. A charged particle moves through a magnetic field perpendicular to **B**. Is the particle's energy affected? Is its momentum?

27.23. In a sketch of magnetic lines of force, what is the significance of lines of force that are closer together in a particular region than they are elsewhere?

27.24. A negative charge is moving west when it enters a magnetic field directed vertically downward. In what direction is the force on the charge?

27.25. A wire carrying a current is placed in a magnetic field **B**. (a) Under what circumstances, if any, will the force on the wire be 0? (b) Under what circumstances will the force on the wire be a maximum?

27.26. Under what circumstances, if any, does a current-carrying wire loop not tend to rotate in a magnetic field?

27.27. The magnetic field 5 cm from a certain straight wire is 10^{-4} T. Find the current in the wire.

27.28. How far away from a compass should a wire carrying a 1-A current be located if its magnetic field at the compass is not to exceed 1 percent of the earth's magnetic field, which is typically 3×10^{-5} T?

27.29. Two parallel wires 20 cm apart carry currents of 5 A in the same direction. Find the magnetic field between the wires 5 cm from one of them and 15 cm from the other.

27.30. What should the current be in a wire loop 1 cm in radius if the magnetic field in the center of the loop is to be 0.01 T?

27.31. What is the magnetic field inside a solenoid wound with 20 turns/cm when the current in it is 5 A?

27.32. An electron in a television picture tube travels at 3×10^7 m/s. Does the earth's gravitational field or its magnetic field exert the greater force on the electron? Assume v is perpendicular to **B**.

27.33. (a) A wire 1 m long is perpendicular to a magnetic field of 0.01 T. What is the force on the wire when it carries a current of 10 A? (b) What is the force on the wire when it is parallel to the magnetic field?

27.34. A horizontal wire 10 cm long whose mass is 1 g is to be supported magnetically against the force of gravity. The current in the wire is 10 A and goes from north to south. (a) What should be the direction of the magnetic field? (b) What should its magnitude be?

27.35. The starting motor of a certain car is connected to the battery by a pair of cables that are 8 mm apart for a distance of 50 cm. Find the force between the cables when the current in them is 100 A.

27.36. An alternating current which varies with time according to the formula $I = I_{max} \sin \omega t$ is sent through a solenoid with an iron core. (a) Does the magnetic intensity in the core also vary sinusoidally with time? (b) Does the magnetic field?

27.37. The magnetic intensity H inside an air-core solenoid 25 cm long that is wound with 300 turns of wire is 600 A/m. (a) What is the current in the solenoid? (b) What is the magnetic field B inside the solenoid?

27.38. An air-core solenoid wound with 20 turns/cm carries a current of 0.1 A. (a) Find H and B inside the solenoid. (b) An iron core whose permeability is 6×10^{-3} T-m/A is inserted in the solenoid. Find H and B now.

27.39. A sample of carbon steel has a permeability of 0.01 T-m/A when the magnetic intensity is 75 A/m. (a) Find the magnetic field in the sample at this value of H. (b) Find the field in air at this value of H.

Answers to Supplementary Problems

27.19. All three fields are detected in both cases.

27.20. At first the mutual electric repulsion of the electrons causes the beam diameter to increase, but as they go faster the magnetic attraction becomes more significant and the beam diameter decreases.

27.21. west; east

27.22. The particle's energy is not changed since the magnetic force on it is perpendicular to its direction of motion and so no work is done on it by the field. The particle's direction changes, however, and hence its momentum, which is a vector quantity, also changes.

27.23. The closer together the lines of force are drawn in a particular region, the stronger the field is there.

27.24. To the north.

27.25. (a) When the wire is parallel to **B** (b) when it is perpendicular to **B**.

27.26. Such a loop does not tend to rotate when its plane is perpendicular to the direction of the magnetic field.

27.27. 25 A

27.28. 67 cm

27.29. Since the fields are in opposite directions,

$$B = 2 \times 10^{-5} \text{ T} - 0.67 \times 10^{-5} \text{ T} = 1.33 \times 10^{-5} \text{ T}$$

27.30. 159 A

27.31. 1.26×10^{-2} T

27.32. The magnetic force evB is more than 10^{13} times greater than the gravitational force mg.

27.33. (a) 0.1 N　(b) 0

27.34. (a) The direction of the field should be toward the west in order that the force on the wire be upward　(b) 9.8×10^{-3} T

27.35. 0.125 N; the force is repulsive

27.36. (a) yes　(b) no

27.37. (a) 0.5 A　(b) 7.54×10^{-4} T

27.38. (a) 200 A/m, 2.51×10^{-4} T　(b) 200 A/m, 1.2 T

27.39. (a) 0.75 T　(b) 9.4×10^{-5} T

Chapter 28

Electromagnetic Induction

ELECTROMAGNETIC INDUCTION

A current is produced in a conductor whenever it cuts across magnetic lines of force, a phenomenon known as *electromagnetic induction*. If the motion is parallel to the lines of force, there is no effect. Electromagnetic induction originates in the force a magnetic field exerts on a moving charge: when a wire moves across a magnetic field, the electrons it contains experience sideways forces which push them along the wire to cause a current. It is not even necessary for there to be relative motion of a wire and a source of magnetic field, since a magnetic field whose strength is changing has moving lines of force associated with it and a current will be induced in a conductor that is in the path of these moving lines of force.

When a straight conductor of length l is moving across a magnetic field $\mathbf{B}$ with the velocity $\mathbf{v}$, the emf induced in the conductor is given by

$$\text{Induced emf} = \mathcal{E} = Blv$$

when $\mathbf{B}$, $\mathbf{v}$, and the conductor are all perpendicular to one another.

FARADAY'S LAW

Figure 28-1 shows a coil of N turns that encloses an area of A. The axis of the coil is parallel to a magnetic field $\mathbf{B}$. According to *Faraday's law of electromagnetic induction*, the emf induced in the coil when the product BA changes by $\Delta(BA)$ in the time Δt is given by

$$\text{Induced emf} = \mathcal{E} = -N\frac{\Delta(BA)}{\Delta t}$$

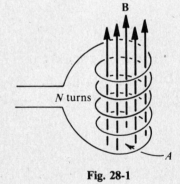

Fig. 28-1

The quantity BA is called the *magnetic flux* enclosed by the coil and is denoted by the symbol Φ (Greek capital letter *phi*):

$$\Phi = BA$$
$$\text{Magnetic flux} = \text{magnetic field} \times \text{cross-sectional area}$$

The unit of magnetic flux is the *weber* (Wb), where 1 Wb = 1 T-m^2. Thus Faraday's law can be written

$$\mathcal{E} = -N\frac{\Delta\Phi}{\Delta t}$$

LENZ'S LAW

The minus sign in Faraday's law is a consequence of *Lenz's law*: an induced current is always in such a direction that its own magnetic field acts to oppose the effect that brought it about. For example, if **B** is decreasing in magnitude in the situation of Fig. 28-1, the induced current in the coil will be counterclockwise in order that its own magnetic field will add to **B** and so reduce the rate at which **B** is decreasing.   Similarly, if **B** is increasing, the induced current in the coil will be clockwise so that its own magnetic field will subtract from **B** and thus reduce the rate at which **B** is increasing.

THE TRANSFORMER

A *transformer* consists of two coils of wire, usually wound on an iron core.   When an alternating current is passed through one of the windings, the changing magnetic field it gives rise to induces an alternating current in the other winding.   The potential difference per turn is the same in both primary and secondary windings, so the ratio of turns in the windings determines the ratio of voltages across them:

$$\frac{V_1}{V_2} = \frac{N_1}{N_2}$$

$$\frac{\text{Primary voltage}}{\text{Secondary voltage}} = \frac{\text{primary turns}}{\text{secondary turns}}$$

Since the power $I_1 V_1$ going into a transformer must equal the power $I_2 V_2$ going out, where I_1 and I_2 are the primary and secondary currents respectively, the ratio of currents is inversely proportional to the ratio of turns:

$$\frac{I_1}{I_2} = \frac{N_2}{N_1}$$

$$\frac{\text{Primary current}}{\text{Secondary current}} = \frac{\text{secondary turns}}{\text{primary turns}}$$

SELF-INDUCTANCE

When the current in a circuit changes, the magnetic field enclosed by the circuit also changes, and the resulting change in flux leads to a *self-induced emf* of

$$\text{Self-induced emf} = \mathcal{E} = -L\frac{\Delta i}{\Delta t}$$

Here $\Delta i / \Delta t$ is the rate of change of the current and L is a property of the circuit called its *self-inductance*, or, more commonly, its *inductance*.   The minus sign indicates that the direction of $\mathcal{E}$ is such as to oppose the change in current Δi that caused it.

The unit of inductance is the *henry* (H).   A circuit or circuit element (such as a solenoid) that has an inductance of 1 H will have a self-induced emf of 1 V when the current through it changes at the rate of 1 A/s.   Because the henry is a rather large unit, the *millihenry* and *microhenry* are often used, where

$$1 \text{ millihenry} = 1 \text{ mH} = 10^{-3} \text{ H}$$

$$1 \text{ microhenry} = 1 \text{ } \mu\text{H} = 10^{-6} \text{ H}$$

The inductance of a solenoid is

$$L = \frac{\mu N^2 A}{l}$$

where μ is the permeability of the core material, N is the number of turns, A is the cross-sectional area, and l is the length of the solenoid.

ENERGY OF A CURRENT-CARRYING INDUCTOR

Because a self-induced emf opposes any change in current in an inductor, work has to be done against this emf in order to establish a current in the inductor. This work is stored as magnetic potential energy. If L is the inductance of an inductor, its potential energy when it carries the current I is

$$W = \tfrac{1}{2}LI^2$$

It is this energy which powers the self-induced emf that opposes any decrease in the current through the inductor.

TIME CONSTANT

Because the self-induced emf in a circuit is always such as to oppose any changes in the current in the circuit, the current does not rise instantaneously to its final value of $\mathcal{E}/R$ when an external source of emf $\mathcal{E}$ is applied. In a circuit that contains an inductance L and a resistance R, the ratio L/R governs the rate at which a current can be established in the circuit. As shown in Fig. 28-2, the current rises gradually in such a manner that after a time equal to L/R it reaches 63% of its final value. If the source of emf is short-circuited, the current decreases in such a manner that after $t = L/R$ it has fallen to 37% of its original value. The quantity L/R is called the *time constant* of the circuit; the smaller the time constant, the more rapidly the current can change.

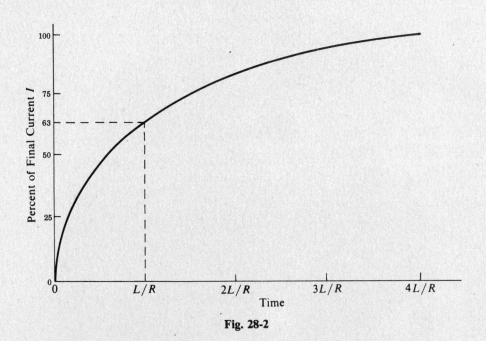

Fig. 28-2

Solved Problems

28.1. The vertical component of the earth's magnetic field in a certain region is 3×10^{-5} T. What is the potential difference between the rear wheels of a car, which are 5 ft apart, when the car's velocity is 40 mi/hr?

The rear axle of the car may be considered as a rod 5 ft long moving perpendicular to the magnetic field's vertical component. First we convert 5 ft and 40 mi/hr to their SI equivalents:

$$l = 5 \text{ ft} \times 0.305 \text{ m/ft} = 1.53 \text{ m}$$

$$v = 40 \text{ mi/hr} \times 0.447 \frac{\text{m/s}}{\text{mi/hr}} = 17.9 \text{ m/s}$$

The potential difference between the wheels is

$$\mathcal{E} = Blv = 3 \times 10^{-5} \text{ T} \times 1.53 \text{ m} \times 17.9 \text{ m/s} = 8.22 \times 10^{-4} \text{ V}$$

28.2. A square wire loop 8 cm on a side is perpendicular to a magnetic field of 5×10^{-3} T. (*a*) What is the magnetic flux through the loop? (*b*) If the field drops to 0 in 0.1 s, what average emf is induced in the loop during this period of time?

(*a*) The area of the loop is

$$A = \frac{8 \text{ cm} \times 8 \text{ cm}}{(100 \text{ cm/m})^2} = 0.0064 \text{ m}^2 = 6.4 \times 10^{-3} \text{ m}^2$$

The flux through the loop is

$$\Phi = BA = 5 \times 10^{-3} \text{ T} \times 6.4 \times 10^{-3} \text{ m}^2 = 3.26 \times 10^{-5} \text{ Wb}$$

(*b*) Since $N = 1$ for a single turn, we have from Faraday's law (disregarding the minus sign)

$$\mathcal{E} = N\frac{\Delta\Phi}{\Delta t} = 1 \times \frac{3.26 \times 10^{-5} \text{ Wb}}{0.1 \text{ s}} = 3.26 \times 10^{-4} \text{ V}$$

28.3. A 100-turn coil whose resistance is 6 Ω encloses an area of 80 cm². How rapidly should a magnetic field parallel to its axis change in order to induce a current of 1 mA in the coil?

The required emf here is

$$\mathcal{E} = IR = 10^{-3} \text{ A} \times 6 \text{ Ω} = 6 \times 10^{-3} \text{ V}$$

and the coil's area is

$$A = \frac{80 \text{ cm}^2}{(100 \text{ cm/m})^2} = 8 \times 10^{-3} \text{ m}^2$$

Since A is constant here,

$$\mathcal{E} = N\frac{\Delta(BA)}{\Delta t} = NA\frac{\Delta B}{\Delta t}$$

and

$$\frac{\Delta B}{\Delta t} = \frac{\mathcal{E}}{NA} = \frac{6 \times 10^{-3} \text{ V}}{100 \times 8 \times 10^{-3} \text{ m}^2} = .0075 \text{ T/s}$$

28.4. Alternating current is in wide use chiefly because its voltage can be so easily changed by a transformer. Since $P = IV$, the higher the voltage, the lower the current, and vice versa. In transmitting electrical energy through long distances, a small current is desirable in order to minimize energy loss to heat, which is equal to I^2R where R is the resistance of the transmission line. On the other hand, both the generation and final use of electrical energy is best accomplished at moderate potential differences. Hence electricity is typically generated at 10,000 V or so, stepped up by transformers at the power station to 500,000 V or even more for transmission, and near the point of consumption other transformers reduce the potential difference to 240 V or 120 V. To verify the advantage of high-voltage transmission, find the rate of energy loss to heat when a 5-Ω cable is used to transmit 1 kW of electricity at 100 V and at 100,000 V.

Since $P = IV$, the currents in the cable are respectively

$$I_A = \frac{P}{V_A} = \frac{1000 \text{ W}}{100 \text{ V}} = 10 \text{ A} \qquad I_B = \frac{P}{V_B} = \frac{1000 \text{ W}}{100,000 \text{ V}} = 0.01 \text{ A}$$

The rates of heat production per kW are respectively

$$I_A^2 R = (10 \text{ A})^2 \times 5 \text{ } \Omega = 500 \text{ W}$$

$$I_B^2 R = (0.01 \text{ A})^2 \times 5 \text{ } \Omega = 0.0005 \text{ W} = 5 \times 10^{-4} \text{ W}$$

Transmission at 100 V therefore means 10^6—a million—times more energy lost as heat than does transmission at 100,000 V.

28.5. A transformer has 100 turns in its primary winding and 500 turns in its secondary winding. If the primary voltage and current are respectively 120 V and 3 A, what are the secondary voltage and current?

$$V_2 = \frac{N_2}{N_1} V_1 = \frac{500 \text{ turns}}{100 \text{ turns}} \times 120 \text{ V} = 600 \text{ V}$$

$$I_2 = \frac{N_1}{N_2} I_1 = \frac{100 \text{ turns}}{500 \text{ turns}} \times 3 \text{ A} = 0.6 \text{ A}$$

28.6. A transformer rated at a maximum power of 10 kW is used to connect a 5000-V transmission line to a 240-V circuit. (*a*) What is the ratio of turns in the windings of the transformer? (*b*) What is the maximum current in the 240-V circuit?

(*a*)
$$\frac{N_1}{N_2} = \frac{V_1}{V_2} = \frac{5000 \text{ V}}{240 \text{ V}} = 20.8$$

(*b*) Since $P = IV$, here

$$I_2 = \frac{P}{V_2} = \frac{10,000 \text{ W}}{240 \text{ V}} = 41.7 \text{ A}$$

28.7. A transformer connected to a 120-V ac power line has 200 turns in its primary winding and 50 turns in its secondary winding. The secondary is connected to a 100-Ω light bulb. How much current is drawn from the 120-V power line?

The voltage across the secondary is

$$V_2 = \frac{N_2}{N_1} V_1 = \frac{50 \text{ turns}}{200 \text{ turns}} \times 120 \text{ V} = 30 \text{ V}$$

and so the current in the secondary circuit is

$$I_2 = \frac{V_2}{R} = \frac{30 \text{ V}}{100 \text{ } \Omega} = 0.3 \text{ A}$$

Hence the current in the primary circuit is

$$I_1 = \frac{N_2}{N_1} I_2 = \frac{50 \text{ turns}}{200 \text{ turns}} \times 0.3 \text{ A} = 0.075 \text{ A}$$

28.8. A solenoid 20 cm long and 2 cm in diameter has an inductance of 0.178 mH. How many turns of wire does it have?

The radius of the solenoid is $r = 1 \text{ cm} = 0.01 \text{ m}$ and so its cross-sectional area is

$$A = \pi r^2 = \pi \times (0.01 \text{ m})^2 = 3.14 \times 10^{-4} \text{ m}^2$$

Since $L = \mu_0 N^2 A / l$ in air,

$$N = \sqrt{\frac{Ll}{\mu_0 A}} = \sqrt{\frac{0.178 \times 10^{-3} \text{ H} \times 0.2 \text{ m}}{4\pi \times 10^{-7} \text{ T-m/A} \times 3.14 \times 10^{-4} \text{ m}^2}} = 300 \text{ turns}$$

28.9. An inductor consists of an iron ring 5 cm in diameter and 1 cm^2 in cross-sectional area that is wound with 1000 turns of wire. If the permeability of the iron is constant at 400 times that of free space at the magnetic intensities at which the inductor will be used, find its inductance.

An inductor of this kind is essentially a solenoid bent into a circle. The length of the equivalent solenoid is therefore

$$l = \pi d = \pi \times 0.05 \text{ m} = 0.157 \text{ m}$$

and its cross-sectional area is $A = 1 \text{ cm}^2/(100 \text{ cm/m})^2 = 10^{-4} \text{ m}^2$. The permeability of the core is $\mu = 400\mu_0$. The inductance of this inductor is therefore

$$L = \frac{\mu N^2 A}{l} = \frac{400 \times 4\pi \times 10^{-7} \text{ T-m/A} \times (10^3)^2 \times 10^{-4} \text{ m}^2}{0.157 \text{ m}} = 0.32 \text{ H}$$

28.10. The current in a circuit falls from 5 A to 1 A in 0.1 s. If an average emf of 2 V is induced in the circuit while this is happening, find the inductance of the circuit.

Since $\mathcal{E} = -L \, \Delta I/\Delta t$, we have (disregarding the minus sign)

$$L = \frac{\mathcal{E} \, \Delta t}{\Delta I} = \frac{2 \text{ V} \times 0.1 \text{ s}}{5 \text{ A} - 1 \text{ A}} = 0.05 \text{ H}$$

28.11. A 0.1-H inductor whose resistance is 20 Ω is connected to a 12-V battery of negligible internal resistance. (a) What is the initial rate at which the current increases? (b) What happens to the rate of current increase? (c) What is the final current?

(a) According to Lenz's law, the induced emf $-L \, \Delta I/\Delta t$ is in the opposite direction to the battery's emf $\mathcal{E}$. At any time the potential difference IR across the resistance in the inductor equals the net emf $\mathcal{E} - L \, \Delta I/\Delta t$ in the circuit:

$$\mathcal{E} - L \frac{\Delta I}{\Delta t} = IR$$

Impressed emf − induced emf = potential difference across resistance

At the moment the connection is made, $I = 0$, and so

$$\mathcal{E} - L \frac{\Delta I}{\Delta t} = 0 \qquad \frac{\Delta I}{\Delta t} = \frac{\mathcal{E}}{L} = \frac{12 \text{ V}}{0.1 \text{ H}} = 120 \text{ A/s}$$

The initial rate at which the current increases is 120 A/s.

(b) Since $\Delta I/\Delta t = (\mathcal{E} - IR)/L$, as the current I increases, its rate of change $\Delta I/\Delta t$ decreases.

(c) When the current has reached its final value, $\Delta I/\Delta t = 0$ and

$$I = \frac{\mathcal{E}}{R} = \frac{12 \text{ V}}{20 \text{ } \Omega} = 0.6 \text{ A}$$

28.12. What period of time is required for the current in the inductor of Problem 28.11 to reach 63% of its final value?

The time constant of the circuit is

$$t = \frac{L}{R} = \frac{0.1 \text{ H}}{20 \text{ } \Omega} = 0.005 \text{ s}$$

The current will reach 63% of its final value in this period of time.

28.13. (a) How much magnetic potential energy is stored in a 20-mH coil when it carries a current of 0.2 A? (b) What should the current in the coil be in order that it contain 1 J of energy?

(a)
$$W = \tfrac{1}{2}LI^2 = \tfrac{1}{2} \times 20 \times 10^{-3} \text{ H} \times (0.2 \text{ A})^2 = 4 \times 10^{-4} \text{ J}$$

(b)
$$I = \sqrt{\frac{2W}{L}} = \sqrt{\frac{2 \times 1 \text{ J}}{20 \times 10^{-3} \text{ H}}} = 10 \text{ A}$$

Supplementary Problems

28.14. What would happen if the primary winding of a transformer were connected to a battery?

28.15. What is it whose action on the secondary winding of a transformer causes an alternating potential difference to occur across its ends even though there is no connection between the primary and secondary windings?

28.16. Verify that L/R has the dimensions of time.

28.17. How fast should a wire 20 cm long be moved through a 0.05-T magnetic field for an emf of 1 V to appear across its ends?

28.18. The magnetic flux through a 50-turn coil increases at the rate of 0.05 Wb/s. (a) What is the induced emf between the ends of the coil? (b) If the coil's resistance is 2 Ω and it is connected to an external circuit whose resistance is 10 Ω, how much current will flow?

28.19. A 30-turn coil 8 cm in diameter is in a magnetic field of 0.1 T that is parallel to its axis. (a) What is the magnetic flux through the coil? (b) In how much time should the field drop to 0 to induce an average emf of 0.7 V in the coil?

28.20. A transformer has 50 turns in its primary winding and 100 turns in its secondary. (a) If a 60-Hz, 3-A current passes through the primary winding, what is the nature and magnitude of the current in the secondary? (b) If a 3-A direct current passes through the primary winding, what is the nature and magnitude of the current in the secondary?

28.21. The transformer in an electric welding machine draws 3 A from a 240-V ac power line and delivers 400 A. What is the potential difference across the secondary of the transformer?

28.22. A 240-V, 400-W electric mixer is connected to a 120-V power line through a transformer. (a) What is the ratio of turns in the transformer? (b) How much current is drawn from the power line?

28.23. A solenoid 2 cm long and 6 mm in diameter has 500 turns of thin wire. What is its inductance?

28.24. Find the emf induced in a 0.1-H coil when the current in it is changing at the rate of 80 A/s.

28.25. A potential difference of 100 V is applied to a 50-mH, 40-Ω inductor. (a) What is the initial rate at which the current increases? (b) What is the rate at which the current is increasing when $I = 1$ A? (c) What is the final current?

28.26. A 5-mH coil carries a current of 2 A. How much energy is stored in it?

28.27. How much current must flow through a 40-mH coil in order that it contain 0.1 J of magnetic potential energy?

28.28. What is the time constant of a 50-mH, 3-Ω inductor?

Answers to Supplementary Problems

28.14. When the connection is made, there will be a momentary current in the secondary winding as the current in the primary builds up to its final value. Afterward, since the primary current will be constant and hence its magnetic field will not change, there will be no current in the secondary.

28.15. The changing magnetic field produced by an alternating current in the primary winding.

28.16. From their definitions $L = Vt/I$ and $R = V/I$. Hence $L/R = (Vt/I)/(V/I) = t$.

28.17. 100 m/s

28.18. (a) 2.5 V   (b) 0.208 A

28.19. (a) 5.03×10^{-4} Wb   (b) 0.0215 s

28.20. (a) 60 Hz, 1.5 A   (b) no current

28.21. 1.8 V

28.22. (a) 2:1   (b) 3.3 A

28.23. 3.53×10^{-8} H

28.24. 8 V

28.25. (a) 2000 A/s   (b) 1200 A/s   (c) 2.5 A

28.26. 0.01 J

28.27. 2.24 A

28.28. 0.0167 s

Chapter 29

Alternating Current Circuits

ALTERNATING CURRENT

The *frequency* of an alternating current is the number of complete back-and-forth cycles it goes through each second (Fig. 29-1). As in the case of harmonic motion, the unit of frequency is the *hertz* (Hz), where 1 Hz = 1 cycle/s.

An alternating emf of frequency f whose maximum value is $\mathcal{E}_{max}$ varies with time according to the formula

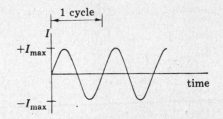

Fig. 29-1

$$\mathcal{E} = \mathcal{E}_{max} \sin 2\pi ft = \mathcal{E}_{max} \sin \omega t$$

The quantity $\omega = 2\pi f$ is the *angular frequency* of the emf in radians/s. Similarly an alternating current of frequency f whose maximum value is I_{max} varies with time according to the formula

$$I = I_{max} \sin 2\pi ft = I_{max} \sin \omega t$$

EFFECTIVE VALUES

Because an alternating current changes continuously, its maximum value $\pm I_{max}$ does not indicate its ability to do work or to produce heat as does the magnitude of a direct current. Instead it is customary to refer to the *effective current*

$$I_{eff} = \frac{I_{max}}{\sqrt{2}} = 0.707 \, I_{max}$$

which is such that a direct current of this value does as much work or produces as much heat as the alternating current whose maximum value is $\pm I_{max}$. Similarly the effective emf in an ac circuit is

$$\mathcal{E}_{eff} = \frac{\mathcal{E}_{max}}{\sqrt{2}} = 0.707 \, \mathcal{E}_{max}$$

Currents and voltages in ac circuits are usually expressed in terms of their effective values. For instance, the potential difference across a 120-V ac power line actually varies between + 170 V and − 170 V since

$$\pm V_{max} = \frac{\pm V_{eff}}{0.707} = \frac{\pm 120 \text{ V}}{0.707} = \pm 170 \text{ V}$$

REACTANCE

The *inductive reactance* of an inductor is a measure of its effectiveness in resisting the flow of an alternating current by virtue of the self-induced back emf that the changing current causes in it. Unlike the case of a resistor, there is no power dissipated in a pure inductor. The inductive reactance X_L of an inductor whose inductance is L (in henries) when the frequency of the current is f (in Hz) is

$$\text{Inductive reactance} = X_L = 2\pi fL$$

When a potential difference V of frequency f is applied across an inductor whose reactance is X_L at the frequency f, the current $I = V/X_L$ will flow. The unit of X_L is the ohm.

The *capacitive reactance* of a capacitor is similarly a measure of its effectiveness in resisting the flow of an alternating current, in this case by virtue of the reverse potential difference across it due to the accumulation of charge on its plates.   No power loss is associated with a capacitor in an ac circuit. The capacitive reactance X_C of a capacitor whose capacitance is C (in farads) when the frequency of the current is f (in Hz) is

$$\text{Capacitive reactance} = X_C = \frac{1}{2\pi f C}$$

When a potential difference of frequency f is applied across a capacitor whose reactance is X_C at the frequency f, the current $I = V/X_C$ will flow.   The unit of X_C is the ohm.

IMPEDANCE

The *impedance* of an ac circuit that contains resistance, inductance, and capacitance is equivalent to the resistance of a dc circuit.   If the resistance is R, the inductive reactance is X_L, and the capacitive reactance is X_C at the frequency f, the impedance Z at that frequency is

$$Z = \sqrt{R^2 + (X_L - X_C)^2}$$

The unit of Z is the ohm.   When a potential difference V whose frequency is f is applied to a circuit whose impedance is Z at that frequency, the result is the current

$$I = \frac{V}{Z}$$

The potential difference V across the entire circuit is related to the potential difference V_R across the resistor, V_L across the inductor, and V_C across the capacitor by the formula

$$V = \sqrt{V_R^2 + (V_L - V_C)^2}$$

where $V_R = IR$, $V_L = IX_L$, and $V_C = IX_C$.

RESONANCE

The impedance in a series ac circuit is a minimum when $X_L = X_C$; under these circumstances $Z = R$ and $I = V/R$.   The *resonant frequency* f_0 of a circuit is that frequency at which $X_L = X_C$:

$$2\pi f_0 L = \frac{1}{2\pi f_0 C} \qquad f_0 = \frac{1}{2\pi\sqrt{LC}}$$

When the potential difference applied to a circuit has the frequency f_0, the current in the circuit will be a maximum.   This condition is known as *resonance*.

PHASE ANGLE

In an ac circuit that contains only resistance, the instantaneous voltage and current are *in phase* with each other; that is, both are 0 at the same time, both reach their maximum values in either direction at the same time, and so on, as shown in Fig. 29-2(a).

In an ac circuit that contains only inductance, the voltage leads the current by $\frac{1}{4}$ cycle.   Since a complete cycle means a change in $2\pi f t$ of $360°$ and $360°/4 = 90°$, it is customary to say that in a pure inductor the voltage leads the current by $90°$.   This situation is shown in Fig. 29-2(b).

In an ac circuit that contains only capacitance, the voltage lags behind the current by $\frac{1}{4}$ cycle, which is $90°$.   This situation is shown in Fig. 29-2(c).

The angle ϕ between voltage and current, called the *phase angle*, in an ac circuit that contains R, L, and C is given by the formula

$$\tan \phi = \frac{X_L - X_C}{R}$$

If $X_L > X_C$, the phase angle ϕ is positive and the voltage leads the current by ϕ. If $X_C > X_L$, the phase angle ϕ is negative and the voltage lags behind the current by ϕ. At resonance $X_L = X_C$ and $\phi = 0$. Another formula for the phase angle is

$$\cos \phi = \frac{R}{Z}$$

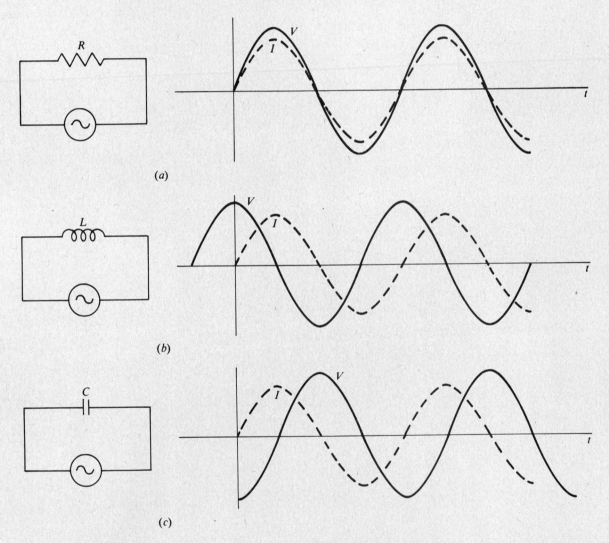

Fig. 29-2

POWER FACTOR

The power absorbed in an ac circuit is given by

$$P = IV \cos \phi$$

where ϕ is the phase angle between voltage and current. The quantity $\cos \phi$ is the *power factor* of the circuit. At resonance $\phi = 0$, $\cos \phi = 1$ and the power absorbed is a maximum. The power factor in an ac circuit is equal to the ratio between its resistance and its impedance:

$$\text{Power factor} = \cos \phi = \frac{R}{Z} = \frac{R}{\sqrt{R^2 + (X_L - X_C)^2}}$$

Power factors are often expressed as percentages, so that a phase angle of, say, 25° would give rise to a power factor of $\cos 25° = 0.906 = 90.6\%$.

Ac power sources are usually rated in *volt-amperes*, the product of V_{eff} and I_{eff}, without regard to the actual power P, because higher values of these quantities must be supplied to a circuit whose power factor is less than 1 than is indicated by its power rating in watts. Thus a power factor of 90.6% means that an apparent power of 1 VA (volt-ampere) must be supplied for each 0.906 W of true power that is consumed by the circuit.

Solved Problems

29.1. (a) What happens to X_L and X_C in the limit of $f = 0$? (b) What is the physical meaning of these results?

(a) When $f = 0$, $X_L = 2\pi f L = 0$ and $X_C = 1/2\pi f C = \infty$.

(b) A current with $f = 0$ is a direct current. When a constant current flows in an inductor, there is no self-induced back emf to hamper the current, and the inductive reactance is accordingly 0. A direct current cannot pass through a capacitor because its plates are insulated from each other, so the capacitive reactance is infinite and $I = V/X_C = 0$ when $f = 0$. (An alternating current does not actually pass *through* a capacitor but surges back and forth in the circuit on both sides of it.)

29.2. The dielectric used in a certain capacitor breaks down at a potential difference of 300 V. Find the maximum effective ac potential difference that can be applied to it.

$$V_{eff} = 0.707 \, V_{max} = 0.707 \times 300 \text{ V} = 212 \text{ V}$$

29.3. Alternating current with a maximum value of 10 A is passed through a 20-Ω resistor. At what rate does the resistor dissipate energy?

The effective current is

$$I_{eff} = 0.707 \, I_{max} = 0.707 \times 10 \text{ A} = 7.07 \text{ A}$$

and so the power dissipated is

$$P = I_{eff}^2 R = (7.07 \text{ A})^2 \times 20 \, \Omega = 1000 \text{ W}$$

29.4. A 10-μF capacitor is connected to a 15-V, 5-kHz power source. Find (a) the reactance of the capacitor and (b) the current that flows.

(a)
$$X_C = \frac{1}{2\pi f C} = \frac{1}{2\pi \times 5 \times 10^3 \text{ Hz} \times 10 \times 10^{-6} \text{ F}} = 3.18 \, \Omega$$

(b)
$$I = \frac{V}{X_C} = \frac{15 \text{ V}}{3.18 \, \Omega} = 4.72 \text{ A}$$

29.5. The reactance of an inductor is 80 Ω at 500 Hz. Find its inductance.

Since $X_L = 2\pi fL$,

$$L = \frac{X_L}{2\pi f} = \frac{80 \ \Omega}{2\pi \times 500 \text{ Hz}} = 0.0255 \text{ H} = 25.5 \text{ mH}$$

29.6. A 5-mH, 20-Ω inductor is connected to a 28-V, 400-Hz power source. Find (a) the current in the inductor, (b) the power dissipated in it, and (c) the phase angle.

(a) The reactance of the inductor is

$$X_L = 2\pi fL = 2\pi \times 400 \text{ Hz} \times 5 \times 10^{-3} \text{ H} = 12.6 \ \Omega$$

and its impedance is

$$Z = \sqrt{R^2 + X_L^2} = \sqrt{(20 \ \Omega)^2 + (12.6 \ \Omega)^2} = 23.6 \ \Omega$$

Hence the current is

$$I = \frac{V}{Z} = \frac{28 \text{ V}}{23.6 \ \Omega} = 1.18 \text{ A}$$

(b) $$P = I^2R = (1.18 \text{ A})^2 \times 20 \ \Omega = 28 \text{ W}$$

The reactance of the inductor does not contribute to the power dissipation even though it must be taken into account in determining the current that flows in the circuit.

(c) $$\tan \phi = \frac{X_L - X_C}{R} = \frac{12.6 \ \Omega - 0}{20 \ \Omega} = 0.63$$

$$\phi = 32°$$

29.7. A coil of unknown resistance and inductance draws 4 A when connected to a 12-V dc power source and 3 A when connected to a 12-V, 100-Hz power source. (a) Find the values of R and L. (b) How much power is dissipated when the coil is connected to the dc source? (c) When it is connected to the ac source?

(a) There is no inductive reactance when direct current passes through the coil, so its resistance is

$$R = \frac{V_1}{I_1} = \frac{12 \text{ V}}{4 \text{ A}} = 3 \ \Omega$$

At $f = 100$ Hz the impedance of the circuit is

$$Z = \frac{V_2}{I_2} = \frac{12 \text{ V}}{3 \text{ A}} = 4 \ \Omega$$

and so, since $Z = \sqrt{R^2 + (X_L^2 - X_C^2)}$ and $X_C = 0$ here,

$$X_L = \sqrt{Z^2 - R^2} = \sqrt{(4 \ \Omega)^2 - (3 \ \Omega)^2} = 2.65 \ \Omega$$

Hence the inductance of the coil is

$$L = \frac{X_L}{2\pi f} = \frac{2.65 \ \Omega}{2\pi \times 100 \text{ Hz}} = 4.22 \text{ mH}$$

(b) $$P_1 = I_1^2 R = (4 \text{ A})^2 \times 3 \ \Omega = 48 \text{ W}$$

(c) $$P_2 = I_2^2 R = (3 \text{ A})^2 \times 3 \ \Omega = 27 \text{ W}$$

29.8. In the antenna circuit of a radio receiver that is tuned to a particular station, $R = 5\ \Omega$, $L = 5$ mH, and $C = 5$ pF. (a) Find the frequency of the station. (b) If the potential difference applied to the circuit is 5×10^{-4} V, find the current that flows.

(a)
$$f_0 = \frac{1}{2\pi\sqrt{LC}} = \frac{1}{2\pi\sqrt{5 \times 10^{-3}\ \text{H} \times 5 \times 10^{-12}\ \text{F}}} = 1006\ \text{kHz}$$

(b) At resonance, $X_L = X_C$ and $Z = R$. Hence
$$I = \frac{V}{R} = \frac{5 \times 10^{-4}\ \text{V}}{5\ \Omega} = 10^{-4}\ \text{A} = 0.1\ \text{mA}$$

29.9. In the antenna circuit of Problem 29.8, the inductance is fixed but the capacitance can be varied. What should the capacitance be in order to receive an 800-kHz radio signal?

$$C = \frac{1}{(2\pi f)^2 L} = \frac{1}{(2\pi \times 800 \times 10^3\ \text{Hz})^2 \times 5 \times 10^{-3}\ \text{H}}$$
$$= 7.9 \times 10^{-12}\ \text{F} = 7.9\ \text{pF}$$

29.10. A 50-μF capacitor, a 0.3-H inductor, and an 80-Ω resistor are connected in series with a 120-V, 60-Hz power source (Fig. 29-3). (a) What is the impedance of the circuit? (b) How much current flows in it? (c) What is the power factor? (d) How much power is dissipated by the circuit? (e) What must be the minimum rating in volt-amperes of the power source?

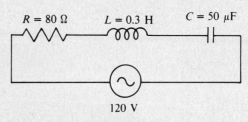

Fig. 29-3

(a)
$$X_L = 2\pi f L = 2\pi \times 60\ \text{Hz} \times 0.3\ \text{H} = 113\ \Omega$$
$$X_C = \frac{1}{2\pi f C} = \frac{1}{2\pi \times 60\ \text{Hz} \times 50 \times 10^{-6}\ \text{F}} = 53\ \Omega$$
$$Z = \sqrt{R^2 + (X_L - X_C)^2} = \sqrt{80^2 + (113\ \Omega - 53\ \Omega)^2} = 100\ \Omega$$

(b)
$$I = \frac{V}{Z} = \frac{120\ \text{V}}{100\ \Omega} = 1.2\ \text{A}$$

(c)
$$\cos\phi = \frac{R}{Z} = \frac{80\ \Omega}{100\ \Omega} = 0.8 = 80\%$$

(d)
$$\text{True power} = P = IV\cos\phi = 1.2\ \text{A} \times 120\ \text{V} \times 0.8 = 115\ \text{W}$$

Alternatively,
$$P = I^2 R = (1.2\ \text{A})^2 \times 80\ \Omega = 115\ \text{W}$$

(e)
$$\text{Apparent power} = IV = 1.2\ \text{A} \times 120\ \text{V} = 144\ \text{VA}$$

29.11. (a) Find the potential differences across the resistor, the inductor, and the capacitor in the circuit of Problem 29.10. (b) Are these values in accord with the applied potential difference of 120 V?

(a)
$$V_R = IR = 1.2 \text{ A} \times 80 \text{ } \Omega = 96 \text{ V}$$

$$V_L = IX_L = 1.2 \text{ A} \times 113 \text{ } \Omega = 136 \text{ V}$$

$$V_C = IX_C = 1.2 \text{ A} \times 53 \text{ } \Omega = 64 \text{ V}$$

(b) The sum of these potential difference is 296 V, more than twice the 120 V applied to the circuit. However, this is a meaningless way to combine the potential differences since they are not in phase with one another: V_L is 90° ahead of V_R and V_C is 90° behind V_R. The correct way to find the total potential difference across the circuit is as follows:

$$V = \sqrt{V_R{}^2 + (V_L - V_C)^2} = \sqrt{(96 \text{ V})^2 + (136 \text{ V} - 64 \text{ V})^2} = 120 \text{ V}$$

This result is in agreement with the applied potential difference of 120 V. We note that the voltage across an inductor or capacitor in an ac circuit can be greater than the voltage applied to the circuit.

29.12. (a) Find the resonant frequency f_0 of the circuit of Problem 29.10. (b) What current will flow in the circuit if it is connected to a 120-V power source whose frequency is f_0? (c) What will be the power factor in this case? (d) How much power will be dissipated by the circuit? (e) What must be the minimum rating in volt-amperes of the power source now?

(a)
$$f_0 = \frac{1}{2\pi\sqrt{LC}} = \frac{1}{2\pi\sqrt{0.3 \text{ H} \times 50 \times 10^{-6} \text{ F}}} = 41 \text{ Hz}$$

(b) At the resonant frequency, $X_L = X_C$ and $Z = R$. Hence

$$I = \frac{V}{R} = \frac{120 \text{ V}}{80 \text{ } \Omega} = 1.5 \text{ A}$$

(c)
$$\cos \phi = \frac{R}{Z} = \frac{R}{R} = 1 = 100\%$$

(d)
$$\text{True power} = IV \cos \phi = 1.5 \text{ A} \times 120 \text{ V} \times 1 = 180 \text{ W}$$

(e)
$$\text{Apparent power} = IV = 1.5 \text{ A} \times 120 \text{ V} = 180 \text{ VA}$$

29.13. (a) Find the potential difference across the resistor, the inductor, and the capacitor in the circuit of Problem 29.10 when it is connected to a 120-V ac source whose frequency is equal to the circuit's resonant frequency of 41 Hz. (b) Are these values in accord with the applied potential difference of 120 V?

(a) At the resonant frequency of $f_0 = 41$ Hz the inductive and capacitive reactances are respectively
$$X_L = 2\pi f_0 L = 2\pi \times 41 \text{ Hz} \times 0.3 \text{ H} = 77 \text{ } \Omega$$
$$X_C = \frac{1}{2\pi f_0 C} = \frac{1}{2\pi \times 41 \text{ Hz} \times 50 \times 10^{-6} \text{ F}} = 77 \text{ } \Omega$$

The various potential differences are therefore
$$V_R = IR = 1.5 \text{ A} \times 80 \text{ } \Omega = 120 \text{ V}$$
$$V_L = IX_L = 1.5 \text{ A} \times 77 \text{ } \Omega = 116 \text{ V}$$
$$V_C = IX_C = 116 \text{ V}$$

(b) The total potential difference across the circuit is

$$V = \sqrt{V_R^2 + (V_L - V_C)^2} = \sqrt{(120 \text{ V})^2 + (0)^2} = 120 \text{ V}$$

which is equal to the applied potential difference.

29.14. In a series ac circuit $R = 20 \text{ } \Omega$, $X_L = 10 \text{ } \Omega$, and $X_C = 25 \text{ } \Omega$ when the frequency is 400 Hz. (a) Find the impedance of the circuit. (b) Find the phase angle. (c) Is the resonant frequency of the circuit greater than or less than 400 Hz? (d) Find the resonant frequency.

(a)
$$Z = \sqrt{R^2 + (X_L - X_C)^2} = \sqrt{(20\ \Omega)^2 + (10\ \Omega - 25\ \Omega)^2} = 25\ \Omega$$

(b)
$$\tan \phi = \frac{X_L - X_C}{R} = \frac{10\ \Omega - 25\ \Omega}{20\ \Omega} = -0.75 \qquad \phi = -37°$$

A negative phase angle signifies that the voltage lags behind the current.

(c) At resonance $X_L = X_C$. At 400 Hz, $X_L < X_C$, so the frequency must be changed in such a way as to increase X_L and decrease X_C. Since $X_L = 2\pi f L$ and $X_C = 1/2\pi f C$, it is clear that increasing the frequency will have this effect. Hence the resonant frequency must be greater than 400 Hz.

(d) Since $X_L = 10\ \Omega$ and $X_C = 25\ \Omega$ when $f = 400$ Hz,

$$L = \frac{X_L}{2\pi f} = \frac{10\ \Omega}{2\pi \times 400\ \text{Hz}} = 4 \times 10^{-3}\ \text{H}$$

$$C = \frac{1}{2\pi f X_C} = \frac{1}{2\pi \times 400\ \text{Hz} \times 25\ \Omega} = 1.6 \times 10^{-5}\ \text{F}$$

Hence

$$f_0 = \frac{1}{2\pi\sqrt{LC}} = \frac{1}{2\pi\sqrt{4 \times 10^{-3}\ \text{H} \times 1.6 \times 10^{-5}\ \text{F}}} = 629\ \text{Hz}$$

29.15. A 5-hp electric motor is 80% efficient and has an inductive power factor of 75%. (a) What minimum rating in kVA must its power source have? (b) A capacitor is connected in series with the motor to raise the power factor to 100%. What minimum rating in kVA must the power source now have?

(a) The power required by the motor is
$$P = \frac{5\ \text{hp} \times 0.746\ \text{kW/hp}}{0.8} = 4.66\ \text{kW}$$

Since $P = IV \cos \phi$, the power source must have the minimum rating

$$IV = \frac{P}{\cos \phi} = \frac{4.66\ \text{kW}}{0.75} = 6.22\ \text{kVA}$$

(b) When $\cos \phi = 1$, $IV = P = 4.66$ kVA.

29.16. A coil connected to a 120-V, 25-Hz power line draws a current of 0.5 A and dissipates 50 W. (a) What is its power factor? (b) What capacitance should be connected in series with the coil to increase the power factor to 100%? (c) What would the current in the circuit be then? (d) How much power would the circuit then dissipate?

(a) Since $P = IV \cos \phi$,

$$\cos \phi = \frac{P}{IV} = \frac{50\ \text{W}}{0.5\ A \times 120\ V} = 0.833 = 83.3\%$$

(b) The power factor will be 100% at resonance, when $X_L = X_C$. The first step is to find X_L, which can be done from the formula $\tan \phi = (X_L - X_C)/R$. Here $X_C = 0$ and, since $\cos \phi = 0.833$, $\phi = 34°$ and $\tan \phi = 0.663$. Since $P = I^2 R$,

$$R = \frac{P}{I^2} = \frac{50\ \text{W}}{(0.5\ A)^2} = 200\ \Omega$$

Hence

$$X_L = R \tan \phi + X_C = 200\ \Omega \times 0.663 + 0 = 133\ \Omega$$

This must also be the value of X_C when $f = 25$ Hz, and so

$$C = \frac{1}{2\pi f X_C} = \frac{1}{2\pi \times 25\ \text{Hz} \times 133\ \Omega} = 4.8 \times 10^{-5}\ \text{F} = 48\ \mu\text{F}$$

(c) $Z = R$ at resonance, so

$$I = \frac{V}{Z} = \frac{V}{R} = \frac{120\text{ V}}{200\ \Omega} = 0.6\text{ A}$$

(d) $P = I^2R = (0.6\text{ A})^2 \times 200\ \Omega = 72\text{ W}$

Supplementary Problems

29.17. The frequency of the alternating emf applied to a series circuit containing resistance, inductance, and capacitance is doubled. What happens to R, X_L, and X_C?

29.18. (a) What is the minimum value the power factor of a circuit can have? Under what circumstances can this occur? (b) What is the maximum value the power factor can have? Under what circumstances can this occur?

29.19. A voltmeter across an ac circuit reads 40 V and an ammeter in series with the circuit reads 6 A. (a) What is the maximum potential difference across the circuit? (b) What is the maximum current in the circuit? (c) What other information is needed to determine the power consumed in the circuit?

29.20. Find the reactance of a 10-mH coil when it is in a 500-Hz circuit.

29.21. A capacitor has a reactance of 200 Ω at 1000 Hz. What is its capacitance?

29.22. A 5-μF capacitor is connected to a 6-kHz alternating emf, and a current of 2 A flows. Find the effective magnitude of the emf.

29.23. A 5-μF capacitor draws 1 A when connected to a 60-V ac source. What is the frequency of the source?

29.24. A 25-μF capacitor is connected in series with a 50-Ω resistor and a 12-V, 60-Hz potential difference is applied. Find the current in the circuit and the power dissipated in it.

29.25. A capacitor is connected in series with an 8-Ω resistor and the combination is placed across a 24-V, 1000-Hz power source. A current of 2 A flows. Find (a) the capacitance of the capacitor, (b) the phase angle, and (c) the power dissipated by the circuit.

29.26. The current in a resistor is 1 A when it is connected to a 50-V, 100-Hz power source. (a) How much inductive reactance is required to reduce the current to 0.5 A? What value of L will accomplish this? (b) How much capacitive reactance is required to reduce the current to 0.5 A? What value of C will accomplish this? (c) What will the current be if the above inductance and capacitance are both placed in series with the resistor?

29.27. A pure resistor, a pure capacitor, and a pure inductor are connected in series across an ac power source. An ac voltmeter placed in turn across these circuit elements reads 10 V, 20 V, and 30 V. What is the potential difference of the source?

29.28. An inductive circuit with a power factor of 80% consumes 750 W of power. What minimum rating in VA must its power source have?

29.29. A 10-μF capacitor, a 10-mH inductor, and a 10-Ω resistor are connected in series with a 45-V, 400-Hz power source. Find (a) the impedance of the circuit, (b) the current in it, (c) the power it dissipates, and (d) the minimum rating in VA of the power source.

29.30. (*a*) Find the resonant frequency f_0 of the circuit of Problem 29.29.   (*b*) What current will flow in the circuit if it is connected to a 45-V power source whose frequency is f_0?   (*c*) How much power will be dissipated by the circuit?   (*d*) What must be the minimum rating of the power source in VA?

29.31. In a series circuit $R = 100\ \Omega$, $X_L = 120\ \Omega$, and $X_C = 60\ \Omega$ when it is connected to an 80-V ac power source. Find (*a*) the current in the circuit, (*b*) the phase angle, and (*c*) the current if the frequency of the power source were changed to be equal to the resonant frequency of the circuit.

29.32. A circuit that consists of a capacitor in series with a 50-Ω resistor draws 4 A from a 250-V, 200-Hz power source.   (*a*) How much power is dissipated?   (*b*) What is the power factor?   (*c*) What inductance should be connected in series in the circuit to increase the power factor to 100%?   (*d*) What would the current then be?   (*e*) How much power would the circuit then dissipate?

Answers to Supplementary Problems

29.17. R is unchanged, X_L is doubled, and X_C is halved.

29.18. (*a*) The minimum value is 0, which can occur only if $R = 0$ in a circuit in which X_L is not equal to X_C. (*b*) The maximum value is 1, which occurs at resonance when $X_L = X_C$.

29.19. (*a*) 56.6 V   (*b*) 8.5 A   (*c*) the phase angle

29.20. 31.4 Ω

29.21. 0.796 μF

29.22. 10.6 V

29.23. 531 Hz

29.24. 0.102 A; 0.524 W

29.25. (*a*) 17.8 μF   (*b*) 48°   (*c*) 32 W

29.26. (*a*) 50 Ω, 79.6 mH   (*b*) 50 Ω, 31.8 μF   (*c*) 1 A

29.27. 14 V

29.28. 938 VA

29.29. (*a*) 18 Ω (*b*) 2.5 A   (*c*) 62.5 W   (*d*) 112.5 VA

29.30. (*a*) 503 Hz   (*b*) 4.5 A   (*c*) 202.5 W   (*d*) 202.5 VA

29.31. (*a*) 0.69 A   (*b*) 31°   (*c*) 0.80 A

29.32. (*a*) 800 W   (*b*) 80%   (*c*) 30 mH   (*d*) 5 A   (*e*) 1250 W

Chapter 30

Light

ELECTROMAGNETIC WAVES

Electromagnetic waves consist of coupled electric and magnetic fields that vary periodically as they move through space. The electric and magnetic fields are perpendicular to each other and to the direction in which the waves travel (Fig. 30-1), so the waves are transverse, and the variations in **E** and **B** occur simultaneously. Electromagnetic waves transport energy and require no material medium for their passage. Radio waves, light waves, X-rays, and gamma rays are examples of electromagnetic waves, and differ only in frequency. The color of light waves depends upon their frequency, with red light having the lowest visible frequencies and violet light the highest. White light contains light waves of all frequencies.

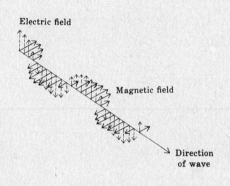

Fig. 30-1

Electromagnetic waves are generated by accelerated electric charges, usually electrons. Electrons oscillating back and forth in an antenna give off radio waves, for instance, and accelerated electrons in atoms give off light waves.

In free space all electromagnetic waves have the *velocity of light*, which is

$$\text{Velocity of light} = c = 3.00 \times 10^8 \text{ m/s} = 186,000 \text{ mi/s}$$

LUMINOUS INTENSITY AND FLUX

The eye responds to only part of the electromagnetic radiation emitted by most light sources (about 10% in the case of an ordinary light bulb), and in addition is not equally sensitive to light of different colors (the greatest sensitivity is to yellow-green light). For these reasons the watt is not a useful unit for comparing light sources and the illumination they provide, and other units, more closely based upon the visual response of the eye, are necessary.

The unit of *luminous intensity I* is the *candela* (cd), also known as the *new international candle*. The candela is defined in terms of the light emitted by a blackbody (see Chapter 22) at the freezing temperature of platinum, 1773 °C. The intensity of a light source is commonly referred to as its *candlepower*.

The unit of *luminous flux F* is the *lumen* (lm). One lumen is equal to the luminous flux which falls on each m² of a sphere 1 m in radius when a 1-candela isotropic light source (one that radiates equally in all directions) is at the center of the sphere. Since the area of a sphere of radius r is $4\pi r^2$, a sphere whose radius is 1 m has 4π m² of area, and the total luminous flux emitted by a 1-cd source is therefore 4π lm. Thus the luminous flux emitted by an isotropic light source of intensity I is given by

$$F = 4\pi I$$

Luminous flux = $4\pi \times$ luminous intensity

This formula does not apply to a light source that radiates different fluxes in different directions.

The *luminous efficiency* of a light source is the amount of luminous flux it radiates per watt of power input.   The luminous efficiency of ordinary tungsten-filament lamps increases with their power, because the higher the power of such a lamp, the greater its temperature and the more of its radiation is in the visible part of the spectrum.   The efficiencies of such lamps range from about 8 lm/W for a 10-W lamp to 22 lm/W for a 1000-W lamp.   Fluorescent lamps have efficiencies from 40 to 75 lm/W.

ILLUMINATION

The *illumination* (or *illuminance*) E of a surface is the luminous flux per unit area that reaches the surface:

$$E = \frac{F}{A} \qquad \text{Illumination} = \frac{\text{luminous flux}}{\text{area}}$$

In the SI system the unit of illumination is the lumen/m², or *lux*; in the British system it is the lumen/ft², or *footcandle*.

The illumination on a surface a distance R away from an isotropic source of light of intensity I is

$$E = \frac{I \cos \theta}{R^2}$$

where θ is the angle between the direction of the light and the normal to the surface (Fig. 30-2). Thus the illumination from such a source varies inversely as R^2, just as in the case of sound waves; doubling the distance means reducing the illumination to $(1/2)^2 = \frac{1}{4}$ its former value.   For light perpendicularly incident on a surface, $\theta = 0$ and $\cos \theta = 1$, so in this situation $E = I/R^2$.

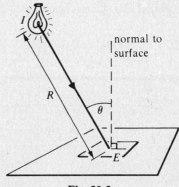

Fig. 30-2

REFLECTION OF LIGHT

When light is reflected from a smooth, plane surface, the angle of reflection equals the angle of incidence (Fig. 30-3). The image of an object in a plane mirror has the same size and shape as the object but with left and right reversed; the image is the same distance behind the mirror as the object is in front of it.

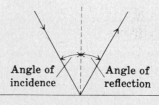

Fig. 30-3

REFRACTION OF LIGHT

When light passes obliquely from one medium to another in which its velocity is different, its direction changes (Fig. 30-4).   The greater the ratio between the two velocities, the greater the deflection.   If the light goes from the medium of high velocity to the one of low velocity, it is bent toward the normal to the surface; if the light goes the other way, it is bent away from the normal.   Light moving along the normal is not deflected.

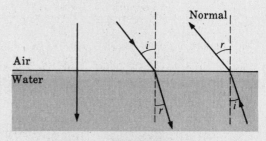

Fig. 30-4

The *index of refraction* of a transparent medium is the ratio between the velocity of light in free space and its velocity in the medium:

$$\text{Index of refraction} = n = \frac{c}{v}$$

The greater its index of refraction, the more light is deflected on entering a medium from air. The index of refraction of air is about 1.0003, so for most purposes it can be considered as equal to 1.

According to *Snell's law*, the angles of incidence i and refraction r shown in Fig. 30-4 are related by the formula

$$\frac{\sin i}{\sin r} = \frac{v_1}{v_2} = \frac{n_2}{n_1}$$

where v_1, n_1 are respectively the velocity of light and index of refraction of the first medium and v_2, n_2 are the corresponding quantities in the second medium. Snell's law is often written

$$n_1 \sin i = n_2 \sin r$$

In general the index of refraction of a medium increases with increasing frequency of the light. For this reason a beam of white light is separated into its component frequencies, each of which produces the sensation of a particular color, when it passes through an object whose sides are not parallel, for instance a glass prism. The resulting band of color is called a *spectrum*.

APPARENT DEPTH

An object submerged in water or other transparent liquid appears closer to the surface than it actually is. As Fig. 30-5 shows, light leaving the object is bent away from the normal to the water-air surface as it leaves the water. Since an observer interprets what he sees in terms of the straight-line propagation of light, the object seems at a shallower depth than its true one. The ratio between apparent and true depths is

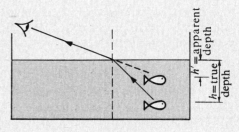

Fig. 30-5

$$\frac{\text{Apparent depth}}{\text{True depth}} = \frac{h'}{h} = \frac{n_2}{n_1}$$

where n_1 is the index of refraction of the liquid and n_2 is the index of refraction of air.

Solved Problems

30.1. Why can light waves travel through a vacuum whereas sound waves cannot?

Light waves consist of coupled fluctuations in electric and magnetic fields and hence require no material medium for their passage. Sound waves, on the other hand, are pressure fluctuations and cannot occur without a material medium to transmit them.

30.2. A marine radar operates at a wavelength of 3.2 cm. What is the frequency of the radar waves?

Radar waves are electromagnetic and hence travel with the velocity of light c. Therefore

$$f = \frac{c}{\lambda} = \frac{3 \times 10^8 \text{ m/s}}{3.2 \times 10^{-2} \text{ m}} = 9.4 \times 10^9 \text{ Hz}$$

30.3. A 10-W fluorescent lamp has a luminous intensity of 35 candelas.   Find (a) the luminous flux it emits, and (b) its luminous efficiency.

(a) $F = 4\pi I = 4\pi \times 35 \text{ cd} = 440 \text{ lm}$

(b)   Luminous efficiency $= \dfrac{F}{P} = \dfrac{440 \text{ lm}}{10 \text{ W}} = 44 \text{ lm/W}$

30.4. A spotlight concentrates all the light from a 100-cd bulb in a circle 3 ft in radius on a wall.   If the spotlight beam is perpendicular to the wall, find the illumination it produces.

The luminous flux emitted by the bulb is

$$F = 4\pi I = 4\pi \times 100 \text{ cd} = 400\pi \text{ lm}$$

The area of a circle 3 ft in radius is $A = \pi r^2 = \pi(3 \text{ ft})^2 = 9\pi \text{ ft}^2$.   Hence the illumination is

$$E = \frac{F}{A} = \frac{400\pi \text{ lm}}{9\pi \text{ ft}^2} = 44 \text{ lm/ft}^2 = 44 \text{ footcandles}$$

30.5. An illumination of about 20 footcandles is recommended for reading.   How far away from a book should a 75-W lamp of intensity 90 cd be located if the angle between the light rays and the plane of the opened book is 60°?

The angle between the light rays and the normal to the book is $\theta = 30°$.   Since $E = (I \cos \theta)/R^2$,

$$R = \sqrt{\frac{I \cos \theta}{E}} = \sqrt{\frac{90 \text{ cd} \times \cos 30°}{20 \text{ lm/ft}^2}} = 2.0 \text{ ft}$$

30.6. A 60-W light bulb whose luminous efficiency is 14 lm/W is suspended 2 m over a table.   (a) What is the illumination on the table directly under the bulb?   (b) How high over the table should the bulb be in order to double that illumination?

(a)   The luminous flux emitted by the bulb is
$$F = 60 \text{ W} \times 14 \text{ lm/W} = 840 \text{ lm}$$

Since $F = 4\pi I$ the intensity of the bulb is

$$I = \frac{F}{4\pi} = \frac{840 \text{ lm}}{4\pi} = 66.8 \text{ cd}$$

The illumination at a distance of $R = 2$ m is

$$E = \frac{I}{R^2} = \frac{66.8 \text{ cd}}{(2 \text{ m})^2} = 16.7 \text{ lm/m}^2 = 16.7 \text{ luxes}$$

(b)   Since $E = I/R^2$,

$$\frac{E_2}{E_1} = \frac{I_2 R_1^2}{I_1 R_2^2}$$

Here $I_1 = I_2$ and $E_2/E_1 = 2$, so

$$R_2 = R_1 \sqrt{\frac{E_1}{E_2}} = 2 \text{ m} \times \sqrt{\frac{1}{2}} = 1.41 \text{ m}$$

30.7. A lamp is suspended 6 ft over a large table.   How much greater is the illumination directly under the lamp as compared with the illumination at a point on the table 4 ft to one side?

Directly under the lamp the illumination is

$$E_1 = \frac{I}{R_1^2}$$

and at a distance d away on the table it is

$$E_2 = \frac{I \cos \theta}{R_2^2}$$

With the help of Fig. 30-6 we have

$$R_2 = \sqrt{R_1^2 + d^2} = \sqrt{(6 \text{ ft})^2 + (4 \text{ ft})^2} = 7.2 \text{ ft}$$

and

$$\cos \theta = \frac{R_1}{R_2} = \frac{6 \text{ ft}}{7.2 \text{ ft}} = 0.83$$

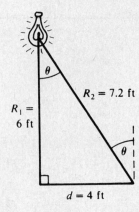

Fig. 30-6

Hence

$$\frac{E_1}{E_2} = \frac{I/R_1^2}{(I \cos \theta)/R_2^2} = \frac{R_2^2}{R_1^2 \cos \theta} = \frac{(7.2 \text{ ft})^2}{(6 \text{ ft})^2 \times 0.83} = 1.7$$

The illumination is 1.7 times greater directly under the lamp.

30.8. The index of refraction of diamond is 2.42. What is the velocity of light in diamond?

Since $n = c/v$, here

$$v = \frac{c}{n} = \frac{3 \times 10^8 \text{ m/s}}{2.42} = 1.24 \times 10^8 \text{ m/s}$$

30.9. Why is a beam of white light that passes perpendicularly through a flat pane of glass not dispersed into a spectrum?

Light incident perpendicularly on a surface is not deflected, so light of the various frequencies in white light stays together despite the different velocities in the glass.

30.10. A woman 5 ft 6 in. tall wishes to buy a mirror in which she can see herself at full length. What is the minimum height of such a mirror? How far from the mirror should she stand?

Since the angle of reflection equals the angle of incidence, the mirror should be half her height (2 ft 9 in.) and placed so its top is level with the middle of her forehead (Fig. 30-7). The distance between the mirror and the woman does not matter.

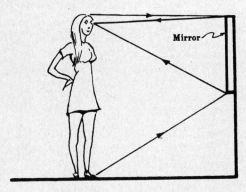

Fig. 30-7

30.11. A beam of light enters a lake at an angle of incidence of 40°. Find the angle of refraction. The index of refraction of water is 1.33.

Here medium 1 is air and medium 2 is water. From Snell's law,

$$\sin r = \frac{n_1}{n_2} \sin i = \frac{1.00}{1.33} \sin 40° = 0.483 \qquad r = 29°$$

The angle of refraction is less than the angle of incidence because $n_2 > n_1$.

30.12. A lantern held by a submerged skin diver directs a beam of light at the surface of a lake at an angle of incidence of 40°. Find the angle of refraction.

Now medium 1 is water and medium 2 is air. Hence

$$\sin r = \frac{n_1}{n_2} \sin i = \frac{1.33}{1.00} \sin 40° = 0.855$$

$$r = 59°$$

The angle of refraction is greater than the angle of incidence because $n_2 < n_1$.

30.13. A beam of light strikes a pane of glass at an angle of incidence of 50°. If the angle of refraction is 30°, find the index of refraction of the glass.

According to Snell's law, $n_1 \sin i = n_2 \sin r$. Here air is medium 1, so $n_1 = 1.00$ and

$$n_2 = n_1 \frac{\sin i}{\sin r} = 1.00 \times \frac{\sin 50°}{\sin 30°} = 1.53$$

30.14. The phenomenon of *total internal reflection* can occur when light goes from a medium of high index of refraction to one of low index of refraction, for example from glass or water to air. The angle of refraction in this situation is greater than the angle of incidence, and a light ray is bent away from the normal, as in Fig. 30-8, at the interface between the two mediums. At the *critical angle* of incidence, the angle of refraction is 90°, and at angles of incidence greater than this the refracted rays are reflected back into the original medium. (*a*) Derive a formula for the critical angle. (*b*) Find the critical angle for light going from crown glass ($n = 1.52$) to air ($n = 1.00$) and for light going from crown glass to water ($n = 1.33$).

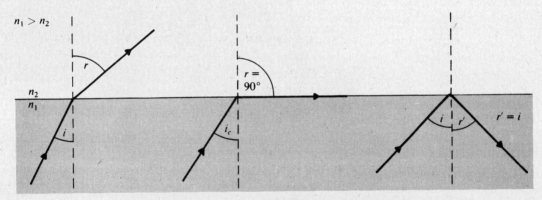

Fig. 30-8

(*a*) At the critical angle i_c the angle of refraction is 90°, so that $\sin r = \sin 90° = 1$ when $i = i_c$. Substituting $\sin r = 1$ into Snell's law

$$n_1 \sin i = n_2 \sin r$$

yields

$$n_1 \sin i_c = n_2 \qquad \sin i_c = \frac{n_2}{n_1}$$

The critical angle depends upon the ratio n_2/n_1.

(*b*) For light going from glass into air,

$$\sin i_c = \frac{1.00}{1.52} = 0.658 \qquad i_c = 41°$$

For light going from glass into water,

$$\sin i_c = \frac{1.33}{1.52} = 0.875 \qquad i_c = 61°$$

30.15. Prisms are used in optical instruments instead of mirrors to change the direction of light beams by 90° because total internal reflection better preserves the sharpness and brightness of light beams. What is the minimum index of refraction of the glass used in such prisms?

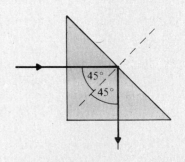

> As Fig. 30-9 shows, the angle of incidence must be 45°, so the critical angle must be at least 45°. Hence the minimum value of n is
>
> $$n = \frac{1}{\sin i_c} = \frac{1}{\sin 45°} = 1.41$$

Fig. 30-9

30.16. The water in a swimming pool is 2 m deep. How deep does it appear to be to someone looking down into it?

> Here $n_1 = 1.33$ and $n_2 = 1.00$, so
>
> $$h' = h\,\frac{n_2}{n_1} = 2 \text{ m} \times \frac{1.00}{1.33} = 1.5 \text{ m}$$

Supplementary Problems

30.17. When a light beam enters one medium from another, which (if any) of the following quantities never changes? The direction of the beam, its velocity, its frequency, its wavelength.

30.18. What is the relationship between the direction of an electromagnetic wave and the directions of its electric and magnetic fields?

30.19. What is the frequency of radio waves whose wavelength is 20 m?

30.20. A certain radio station transmits at a frequency of 1050 kHz. What is the wavelength of these waves?

30.21. Is it possible for the index of refraction of a substance to be less than 1?

30.22. A 100-W tungsten-filament lamp has a luminous efficiency of 16 lm/W. Find its intensity in candelas.

30.23. A newspaper is held 20 ft from a 2000-cd street lamp at midnight. What is the maximum illumination of the newspaper?

30.24. A 150-cd lamp is to be substituted for a 100-cd lamp over a workbench. If the original lamp was 5 ft above the workbench, how much higher can the new lamp be and still provide the same illumination?

30.25. What intensity should a lamp have if it is to provide an illumination of 400 luxes at a distance of 3 m on a surface whose normal is 20° from the direction of the light rays?

30.26. A playing field 250 ft by 400 ft is to have an average illumination at night of 30 footcandles. How many 4000-cd lamps are needed if reflectors are used that enable 40% of their luminous flux to reach the field?

30.27. The index of refraction of benzene is 1.50. Find the velocity of light in benzene.

30.28. The velocity of light in ice is 2.3×10^8 m/s. What is its index of refraction?

30.29. A beam of light enters a plate of flint glass ($n = 1.63$) at an angle of incidence of 40°. Find the angle of refraction.

30.30. A beam of light enters a tank of glycerin at an angle of incidence of 45°. The angle of refraction is 29°. Find the index of refraction of glycerin.

30.31. Find the critical angle for total internal reflection for light going from ice ($n = 1.31$) into air.

30.32. A stick frozen into a pond in winter appears to be 3 in. below the surface. What is its actual depth in the ice?

Answers to Supplementary Problems

30.17. Only the frequency never changes.

30.18. The electric and magnetic fields are perpendicular to the direction of the wave and to each other.

30.19. 1.5×10^7 Hz = 15 MHz

30.20. 286 m

30.21. No, because this would mean a velocity of light greater than its velocity in free space, which is not possible.

30.22. 127 cd

30.23. 5 footcandles

30.24. 1.12 ft higher for a total height of 6.12 ft

30.25. 3831 cd

30.26. 149 lamps

30.27. 2×10^8 m/s

30.28. 1.3

30.29. 23°

30.30. 1.46

30.31. 50°

30.32. 3.93 in.

Chapter 31

Spherical Mirrors

FOCAL LENGTH

Figure 31-1 shows how a concave mirror converges a parallel beam of light to a real focal point F, and Fig. 31-2 shows how a convex mirror diverges a parallel beam of light so that the reflected rays appear to come from a virtual focal point F behind it. In either case, if the radius of curvature of the mirror is R, the focal length f is $R/2$. For a concave mirror, f is positive and for a convex mirror f is negative. Thus

$$\text{Concave mirror:} \quad f = + \frac{R}{2}$$

$$\text{Convex mirror:} \quad f = - \frac{R}{2}$$

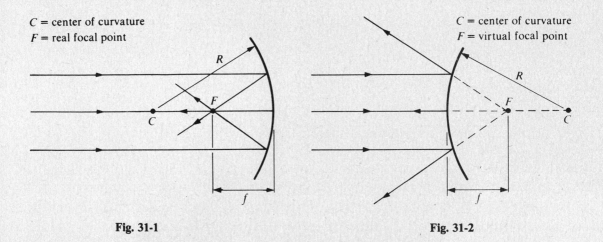

C = center of curvature	C = center of curvature
F = real focal point	F = virtual focal point

Fig. 31-1   **Fig. 31-2**

RAY TRACING

The position and size of the image formed by a spherical mirror of an object in front of it can be found by constructing a scale drawing. What is done is to trace two different light rays from each point of interest in the object to where they (or their extensions in the case of a virtual image) intersect after being reflected by the mirror. Three rays that are especially useful for this purpose are shown in Fig. 31-3; any two of these are sufficient. They are:

1. A ray that leaves the object parallel to the axis of the mirror. After reflection, this ray passes through the focal point of a concave mirror or seems to come from the focal point of a convex mirror.

2. A ray that passes through the focal point of a concave mirror or is directed toward the focal point of a convex mirror. After reflection, this ray travels parallel to the axis of the mirror.

3. A ray that leaves the object along a radius of the mirror. After reflection, this ray returns along the same radius.

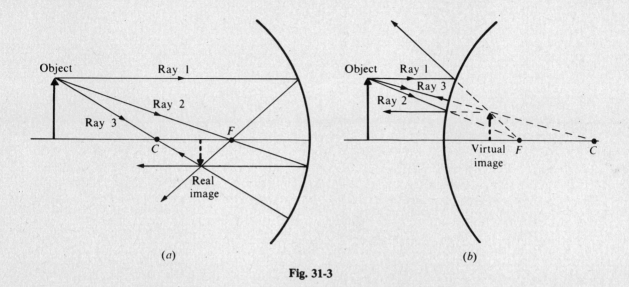

Fig. 31-3

MIRROR EQUATION

When an object is the distance p from a mirror of focal length f, the image will be located the distance q from the mirror, where

$$\frac{1}{p} + \frac{1}{q} = \frac{1}{f}$$

$$\frac{1}{\text{Object distance}} + \frac{1}{\text{image distance}} = \frac{1}{\text{focal length}}$$

This equation holds for both concave and convex mirrors (see Fig. 31-4). The mirror equation is readily solved for p, q, or f:

$$p = \frac{qf}{q-f} \qquad q = \frac{pf}{p-f} \qquad f = \frac{pq}{p+q}$$

f = focal length
p = object distance
q = image distance

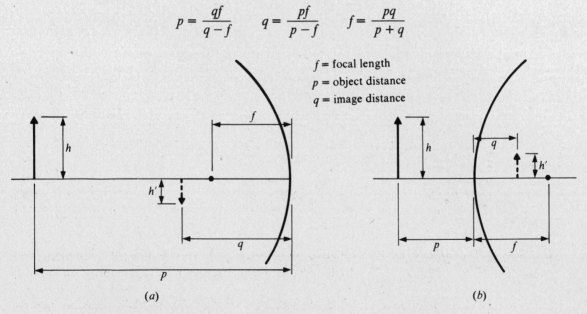

Fig. 31-4

A positive value of p or q denotes a real object or image, and a negative value denotes a virtual object or image. A *real object* is one that is in front of a mirror; a *virtual object* appears to be located behind the mirror, and must itself be an image produced by another mirror or lens. A *real image* is formed by light rays that actually pass through the image, so a real image will appear on a screen placed at the position of the image. On the other hand, a *virtual image* can only be seen by the eye since the light rays that appear to come from the image actually do not pass through it. Real images are located in front of a mirror, virtual images behind it.

MAGNIFICATION

The *linear magnification m* of any optical system is the ratio between the size (height or width or other transverse linear dimension) of the image and the size of the object. In the case of a mirror,

$$m = \frac{h'}{h} = -\frac{q}{p}$$

$$\text{Linear magnification} = \frac{\text{image height}}{\text{object height}} = -\frac{\text{image distance}}{\text{object distance}}$$

A positive magnification signifies an erect image, as in Fig. 31-4(b); a negative one signifies an inverted image, as in Fig. 31-4(a). Table 31-1 is a summary of the sign conventions used in connection with spherical mirrors.

Table 31-1

Quantity	Positive	Negative
Focal length f	Concave mirror	Convex mirror
Object distance p	Real object	Virtual object
Image distance q	Real image	Virtual image
Magnification m	Erect image	Inverted image

Solved Problems

31.1. What is *spherical aberration*?

Spherical aberration refers to the fact that light rays from a point on an object that are reflected at different distances from the axis of a spherical mirror do not converge to (or appear to diverge from) a single point. This effect is shown in Fig. 31-5 for parallel rays reaching a concave mirror: rays reflected from the outer parts of the mirror converge at focal points closer to the mirror than those reflected near the mirror's axis. As a result a spherical mirror produces sharp images only when its diameter is small compared with its focal length.

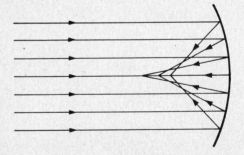

Fig. 31-5

31.2. What is the nature of the image of a real object formed by a convex mirror?

The image is virtual, erect, and smaller in size than the object, as in Fig. 31-3(b).

31.3. Describe the image formed by a concave mirror of an object placed at the focal point of the mirror.

Here $p = f$, and so

$$q = \frac{pf}{p - f} = \infty$$

As Fig. 31-6 shows, the reflected rays are parallel to each other and so no image is formed.

31.4. A candle 5 in. high is placed 20 in. in front of a concave mirror whose focal length is 15 in. Find the location, size, and nature of the image.

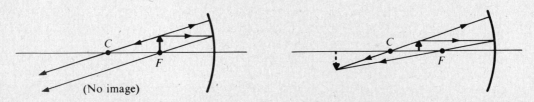

Fig. 31-6   **Fig. 31-7**

Here $p = 20$ in. and $f = 15$ in., so the image distance is

$$q = \frac{pf}{p - f} = \frac{20 \text{ in.} \times 15 \text{ in.}}{20 \text{ in.} - 15 \text{ in.}} = 60 \text{ in.}$$

The image is real and on the same side of the mirror as the candle (Fig. 31-7). The height of the image is

$$h' = -h\frac{q}{p} = -5 \text{ in.} \times \frac{60 \text{ in.}}{20 \text{ in.}} = -15 \text{ in.}$$

which is greater than the height of the candle. The minus sign indicates an inverted image.

In general, an object placed between the focal point and the center of curvature C of a concave mirror (that is, with p greater than f but less than $2f$) will have a real, inverted image that is larger than the object.

31.5. A pencil 12 cm long is placed at the center of curvature of a concave mirror whose focal length is 40 cm. Find the location, size, and nature of the image.

Since $f = R/2$ for a concave mirror, $p = R = 2f = 80$ cm here. The image distance is therefore

$$q = \frac{pf}{p - f} = \frac{80 \text{ cm} \times 40 \text{ cm}}{80 \text{ cm} - 40 \text{ cm}} = 80 \text{ cm}$$

The image is real and at the same distance from the mirror as the object (Fig. 31-8). The height of the image is

$$h' = -h\frac{q}{p} = -12 \text{ cm} \times \frac{80 \text{ cm}}{80 \text{ cm}} = -12 \text{ cm}$$

which is the same as the height of the object. The minus sign indicates an inverted image.

In general, an object placed at the center of curvature of a concave mirror will have a real, inverted image the same size as the object and at the same distance from the mirror.

31.6. A cigar 6 in. long is placed 30 in. in front of a concave mirror whose focal length is 12 in. Find the location, size, and nature of the image.

Here $p = 30$ in. and $f = 12$ in., so the image distance is

$$q = \frac{pf}{p - f} = \frac{30 \text{ in.} \times 12 \text{ in.}}{30 \text{ in.} - 12 \text{ in.}} = 20 \text{ in.}$$

The image is real and on the same side of the mirror as the cigar (Fig. 31-9).   The length of the image is

$$h' = -h\frac{q}{p} = -6 \text{ in.} \times \frac{20 \text{ in.}}{30 \text{ in.}} = -4 \text{ in.}$$

which is smaller than the length of the cigar.   The minus sign indicates an inverted image.

In general, an object placed beyond the center of curvature of a concave mirror (that is, with p greater than $2f$) will have a real, inverted image that is smaller than the object.

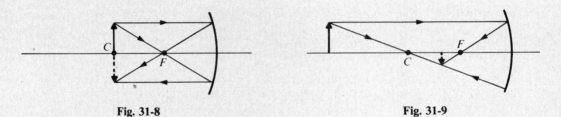

Fig. 31-8   **Fig. 31-9**

31.7. A concave mirror whose radius of curvature is 4 m is used to produce an image of the moon on a photographic plate.   The moon's diameter is approximately 3500 km and it is about 384,000 km from the earth.   (*a*) How far from the mirror should the photographic plate be placed?   (*b*) What will the size and nature of the moon's image be?

(*a*)   The object distance is so much greater than the mirror's focal length of $f = R/2 = 2$ m that we can let $p = \infty$ here.   Substituting $1/p = 0$ and $f = 2$ m into the mirror equation yields

$$\frac{1}{p} + \frac{1}{q} = \frac{1}{f} \qquad 0 + \frac{1}{q} = \frac{1}{2 \text{ m}} \qquad q = 2 \text{ m}$$

The image is real and on the same side of the mirror as the moon.

(*b*)   Since the moon's diameter is $h = 3500 \text{ km} = 3.5 \times 10^6$ m and the object distance is $p = 384,000 \text{ km} = 3.84 \times 10^8$ m, the diameter of the moon's image will be

$$h' = -h\frac{q}{p} = -3.5 \times 10^6 \text{ m} \times \frac{2 \text{ m}}{3.84 \times 10^8 \text{ m}} = -0.018 \text{ m} = -18 \text{ mm}$$

The minus sign indicates an inverted image (Fig. 31-10).

In general, an object very far away from a concave mirror relative to its focal length will have a real, inverted image smaller than the object and located very nearly at the focal point of the mirror.

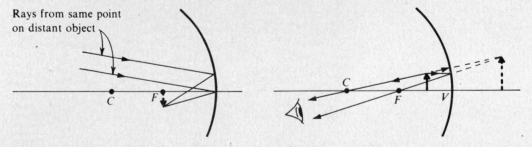

Rays from same point on distant object

Fig. 31-10   **Fig. 31-11**

31.8. A concave mirror has a radius of curvature of 120 cm.   How far from the mirror should one's face be in order that the image be erect and twice the size of the actual face?   Is the image real or virtual?

The focal length of the mirror is $f = R/2 = 60$ cm and is positive since the mirror is **concave**. Because the image is to be erect and twice the size of the object, $m = +2$.   We therefore know m and f and are to find the object distance p.   There are various ways to do this, one of which is as follows.

Since $m = -q/p$ we have $q = -mp$. The negative image distance signifies a virtual image (Fig. 31-11). Substituting $q = -mp$ in the mirror equation enables us to solve for p:

$$\frac{1}{p} + \frac{1}{q} = \frac{1}{f} \qquad \frac{1}{p} - \frac{1}{mp} = \frac{1}{f} \qquad \frac{m-1}{mp} = \frac{1}{f} \qquad \frac{mp}{m-1} = f$$

$$p = \frac{f}{m}(m-1) = \frac{60 \text{ cm}}{2}(2-1) = 30 \text{ cm}$$

In general, an object placed closer to a concave mirror than its focal point (that is, closer than f) will have a virtual, erect image that is larger than the object.

31.9. How far away from a concave mirror of 40 cm focal length should a real object 30 mm long be located in order that its image be 8 mm long?

When a concave mirror forms an image that is smaller than the object, the image is always inverted (see Fig. 31-9). An inverted image is considered to have a negative height, so here the magnification is

$$m = \frac{h'}{h} = \frac{-8 \text{ mm}}{30 \text{ mm}} = -0.267$$

From the solution to Problem 31.8,

$$p = \frac{f}{m}(m-1) = \frac{40 \text{ cm}}{-0.267}(-0.267 - 1) = 190 \text{ cm}$$

If we had not known that the image is inverted and had used a magnification of $+0.267$, the result would have been an object distance of -110 cm. But a negative object distance signifies a virtual object, whereas here we are given a real object, so it would be clear that the wrong sign had been used for the magnification.

31.10. A grasshopper 5 cm long is 25 cm in front of a convex mirror whose radius of curvature is 80 cm. Find the location, size, and nature of the image.

The focal length of the mirror is

$$f = -\frac{R}{2} = -\frac{80 \text{ cm}}{2} = -40 \text{ cm}$$

The image distance is

$$q = \frac{pf}{p-f} = \frac{25 \text{ cm} \times (-40 \text{ cm})}{25 \text{ cm} - (-40 \text{ cm})} = -\frac{25 \text{ cm} \times 40 \text{ cm}}{25 \text{ cm} + 40 \text{ cm}} = -15.4 \text{ cm}$$

The minus sign indicates a virtual image located behind the mirror. The length of the image is

$$h' = -h\frac{q}{p} = -5 \text{ cm} \times \frac{-15.4 \text{ cm}}{25 \text{ cm}} = 3.1 \text{ cm}$$

A positive value for h' signifies an erect image.

Supplementary Problems

31.11. What does a negative magnification signify? A magnification that is less than 1?

31.12. Under what circumstances does a concave mirror produce a real image of a real object? A virtual image?

31.13. A moth flies toward a convex mirror. Does its image become larger or smaller as it approaches the mirror's focal point? What kind of image is it? What happens when the moth is at the focal point?

31.14. (*a*) A concave mirror has a radius of curvature of 30 in.   What is its focal length?   (*b*) A convex mirror has a radius of curvature of 30 in.   What is its focal length?

31.15. A match 6 cm long is placed 30 cm in front of a concave mirror whose focal length is 50 cm.   Find the location, size, and nature of the image.

31.16. The match of Problem 31.15 is placed 50 cm in front of the same mirror.   Find the location, size, and nature of the image.

31.17. The match of Problem 31.15 is placed 80 cm in front of the same mirror.   Find the location, size, and nature of the image.

31.18. The match of Problem 31.15 is placed 100 cm in front of the same mirror.   Find the location, size, and nature of the image.

31.19. The match of Problem 31.15 is placed 3 m in front of the same mirror.   Find the location, size, and nature of the image.

31.20. A pencil 6 in. long is placed 8 in. in front of a mirror whose focal length is -15 in.   Find the location, size, and nature of the image.

31.21. How far away from a concave mirror of 25-in. focal length should a real object be located in order that its image be one-third its actual size?

31.22. A shaving mirror is intended to produce an erect image of a man's face 2.5 times its actual size when the face is 60 cm in front of it.   (*a*) Should the mirror be concave or convex?   (*b*) What should its radius of curvature be?

Answers to Supplementary Problems

31.11. An inverted image; an image smaller than the object.

31.12. When the object distance is greater than the focal length of the mirror.   When the object distance is less than the focal length.

31.13. Larger; real and inverted; there is no image then.

31.14. (*a*) 15 in.   (*b*) -15 in.

31.15. The image is 75 cm behind the mirror, 15 cm long, virtual and erect.

31.16. No image is formed.

31.17. The image is 133 cm in front of the mirror, 10 cm long, real and inverted.

31.18. The image is 100 cm in front of the mirror, 6 cm long, real and inverted.

31.19. The image is 60 cm in front of the mirror, 1.2 cm long, real and inverted.

31.20. The image is 5.22 in. behind the mirror, 3.91 in. long, virtual and erect.

31.21. 100 in.

31.22. (*a*) Concave   (*b*) 200 cm

Chapter 32

Lenses

FOCAL LENGTH

Figure 32-1 shows how a converging lens brings a parallel beam of light to a real focal point *F*, and Fig. 32-2 shows how a diverging lens spreads out a parallel beam of light so that the refracted rays appear to come from a virtual focal point *F*. In this chapter we shall consider only *thin lenses*, whose thickness can be neglected as far as optical effects are concerned. The focal length *f* of a thin lens is given by the *lensmaker's equation*:

$$\frac{1}{f} = (n-1)\left(\frac{1}{R_1} + \frac{1}{R_2}\right)$$

In this equation *n* is the index of refraction of the lens material relative to the medium it is in, and R_1 and R_2 are the radii of curvature of the two surfaces of the lens. Both R_1 and R_2 are considered as + for a convex (curved outward) surface and as − for a concave (curved inward) surface; obviously it does not matter which surface is labeled as 1 and which as 2.

A positive focal length corresponds to a converging lens and a negative focal length to a diverging lens.

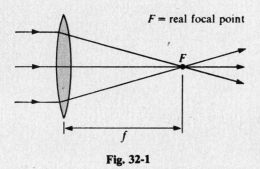

F = real focal point

Fig. 32-1

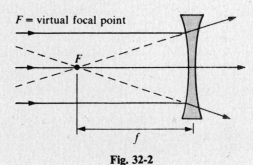

F = virtual focal point

Fig. 32-2

RAY TRACING

As in the case of a spherical mirror, the position and size of the image of an object formed by a lens can be found by constructing a scale drawing. Again, what is done is to trace two different light rays from each point of interest in the object to where they (or their extensions in the case of a virtual image) intersect after being refracted by the lens. Three rays that are especially useful for this purpose are shown in Fig. 32-3; any two of these are sufficient. They are:

1. A ray that leaves the object parallel to the axis of the lens. After refraction, this ray passes through the far focal point of a converging lens or seems to come from the near focal point of a diverging lens.

2. A ray that passes through the near focal point of a converging lens or is directed toward the far focal point of a diverging lens. After refraction, this ray travels parallel to the axis of the lens.

3. A ray that leaves the object and proceeds toward the center of the lens. This ray is not deviated by refraction.

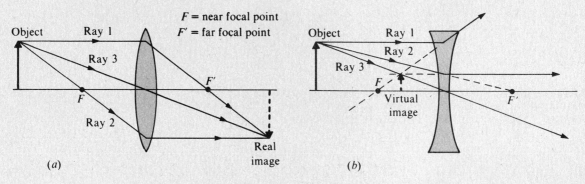

Fig. 32-3

LENS EQUATION

The object distance p, image distance q, and focal length f of a lens (Fig. 32-4) are related by the lens equation:

$$\frac{1}{p} + \frac{1}{q} = \frac{1}{f}$$

$$\frac{1}{\text{Object distance}} + \frac{1}{\text{image distance}} = \frac{1}{\text{focal length}}$$

This equation holds for both converging and diverging lenses. The lens equation is readily solved for p, q, or f:

$$p = \frac{qf}{q-f} \qquad q = \frac{pf}{p-f} \qquad f = \frac{pq}{p+q}$$

As in the case of mirrors, a positive value of p or q denotes a real object or image, and a negative value denotes a virtual object or image. A real image of a real object is always on the opposite side of the lens from the object, and a virtual image is on the same side; thus if a real object is on the left of a lens, a positive image distance q signifies a real image to the right of the lens whereas a negative image distance q denotes a virtual image to the left of the lens.

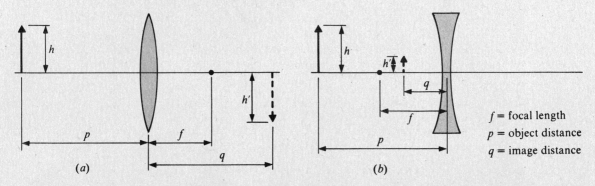

Fig. 32-4

MAGNIFICATION

The linear magnification m produced by a lens is given by the same formula that applies for mirrors:

$$m = \frac{h'}{h} = -\frac{q}{p}$$

$$\text{Linear magnification} = \frac{\text{image height}}{\text{object height}} = -\frac{\text{image distance}}{\text{object distance}}$$

Again, a positive magnification signifies an erect image, a negative one signifies an inverted image. Table 32-1 is a summary of the sign conventions used in connection with lenses.

Table 32-1

Quantity	Positive	Negative
Focal length f	Converging lens	Diverging lens
Object distance p	Real object	Virtual object
Image distance q	Real image	Virtual image
Magnification m	Erect image	Inverted image

LENS SYSTEMS

When a system of lenses is used to produce an image of an object, for instance in a telescope or microscope, the procedure for finding the position and nature of the final image is to let the image formed by each lens in turn be the object for the next lens in the system. Thus to find the image produced by a system of two lenses, the first step is to determine the image formed by the lens nearest the object. This image then serves as the object for the second lens, with the usual sign convention: if the image is on the front side of the second lens, the object distance is considered positive, whereas if the image is on the back side, the object distance is considered negative.

The total magnification produced by a system of lenses is equal to the product of the magnifications of the individual lenses. Thus if the magnification of the objective lens of a microscope or telescope is m_1 and that of the eyepiece is m_2, the total magnification is $m = m_1 m_2$.

Solved Problems

32.1. What is the nature of the image formed by a diverging lens of a real object?

It is virtual, erect, and smaller than the object, as in Fig. 32-3.

32.2. A *plano-convex lens* has one plane surface and one convex surface. If a plano-convex lens of focal length 12 in. is to be ground from glass of index of refraction 1.60, find the radius of curvature of the convex surface.

The radius of curvature of the plane surface is ∞, so if we call this surface 1, $1/R_1 = 0$. From the lensmaker's equation

$$\frac{1}{f} = (n-1)\left(\frac{1}{R_1} + \frac{1}{R_2}\right) = (n-1)\left(\frac{1}{R_2}\right)$$

and so

$$R_2 = (n-1)f = (1.60 - 1.00) \times 12 \text{ in.} = 7.2 \text{ in.}$$

32.3. A *meniscus lens* has one concave and one convex surface. The concave surface of a particular meniscus lens has a radius of curvature of 30 cm and its convex surface has a radius of curvature of 50 cm. The index of refraction of the glass used is 1.50. (*a*) Find the focal length of the lens. (*b*) Is it a converging lens or a diverging lens?

(*a*) Here $R_1 = -30$ cm (since the first surface is concave, its radius is considered negative) and $R_2 = +50$ cm. Hence

$$\frac{1}{f} = (n-1)\left(\frac{1}{R_1} + \frac{1}{R_2}\right) = (1.50 - 1.00)\left(-\frac{1}{30\text{ cm}} + \frac{1}{50\text{ cm}}\right) = -0.00667\text{ cm}^{-1}$$

$$f = -\frac{1}{0.00667\text{ cm}^{-1}} = -150\text{ cm}$$

(*b*) A negative focal length signifies a diverging lens.

32.4. A lens made of glass whose index of refraction is 1.60 has a focal length of $+20$ cm in air. Find its focal length in water, whose index of refraction is 1.33.

Let f be the focal length of the lens in air and f' be its focal length in water. The index of refraction of the glass relative to air is $n = 1.60$ since the index of refraction of air is very nearly equal to 1. The index of refraction of the glass relative to water is

$$n' = \frac{\text{index of refraction of glass}}{\text{index of refraction of water}} = \frac{1.60}{1.33} = 1.20$$

From the lensmaker's equation, since R_1 and R_2 are the same in both air and water,

$$\frac{f'}{f} = \frac{n-1}{n'-1} = \frac{1.60 - 1.00}{1.20 - 1.00} = 3$$

and so

$$f' = 3f = 3 \times (+20\text{ cm}) = +60\text{ cm}$$

The focal length of any lens made of this glass is three times longer in water than in air.

32.5. Both surfaces of a double-concave lens whose focal length is -9 in. have radii of 10 in. Find the index of refraction of the glass.

From the lensmaker's equation

$$n - 1 = \frac{1}{f(1/R_1 + 1/R_2)} = \frac{1}{(-9\text{ in.})\left(-\dfrac{1}{10\text{ in.}} - \dfrac{1}{10\text{ in.}}\right)} = 0.56$$

$$n = 0.56 + 1 = 1.56$$

32.6. Describe the image formed by a converging lens of an object located at the focal point of the lens.

Here $p = f$, and so

$$q = \frac{pf}{p - f} = \infty$$

As Fig. 32-5 shows, the refracted rays are parallel to each other and so no image is formed.

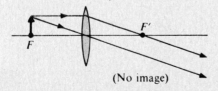

(No image)

Fig. 32-5

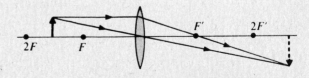

Fig. 32-6

32.7. A doughnut 3 in. in diameter is placed 24 in. from a converging lens whose focal length is 16 in. Find the location, size, and nature of the image.

Here $p = 24$ in. and $f = +16$ in., so the image distance is

$$q = \frac{pf}{p-f} = \frac{24 \text{ in.} \times 16 \text{ in.}}{24 \text{ in.} - 16 \text{ in.}} = 48 \text{ in.}$$

The image is real since q is positive (Fig. 32-6). The diameter of the doughnut's image is, since $m = h'/h = -q/p$,

$$h' = -h\frac{q}{p} = -3 \text{ in.} \times \frac{48 \text{ in.}}{24 \text{ in.}} = -6 \text{ in.}$$

The image is inverted (since h' is negative) and twice as large as the object.

In general, an object that is between f and $2f$ from a converging lens will have a real, inverted image that is larger than the object.

32.8. A sardine 8 cm long is 30 cm from a converging lens whose focal length is 15 cm. Find the location, size and nature of the image.

Here $p = 30$ cm and $f = +15$ cm, so the image distance is

$$q = \frac{pf}{p-f} = \frac{30 \text{ cm} \times 15 \text{ cm}}{30 \text{ cm} - 15 \text{ cm}} = 30 \text{ cm}$$

The image is real since q is positive (Fig. 32-7). The length of the sardine's image is

$$h' = -h\frac{q}{p} = -8 \text{ cm} \times \frac{30 \text{ cm}}{30 \text{ cm}} = -8 \text{ cm}$$

The image is inverted (since h' is negative) and is the same size as the object.

In general, an object that is the distance $2f$ from a converging lens will have a real, inverted image the same size as the object with an image distance equal to $2f$.

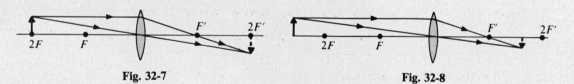

Fig. 32-7   **Fig. 32-8**

32.9. A key 6 cm long is 100 cm from a converging lens whose focal length is 40 cm. Find the location, size, and nature of the image.

Here $p = 100$ cm and $f = +40$ cm, so the image distance is

$$q = \frac{pf}{p-f} = \frac{100 \text{ cm} \times 40 \text{ cm}}{100 \text{ cm} - 40 \text{ cm}} = 66.7 \text{ cm}$$

The image is real since q is positive (Fig. 32-8). The length of the key's image is

$$h' = -h\frac{q}{p} = -6 \text{ cm} \times \frac{66.7 \text{ cm}}{100 \text{ cm}} = -4 \text{ cm}$$

The image is inverted (since h' is negative) and is smaller than the object.

In general, an object that is farther than $2f$ from a converging lens will have a real, inverted image smaller than the object with an image distance between f and $2f$.

32.10. A diverging lens has a focal length of -2 ft. What is the location, size, and nature of the image formed by the lens when it is used to look at an object 12 ft away?

Here $p = 12$ ft and $f = -2$ ft, so the image distance is

$$q = \frac{pf}{p-f} = \frac{12 \text{ ft} \times (-2 \text{ ft})}{12 \text{ ft} - (-2 \text{ ft})} = -1.71 \text{ ft}$$

A negative image distance signifies a virtual image.   The magnification is

$$m = -\frac{q}{p} = -\frac{(-1.71 \text{ ft})}{12 \text{ ft}} = 0.143 = \tfrac{1}{7}$$

The image is erect (since m is positive) and $\tfrac{1}{7}$ the size of the object.

32.11. A double-convex lens has a focal length of 6 cm.   (a) How far from an insect 2 mm long should the lens be held in order to produce an erect image 5 mm long?   (b) What is the image distance?

(a) A double-convex lens is always converging, so the focal length of the lens here is $+6$ cm. An erect image means a positive magnification, which is

$$m = \frac{h'}{h} = \frac{5 \text{ mm}}{2 \text{ mm}} = 2.5$$

Since $m = -q/p$, the image distance is $q = -mp$. We proceed by substituting $q = -mp$ in the lens equation and solving for p:

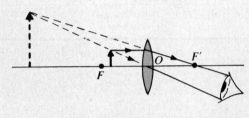

$$\frac{1}{p} = \frac{1}{q} = \frac{1}{f}, \qquad \frac{1}{p} - \frac{1}{mp} = \frac{1}{f}, \qquad \frac{m-1}{mp} = \frac{1}{f}$$

$$p = f\left(\frac{m-1}{m}\right) = +6 \text{ cm} \times \left(\frac{2.5-1}{2.5}\right) = 3.6 \text{ cm}$$

(b) $q = -mp = -2.5 \times 3.6 \text{ cm} = -9 \text{ cm}$

The negative image distance signifies a virtual image (Fig. 32-9).

Fig. 32-9

32.12. Find the focal length of a magnifying glass that produces an erect image magnified three times of an object 1.5 in. away.

An erect image magnified three times means a magnification of $+3$.   Since the object distance is $p = 1.5$ in. and $m = -q/p$,

$$q = -mp = -3 \times 1.5 \text{ in.} = -4.5 \text{ in.}$$

The negative image distance signifies a virtual image.   The required focal length is

$$f = \frac{pq}{p+q} = \frac{1.5 \text{ in.} \times (-4.5 \text{ in.})}{1.5 \text{ in.} + (-4.5 \text{ in.})} = +2.25 \text{ in.}$$

32.13. A 35-mm camera has a telephoto lens whose focal length is 150 mm.   What range of adjustment should the lens have in order to be able to bring to a sharp focus objects as close as 1.5 m from the camera?

The image distance that corresponds to $f = 0.15$ m and $p = 1.5$ m is

$$q = \frac{pf}{p-f} = \frac{1.5 \text{ m} \times 0.15 \text{ m}}{1.5 \text{ m} - 0.15 \text{ m}} = 0.167 \text{ m} = 167 \text{ mm}$$

An object at $p = \infty$ is brought to a focus at $q = f = 150$ mm (see Fig. 32-10).   Hence a range of adjustment of $167 \text{ mm} - 150 \text{ mm} = 17 \text{ mm}$ will permit objects to be photographed at distances from 1.5 m to infinity.

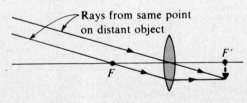

Rays from same point on distant object

Fig. 32-10

32.14. A slide projector uses a converging lens to produce a real, inverted image of a transparent slide on a screen.   What should the focal length of the lens of a slide projector be if a 2 in. × 2 in. slide is to appear 3 ft × 3 ft on a screen 15 ft away?

Here $h = 2$ in. and $h' = -36$ in. (h' is negative because the image is inverted), so the magnification is

$$m = \frac{h'}{h} = \frac{-36 \text{ in.}}{2 \text{ in.}} = -18$$

Since $q = 15$ ft and $m = -q/p$, the object distance is

$$p = -\frac{q}{m} = \frac{-15 \text{ ft}}{-18} = 0.833 \text{ ft}$$

The required focal length is therefore

$$f = \frac{pq}{p+q} = \frac{0.833 \text{ ft} \times 15 \text{ ft}}{0.833 \text{ ft} + 15 \text{ ft}} = 0.789 \text{ ft} = 9.5 \text{ in.}$$

32.15. The focal length f of a combination of two thin lenses in contact whose individual lengths are f_1 and f_2 is given by

$$\frac{1}{f} = \frac{1}{f_1} + \frac{1}{f_2}$$

Use this formula to find the focal length of a combination of a converging lens of $f = +10$ cm and a diverging lens of $f = -20$ cm that are in contact.

The above formula can be rewritten in the more convenient form

$$f = \frac{f_1 f_2}{f_1 + f_2}$$

Since $f_1 = 10$ cm and $f_2 = -20$ cm,

$$f = \frac{10 \text{ cm} \times (-20 \text{ cm})}{10 \text{ cm} + (-20 \text{ cm})} = +20 \text{ cm}$$

The combination acts as a converging lens of focal length $+20$ cm.

32.16. For most distinctness of vision with a normal eye, an object should be about 25 cm (10 in.) away.   (a) Find a formula for the magnification of a converging lens of focal length f when it is used as a magnifying glass with an image distance of -25 cm.   (b) Use this formula to find the magnification of a lens of focal length $+5$ cm.

(a)   From $m = -q/p$ the object distance that corresponds to an image distance of $q = -25$ cm is

$$p = -\frac{q}{m} = \frac{25 \text{ cm}}{m}$$

Substituting $p = (25 \text{ cm})/m$ and $q = -25$ cm in the lens equation yields

$$\frac{1}{f} = \frac{1}{p} + \frac{1}{q} = \frac{m}{25 \text{ cm}} - \frac{1}{25 \text{ cm}} = \frac{m-1}{25 \text{ cm}}$$

$$m = \frac{25 \text{ cm}}{f} - 1$$

(b)
$$m = \frac{25 \text{ cm}}{5 \text{ cm}} - 1 = 5 - 1 = 4$$

32.17. A microscope has an objective of focal length 6 mm and an eyepiece of focal length 25 mm.   If the image distance of the objective is 160 mm and that of the eyepiece is 250 mm (both of which are typical figures), find the magnification produced by (a) the objective, (b) the eyepiece, (c) the entire microscope.

(a)   The object distance of the objective is

$$p = \frac{qf}{q-f} = \frac{160 \text{ mm} \times 6 \text{ mm}}{160 \text{ mm} - 6 \text{ mm}} = 6.23 \text{ mm}$$

and so its magnification is

$$m_1 = -\frac{q}{p} = \frac{-160 \text{ mm}}{6.23 \text{ mm}} = -25.7$$

The minus sign means an inverted image.

(b) From the result of Problem 32.16,

$$m_2 = \frac{250 \text{ mm}}{f} - 1 = \frac{250 \text{ mm}}{25 \text{ mm}} - 1 = 9$$

Alternatively the same figure can be obtained by the procedure of part (a).

(c) The total magnification is the product of m_1 and m_2:

$$m = m_1 m_2 = -25.7 \times 9 = -231$$

32.18. The *angular magnification* of a telescope when used to view a distant object is

$$m_{\text{ang}} = \frac{f_o}{f_e} = \frac{\text{focal length of objective}}{\text{focal length of eyepiece}}$$

(a) The objective of a telescope has a focal length of 4 ft. What should the focal length of the eyepiece be in order to produce an angular magnification of 40? (b) The telescope is used to examine a boat 60 ft long which is 2000 ft away. If the image distance of the telescope eyepiece is 10 in., what is the apparent length of the boat?

(a)

$$f_e = \frac{f_0}{m_{\text{ang}}} = \frac{48 \text{ in.}}{40} = 1.2 \text{ in.}$$

(b) The angle subtended by the boat from the location of the telescope is

$$\theta = \frac{\text{object length}}{\text{object distance}} = \frac{60 \text{ ft}}{2000 \text{ ft}} = 0.03 \text{ radian}$$

If L is the length of the boat's image as seen through the telescope, the angle this image subtends is

$$\theta' = \frac{\text{image length}}{\text{image distance}} = \frac{L}{10 \text{ in.}}$$

Since the angular magnification of the telescope is 40 and $m_{\text{ang}} = \theta'/\theta$, $\theta' = m_{\text{ang}}\theta$ and

$$\frac{L}{10 \text{ in.}} = 40 \times 0.03 \text{ radian} = 1.2 \text{ radians}$$

$$L = 1.2 \text{ radians} \times 10 \text{ in.} = 12 \text{ in.}$$

The boat seems to be 12 in. long and to be located 10 in. from the viewer's eye.

Supplementary Problems

32.19. Can a diverging lens ever form an inverted image of a real object? Can a converging lens?

32.20. If the screen is moved closer to a movie projector, how should the projector's lens be moved to restore the image to a sharp focus?

32.21. A double-convex lens has surfaces whose radii are both 50 cm. The index of the refraction of the glass is 1.52. Find the focal length of the lens.

32.22. A meniscus lens has a concave surface of radius 30 cm and a convex surface of radius 25 cm. The index of refraction of the glass is 1.50. (a) Find the focal length of the lens. (b) Is it a converging or a diverging lens?

32.23. A *plano-concave lens* has one plane surface and one concave surface. If a plano-concave lens of focal length −10 in. is to be ground from optical glass of index of refraction 1.50, find the radius of curvature of the concave surface.

32.24. Glycerin has an index of refraction of 1.47.   Find the focal length in glycerin of a lens made of flint glass ($n = 1.63$) whose focal length in air is $+10$ cm.

32.25. Find the index of refraction of the glass used in a plano-convex lens of focal length 12 in. whose convex surface has a radius of 7 in.

32.26. A button 1 cm in diameter is held 10 cm from a converging lens whose focal length is $+25$ cm.   Find the location, size, and nature of the image.

32.27. The button of Problem 32.26 is held 25 cm from the same lens.   Find the location, size, and nature of the image.

32.28. The button of Problem 32.26 is held 40 cm from the same lens.   Find the location, size, and nature of the image.

32.29. The button of Problem 32.26 is held 50 cm from the same lens.   Find the location, size, and nature of the image.

32.30. The button of Problem 32.26 is held 100 cm from the same lens.   Find the location, size, and nature of the image.

32.31. A diverging lens with a focal length of -1 m is used to examine a man 1.8 m tall who is standing 4 m away.   How tall does the man appear to be?

32.32. A magnifying glass of focal length 3 in. is held 1 in. from a postage stamp.   What magnification does it produce?   What is the nature of the image?

32.33. A converging lens has a focal length of 10 cm.   (a) How far from an object should it be held to produce an erect, virtual image magnified three times?   (b) What is the image distance?

32.34. A slide projector has a lens whose focal length is 8 in.   How far from the lens should the screen be located if a 2 in. $\times$ 2 in. slide is to appear 40 in. $\times$ 40 in.?

32.35. What is the focal length of a combination of two thin lenses in contact both of whose focal lengths are $+8$ in.?

32.36. A microscope has an objective of focal length 4 mm which is used with a $15\times$ eyepiece.   If the image distance for the objective is 160 mm, find the magnification of the instrument.

32.37. A telescope has an objective lens whose focal length is 120 cm and an eyepiece whose focal length is 4 cm. (a) What is the angular magnification of this telescope?   (b) The moon has an angular diameter of about $0.5°$ as seen from the earth.   What is its angular diameter through the telescope?   (c) If the image distance is 25 cm, what is the apparent diameter of the moon as seen through the telescope?

Answers to Supplementary Problems

32.19. no; yes

32.20. The lens should be moved farther away from the film.

32.21. 48 cm

32.22. (a) 300 cm   (b) converging

32.23. -5 in.

32.24. $+58$ cm

32.25. 1.58

32.26. The image is 16.7 cm in front of the lens, 1.67 cm in diameter, virtual and erect.

32.27. No image is formed.

32.28. The image is 66.7 cm behind the lens, 1.67 cm in diameter, real and inverted.

32.29. The image is 50 cm behind the lens, 1 cm in diameter, real and inverted.

32.30. The image is 33.3 cm behind the lens, 3.3 mm in diameter, real and inverted.

32.31. 36 cm

32.32. 1.5; erect and virtual

32.33. (*a*) 6.67 cm   (*b*) 20 cm

32.34. 14 ft

32.35. $+4$ in.

32.36. 585

32.37. (*a*) 30   (*b*) 15°   (*c*) 6.54 cm

Chapter 33

Physical and Quantum Optics

INTERFERENCE

In examining the reflection and refraction of light, it is sufficient to consider light as though it consists of rays that travel in straight lines in a uniform medium. Other phenomena, notably interference, diffraction, and polarization, can only be understood in terms of the wave nature of light, and the study of these phenomena is called *physical optics*.

Interference occurs when waves of the same nature from different sources meet at the same place. In *constructive interference* the waves are in phase ("in step") and reinforce each other; in *destructive* interference the waves are out of phase and partially or completely cancel each other (Fig. 33-1). All types of waves exhibit interference under appropriate circumstances. Thus water waves interfere to produce the irregular surface of the sea, sound waves close together in frequency interfere to produce beats, and light waves interfere to produce the fringes seen around the images formed by optical instruments and the bright colors of soap bubbles and thin films of oil on water.

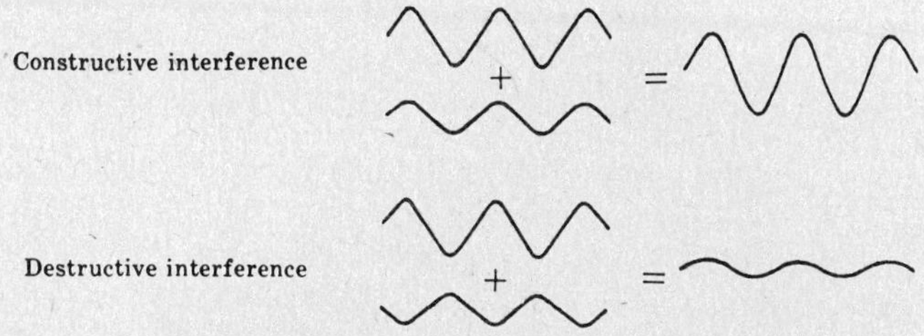

Fig. 33-1

DIFFRACTION

The ability of a wave to bend around the edge of an obstacle is called *diffraction*. Owing to the combined effects of diffraction and interference, the image of a point source of light is always a small disk with bright and dark fringes around it. The smaller the lens or mirror used to form the image, the larger the disk. The angular width in radians of the image disk of a point source is about

$$\theta_0 = 1.22\,\frac{\lambda}{D}$$

where λ is the wavelength of the light and D is the lens or mirror diameter. The images of objects closer together than θ_0 will overlap, and hence cannot be resolved no matter how great the magnification produced by the lens or mirror. In the case of a telescope or microscope, D refers to the diameter of the objective lens. If two objects d_0 apart that can just be resolved are the distance L from the observer, the angle in radians between them is $\theta_0 = d_0/L$, so the above formula can be rewritten in the form

$$\text{Resolving power} = d_0 = 1.22\,\frac{\lambda L}{D}$$

POLARIZATION

A *polarized* beam of light is one in which the electric fields of the waves are all in the same direction. If the electric fields are in random directions (though, of course, always perpendicular to the direction of propagation), the beam is *unpolarized*. Various substances affect differently light with different directions of polarization, and these substances can be used to prepare devices that permit only light polarized in a certain direction to pass through them.

QUANTUM THEORY OF LIGHT

Certain features of the behavior of light can be explained only on the basis that light consists of individual *quanta* or *photons*. The energy of a photon of light whose frequency is f is

$$\text{Quantum energy} = E = hf$$

where h is *Planck's constant*:

$$\text{Planck's constant} = h = 6.63 \times 10^{-34} \text{ joule-second}$$

A photon has most of the properties associated with particles—it is localized in space and possesses energy and momentum—but it has no mass. Photons travel with the velocity of light.

The electromagnetic and quantum theories of light complement each other: under some circumstances light exhibits a wave character, under other circumstances it exhibits a particle character. Both are aspects of the same basic phenomenon.

X-RAYS

X-rays are high-frequency electromagnetic waves produced when fast electrons impinge on a target. If the electrons are accelerated through a potential difference of V, each electron has the energy $\text{KE} = eV$. If all of this energy goes into creating an X-ray photon,

$$eV = hf$$

$$\text{Electron kinetic energy} = \text{X-ray photon energy}$$

and the frequency of the X-rays is $f = eV/h$.

THE ELECTRON VOLT

A common energy unit in atomic and quantum physics is the *electron volt* (eV), which is defined as the energy an electron gains when it moves through a potential difference of 1 volt. Hence

$$1 \text{ eV} = 1.60 \times 10^{-19} \text{ J}$$

Multiples of the eV are the *keV*, *MeV*, and *GeV*, where

$$1 \text{ keV} = 10^3 \text{ eV} \qquad 1 \text{ MeV} = 10^6 \text{ eV} \qquad 1 \text{ GeV} = 10^9 \text{ eV}$$

Solved Problems

33.1 When is it appropriate to think of light as consisting of waves and when as consisting of rays?

When paths or path differences are involved whose lengths are comparable with the wavelengths found in light, the wave nature of light is significant and must be taken into account. Thus diffraction and interference can be understood only on a wave basis. When paths are involved that are many wavelengths long and neither diffraction nor interference occurs, as in reflection and refraction, it is more convenient to consider light as consisting of rays.

33.2. When two light beams of the same wavelength interfere, the result is a pattern of bright and dark lines. What becomes of the energy of the light waves whose destructive interference leads to the dark lines?

The missing energy is found in the bright lines, whose brightness is greater than the simple addition of the two light beams would produce in the absence of interference.   The total energy remains the same.

33.3.   When two trains of waves meet on the surface of a body of water, the resulting interference pattern is obvious.   However, when the light beams from two flashlights overlap on a screen, there is no evidence of an interference pattern.   Why not?   Is there any way in which the interference of light can be demonstrated?

There are two reasons why such an experiment does not yield a conspicuous interference pattern. First, the wavelengths found in light are so short that such a pattern would be on an extremely small scale. Second and more important, all sources of light (except lasers) emit light waves as short trains of random phase and not as continuous trains.   The interference that occurs between light beams from two independent sources is therefore averaged out during all but the briefest of observation times and cannot be seen by eye or recorded on photographic film.   Such light sources are said to be *incoherent*.

To exhibit an interference pattern in light, sources must be used whose waves have fixed phase relationships during the observation period.   The waves from one source can be in step with those from the other when they are produced, or out of step, or something in between, but the essential thing is that the relationship be constant.   Such sources are *coherent*.   Three ways to construct coherent sources are:

1.   Pass light from a single source (such as an illuminated slit or a narrow filament) through two or more other slits.   The waves that emerge from the latter slits are necessarily coordinated and can interfere to produce a visible pattern.

2.   Combine a direct light beam from a source with an indirect beam from the same source produced by refraction or reflection.   This is how the interference patterns produced by thin oil films floating on water are caused.

3.   Coordinate the radiating atoms in each individual source so that the radiating atoms always emit wave trains in step with one another.   This is done in the laser.

33.4.   What is a *diffraction grating*?

A diffraction grating consists of a series of closely spaced parallel slits that diffract light passing through them.   The diffracted wave trains of a particular wavelength interfere constructively in certain directions only (Fig. 33-2).   When light is directed at a grating, each wavelength present undergoes constructive interference in different directions from those of other wavelengths, and the result is a series of spectra.   A grating is better able than a prism to separate nearby wavelengths, and as a result gratings rather than prisms are used in nearly all spectrographs.

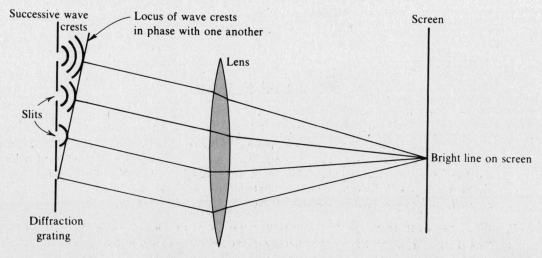

Fig. 33-2

33.5. A pair of "7×50" binoculars has a magnification of 7 and objective lens diameters of 50 mm. Find the length of the smallest detail that can possibly be resolved by such binoculars when examining something 1 km away. Consider the wavelength of the light to be 5×10^{-7} m, which is near the middle of the visible spectrum and corresponds to green.

Since $D = 50$ mm $= 5 \times 10^{-2}$ m and $L = 1$ km $= 10^3$ m,

$$d_0 = 1.22 \frac{\lambda L}{D} = \frac{1.22 \times 5 \times 10^{-7} \text{ m} \times 10^3 \text{ m}}{5 \times 10^{-2} \text{ m}} = 1.22 \times 10^{-2} \text{ m} = 1.22 \text{ cm}$$

33.6. A radar operating at a wavelength of 3 cm is to have a resolving power of 100 ft at a range of 1 mi. Find the minimum width its antenna must have.

The width of a radar antenna corresponds to the diameter of the objective lens of an optical system. Here $L = 1$ mi $= 5280$ ft and

$$\lambda = \frac{3 \text{ cm}}{2.54 \text{ cm/in} \times 12 \text{ in/ft}} = 0.098 \text{ ft}$$

Hence

$$D = 1.22 \frac{\lambda L}{d_0} = \frac{1.22 \times 0.098 \text{ ft} \times 5280 \text{ ft}}{100 \text{ ft}} = 6.34 \text{ ft} = 6 \text{ ft } 4 \text{ in.}$$

33.7. The human eye can respond to as few as three photons of light. If the light is yellow ($f = 5 \times 10^{14}$ Hz), how much energy does this represent?

The energy of each photon is

$$E = hf = 6.63 \times 10^{-34} \text{ J-s} \times 5 \times 10^{14} \text{ Hz} = 3.3 \times 10^{-19} \text{ J}$$

The total energy is $3E = 10^{-18}$ J.

33.8. The average wavelength of the light emitted by a certain 100-W light bulb is 5.5×10^{-7} m. How many photons per second does the light bulb emit?

The frequency of the light is

$$f = \frac{c}{\lambda} = \frac{3 \times 10^8 \text{ m/s}}{5.5 \times 10^{-7} \text{ m}} = 5.5 \times 10^{14} \text{ Hz}$$

and the energy of each photon is

$$E = hf = 6.63 \times 10^{-34} \text{ J-s} \times 5.5 \times 10^{14} \text{ Hz} = 3.6 \times 10^{-19} \text{ J}$$

Since 100 W $= 100$ J/s, the number of photons emitted per second is

$$\frac{100 \text{ J/s}}{3.6 \times 10^{-19} \text{ J/photon}} = 2.8 \times 10^{20} \text{ photons/s}$$

33.9. In the *photoelectric effect*, light directed at the surfaces of certain metals causes electrons to be emitted. In the case of potassium, 2 eV of work must be done to remove an electron from the surface. (*a*) If light of wavelength 5×10^{-7} m falls on a potassium surface, what is the maximum energy of the photoelectrons that emerge? (*b*) If light of wavelength 4×10^{-7} m falls on the same surface, will the photoelectrons have more or less energy?

(*a*) Since $c = f\lambda$, the frequency of the light is

$$f = \frac{c}{\lambda} = \frac{3 \times 10^8 \text{ m/s}}{5 \times 10^{-7} \text{ m}} = 6 \times 10^{14} \text{ Hz}$$

The energy of each photon is therefore

$$E = hf = 6.63 \times 10^{-34} \text{ J-s} \times 6 \times 10^{14} \text{ Hz} = 3.98 \times 10^{-19} \text{ J}$$

Since 1 eV $= 1.6 \times 10^{-19}$ J,

$$E = \frac{3.98 \times 10^{-19} \text{ J}}{1.6 \times 10^{-19} \text{ J/eV}} = 2.49 \text{ eV}$$

This is the maximum energy that can be given to an electron by a photon of this light. Because 2 eV is needed to remove an electron, the maximum energy of the photoelectrons in this situation is 0.49 eV.

(b) A shorter wavelength means a higher frequency and hence more energy to be imparted to the photoelectrons.

33.10. In a certain televeision picture tube, electrons are accelerated through a potential difference of 10,000 V. Find the frequency of the X-rays that are emitted when these electrons strike the screen.

Since $hf = eV$, here we have

$$f = \frac{eV}{h} = \frac{1.6 \times 10^{-19} \text{ C} \times 10^4 \text{ V}}{6.63 \times 10^{-34} \text{ J-s}} = 2.4 \times 10^{18} \text{ Hz}$$

33.11. An X-ray tube emits X-rays whose wavelength is 2×10^{-11} m. What is the operating voltage of the tube?

The frequency of the X-rays is

$$f = \frac{c}{\lambda} = \frac{3 \times 10^8 \text{ m/s}}{2 \times 10^{-11} \text{ m}} = 1.5 \times 10^{19} \text{ Hz}$$

Since $hf = eV$,

$$V = \frac{hf}{e} = \frac{6.63 \times 10^{-34} \text{ J-s} \times 1.5 \times 10^{19} \text{ Hz}}{1.6 \times 10^{-19} \text{ C}} = 62,000 \text{ V}$$

33.12. What is the kinetic energy in eV of an electron whose velocity is 10^7 m/s?

$$\text{KE} = \frac{1}{2} mv^2 = \frac{1}{2} \times 9.1 \times 10^{-31} \text{ kg} \times (10^7 \text{ m/s})^2 = 4.55 \times 10^{-17} \text{ J}$$

Since 1 eV $= 1.6 \times 10^{-19}$ J,

$$\text{KE} = \frac{4.55 \times 10^{-17} \text{ J}}{1.6 \times 10^{-19} \text{ J/eV}} = 284 \text{ eV}$$

33.13. A proton ($m = 1.67 \times 10^{-27}$ kg) is accelerated through a potential difference of 200 V. (a) What is its kinetic energy in eV? (b) What is its velocity?

(a) Since the charge on the proton is $+e$, its kinetic energy is 200 eV.

(b) $$\text{KE} = 200 \text{ eV} \times 1.6 \times 10^{-19} \text{ J/eV} = 3.2 \times 10^{-17} \text{ J}$$

Since KE $= \frac{1}{2} mv^2$,

$$v = \sqrt{\frac{2\text{KE}}{m}} = \sqrt{\frac{2 \times 3.2 \times 10^{-17} \text{ J}}{1.67 \times 10^{-27} \text{ kg}}} = 2 \times 10^5 \text{ m/s}$$

Supplementary Problems

33.14. Which of the following phenomena occur only in transverse waves (such as light) and not in longitudinal waves (such as sound)? Reflection, refraction, interference, diffraction, polarization.

33.15. Which of the following optical phenomena, if any, are independent of the wavelength of the light involved? Interference, diffraction, resolving power, polarization.

33.16. Why does the electric field of an electromagnetic wave determine its direction of polarization rather than its magnetic field?

33.17. State two advantages in having the objective lens or mirror of a telescope be of large diameter.

33.18. Why do radio waves readily diffract around buildings whereas light waves, which are also electromagnetic in nature, do not?

33.19. The energy of a light beam is carried by separate photons, yet we do not perceive light as a series of tiny flashes. Why not?

33.20. If Planck's constant were equal to 6.63 J-s instead of 6.63×10^{-34} J-s, would quantum phenomena be more or less conspicuous in everyday life than they are now?

33.21. When light is directed at a metal surface, upon what property of the light does the maximum energy of the emitted electrons depend?

33.22. A radar whose operating frequency is 9000 mHz has an antenna 1.5 m wide. What is its resolving power at a range of 5 km?

33.23. A telescope has an objective lens 10 cm in diameter. What is the maximum distance at which the telescope can resolve two objects 1 cm apart? Assume that $\lambda = 5 \times 10^{-7}$ m.

33.24. Light from the sun arrives at the earth at the rate of about 1400 W/m^2 of area perpendicular to the direction of the light. Assuming that sunlight consists exclusively of light of wavelength 6×10^{-7} m, find the number of photons per second that fall on each m^2 of the earth's surface directly facing the sun.

33.25. How many photons per second are emitted by a 50-kW radio transmitter that operates at a frequency of 1200 kHz?

33.26. What is the operating voltage of an X-ray tube that produces X-rays of frequency 10^{19} Hz?

33.27. Find the wavelength of the X-rays produced by a 50,000-V X-ray machine.

33.28. The work needed to remove an electron from the surface of sodium is 2.3 eV. Find the maximum wavelength of light that will cause photoelectrons to be emitted from sodium. (See Problem 33.9.)

33.29. Photoelectrons are emitted by a copper surface only when light whose frequency is 1.1×10^{15} Hz or more is directed at it. What is the maximum energy of the photoelectrons when light of frequency 1.5×10^{15} Hz is directed at the surface?

33.30. What is the kinetic energy in eV of a proton ($m = 1.67 \times 10^{-27}$ kg) whose velocity is 5×10^6 m/s?

33.31. Find the velocity of a 50-eV electron.

Answers to Supplementary Problems

33.14. polarization

33.15. polarization

33.16. The interaction between the electric field of an electromagnetic wave and the matter it passes through is responsible for nearly all optical effects, hence it is this field which is used to specify the direction of polarization.

33.17. (*a*) The larger the diameter, the greater the ability to resolve objects close together.   (*b*) A large diameter means that more light reaches the eye from a given object, hence it can be seen even if poorly illuminated.

33.18. The wavelengths of visible light are very short relative to the size of a building, so their diffraction is imperceptible.   The wavelengths of radio waves are more nearly comparable with the size of a building.

33.19. Even a weak light involves many photons per second.   Visual responses persist for a short time, so successive photons give the impression of a continuous transfer of energy.

33.20. more conspicuous	**33.24.** 4.2×10^{21} photons/m²-s	**33.28.** 5.4×10^{-7} m
33.21. frequency	**33.25.** 6.3×10^{31} photons/s	**33.29.** 1.7 eV
33.22. 136 m	**33.26.** 41,400 V	**33.30.** 1.3×10^{5} eV = 0.13 MeV
33.23. 1.64 km	**33.27.** 2.5×10^{-11} m	**33.31.** 4.2×10^{6} m/s

Chapter 34

Atomic Physics

ATOMIC STRUCTURE

An atom consists of a small, central nucleus composed of protons and neutrons that is surrounded by electrons. All the atoms of an element of atomic number Z contain Z protons and Z electrons. Although the actual situation is more complicated, the electrons in an atom may be thought of as moving in orbits around the nucleus. An atomic electron can occupy only certain specific orbits, each of which has a particular energy associated with it. The possible electron orbits of an atom fall into groups called *shells*; the electrons in a shell all have similar energies and are similar distances from the nucleus. The innermost shell of an atom can contain at most 2 electrons; the next shell can contain at most 8 electrons; the third shell can contain at most 18 electrons; and so on. In an atom in its normal, or *ground*, state, the electrons all occupy orbits of lowest energy subject to the limitation on the number possible in each shell. When one of the outer electrons of an atom is temporarily in an orbit of higher energy than usual, the atom is said to be in an *excited* state. The *ionization energy* of an atom is the energy required to detach one of its outer electrons.

ATOMIC SPECTRA

When a gas or vapor is excited by the passage of an electric current, light is given off which consists of certain specific wavelengths only. Every element has a characteristic *emission line spectrum*. The wavelengths in this spectrum fall into definite series whose member wavelengths are related by simple formulas.

When white light is passed through a cool gas or vapor, light of certain specific wavelengths is absorbed. The wavelengths in the resulting *absorption line spectrum* correspond to a number of the wavelengths in the emission spectrum of that element.

Line spectra owe their origin to the presence of energy levels in atoms. An atom in an excited state can remain there only a brief time (normally about 10^{-8} s) before dropping to a lower state. The difference in energy appears as a photon of frequency f, where

$$E_{\text{initial}} - E_{\text{final}} = hf$$

An absorption spectrum is produced by transitions in the opposite direction, from the ground state to excited states. Light of frequencies that correspond to the various energy differences is absorbed by atoms illuminated by light whose spectrum is continuous (that is, which contains all frequencies). These atoms then reradiate light as they fall to their ground states, but the reradiation occurs in random directions and so is much fainter in the direction of the original beam.

CHEMICAL BONDS

When a compound is formed, atoms of the elements present are linked together by *chemical bonds*. It is customary to classify chemical bonds as being *ionic* or *covalent*, although actual bonds are often intermediate between the two extremes. In an ionic bond, one or more electrons from one atom are transferred to another atom, and the resulting positive and negative ions then attract each other. In a covalent bond, one or more pairs of electrons are shared by two adjacent atoms. As these electrons move about, they spend more time between the atoms than elsewhere, which results in an attractive electrical force that holds the atoms together.

A molecule is a group of atoms that are held together tightly enough by covalent bonds to behave as a single particle. A molecule always has a definite composition and structure, and has little tendency to gain or lose atoms. Ionic bonds usually result in crystalline solids, not in molecules; such solids consist of aggregates of positive and negative ions in a stable arrangement characteristic of the compound involved. Some crystalline solids are covalent rather than ionic, as discussed below.

CRYSTALS

Most solids are crystalline, with the ions, atoms, or molecules of which they consist being arranged in a regular pattern. Four kinds of bonds are found in crystals: ionic, covalent, metallic, and van der Waals.

A crystal of ordinary salt, NaCl, is an example of an ionic solid, with Na^+ and Cl^- ions in alternate positions in a simple lattice (Fig. 34-1).

An example of a covalent solid is diamond, each of whose carbon atoms is joined by covalent bonds to four other carbon atoms in a structure that is repeated throughout the crystal (Fig. 34-2). Both ionic and covalent solids are hard and have high melting points, which are reflections of the strength of the bonds. Ionic solids are much more common than covalent ones.

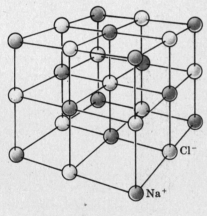

Fig. 34-1

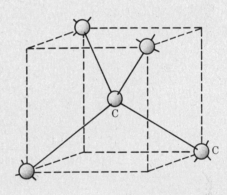

Fig. 34-2

In a metal, the outermost electrons of each atom are shared by the entire assembly, so that a "gas" or "sea" of electrons moves relatively freely throughout it. The interaction between this electron sea and the positive metal ions leads to a cohesive force, much as in the case of the shared electrons in a covalent bond but on a larger scale. The presence of the free electrons accounts for such typical properties of metals as their opacity, surface luster, and high electrical and heat conductivities.

All molecules, and even inert-gas atoms such as those of helium, exhibit weak, short-range attractions for one another due to *van der Waals forces*. These forces are responsible for the condensation of gases into liquids and the freezing of liquids into solids even in the absence of ionic, covalent, or metallic bonds between the atoms or molecules involved. Such familiar aspects of the behavior of matter as friction, viscosity, and adhesion are due to van der Waals forces. Van der Waals forces arise from the lack of symmetry in the momentary distributions of the electrons in a molecule. When two molecules are close together, these momentary charge asymmetries tend to shift together, with the positive part of one molecule always near the negative part of the other even though the locations of these parts are always changing. Van der Waals forces are quite weak, and substances composed of whole molecules, such as water, usually have low melting and boiling points and little mechanical strength in the solid state.

ENERGY BANDS

The atoms in almost all crystalline solids, whether metals or not, are so close together that their outer electrons constitute a single system of electrons common to the entire crystal. In place of each precisely defined characteristic energy level of an individual atom, the entire crystal possesses an *allowed energy band* which spans a range of possible energies. The allowed energy bands in a solid thus correspond to the energy levels in an atom, and an electron in a solid can have only those energies that fall within these energy bands. If adjacent allowed energy bands do not overlap, the intervals between them represent energies which their electrons cannot have. Such intervals are called *forbidden bands*. The electrical behavior of a crystalline solid is determined both by its energy-band structure and by how these bands are normally filled with electrons.

Solved Problems

34.1. Describe two mechanisms by which the atoms of a gas can be excited so that they emit light whose frequencies make up the characteristic line spectrum of the element involved.

(a) One mechanism is a collision with another atom, as a result of which some of their kinetic energy becomes excitation energy within one or both of the atoms. The excited atoms then lose this energy by emitting one or more photons. In an electric discharge in a gas, for instance in a neon sign or a sodium-vapor highway lamp, an electric field accelerates electrons and ions to velocities sufficient for atomic excitation.

(b) Another mechanism is the absorption by an atom of a photon of light for which hf is just right to raise the atom from its ground state to one of its excited states. If white light is directed at the gas, photons of those energies that correspond to such transitions will be absorbed. When the energy is reradiated, all of the possible transitions from the highest excited state reached will show up in the emitted light.

34.2. What is a *laser*?

A laser is a device that produces an intense beam of monochromatic, coherent light from the cooperative radiation of excited atoms. The light waves in a coherent beam are all in phase with one another, which greatly increases their effectiveness. A laser beam is virtually nondivergent and hence remains as a narrow pencil of light even after traveling a large distance.

The word "laser" stands for light amplification by stimulated emission of radiation. In a laser, atoms of a particular kind are raised to *metastable* (temporarily stable) states of energy hf, which are excited states of relatively long lifetimes. Radiation of frequency f induces the excited atoms to emit photons of the same frequency and thereby to return to their ground (normal) states, so that a small amount of initial radiation can be greatly amplified.

34.3. Atoms absorb or radiate photons when the energies of their electrons change. What is the origin of molecular spectra?

Molecular energy states arise from the rotation of a molecule as a whole and from the vibration of its constituent atoms relative to one another as well as from changes in its electron energies. Rotational energy states are separated by quite small energy intervals (0.001 eV is typical), and the spectra that arise from transitions between these states are in the microwave region with wavelengths of 0.1 mm to 1 cm. Vibrational states are separated by somewhat larger energy intervals (0.1 eV is typical), and vibrational spectra are in the infrared region with wavelengths of 0.001–0.1 mm. Molecular electronic states have high energies, with typical separations between the energy levels of outer electrons of several eV and spectra in the visible and ultraviolet regions. Molecular spectra can be used not only to identify particular molecules in a sample of unknown composition but also to obtain a detailed picture of the structure of a molecule.

34.4. Are all solids crystalline?

No. The atoms, ions, or molecules of which a crystalline solid is composed fall into regular, repeated patterns. The presence of such long-range order is the defining property of crystals. Other solids lack long-range order in their structures, and may be regarded as supercooled liquids whose stiffness is due to exceptionally high viscosity. Glass, pitch, and many plastics are examples of such *amorphous* ("without form") solids.

34.5. How are the crystal structures of solids usually determined?

The structure of a crystal is usually determined by means of the interference patterns produced when an X-ray beam passes through it. A crystal consists of a regular array of atoms, each of which is able to scatter an electromagnetic wave that happens to strike it. A beam of X-rays, all of the same wavelength, that falls upon a crystal will be scattered in all directions within it, but, owing to the regular arrangement of the atoms, in certain directions the scattered waves will constructively interfere with one another while in others they will destructively interfere. The phenomenon is known as *X-ray diffraction*. The resulting pattern of high and low X-ray intensities can be analyzed to yield the arrangement in space of the scattering centers, which are the atoms of the crystal.

34.6. The upper energy band of a metal is only partly filled with electrons; that is, the band does not contain the maximum number of electrons that can have energies in its range. How does this fact account for the ability of metals to conduct electric current?

When an electric field is established in a metal, electrons easily acquire additional energy while remaining in their original energy band. The additional energy is in the form of kinetic energy, and the moving electrons constitute an electric current.

34.7. The upper energy band of an insulator is completely filled with electrons, and is separated from the next higher energy band by a forbidden band several eV wide. How does this fact account for the inability of insulators to conduct electric current?

An electron in an insulator must acquire at least as much energy as the width of the forbidden band if it is to have the kinetic energy required to move through the crystal. An energy increment of several eV cannot be readily given to an electron in a solid by an electric field because of the presence of so many other electrons and nuclei, so insulators are very poor conductors of electric current.

34.8. What is the energy-band structure of a semiconductor?

In semiconductors, a very narrow forbidden band separates a filled upper energy band from the next empty one, and some electrons have enough kinetic energy of thermal origin to jump this gap. Such a substance can conduct electric current to a limited extent. Some semiconductors contain small amounts of impurities which provide energy levels in the forbidden band that reduce the width of the energy gap electrons must overcome in order to move freely and thus constitute an electric current.

34.9. Why are metals opaque to visible light, whereas insulators are transparent when in the form of regular crystals?

Photons of visible light have energies of between about 1 and 3 eV. Such amounts of energy are readily absorbed by a "free electron" in a metal, since its allowed energy band is only partly filled, and metals are accordingly opaque. The electrons in an insulator, on the other hand, need more than 3 eV of energy to jump across the forbidden band to the next allowed band. Insulators therefore cannot absorb photons of visible light and so are transparent. Of course, most samples of insulating materials do not appear transparent, but that is because of such other factors as the scattering of light by irregularities in their structures or an amorphous character.

Supplementary Problems

34.10. What relationship would you expect between the chemical activity of a metal and its ionization energy?

34.11. What is the nature of the spectrum found in (*a*) light from the hot filament of a light bulb; (*b*) light from a neon sign; and (*c*) light originating as in (*a*) that has passed through cool neon gas?

34.12. How are the compositions of the sun and stars determined?

34.13. Why does the spectrum of hydrogen consist of many lines even though a hydrogen atom has only a single electron?

34.14. You are given two solids of almost identical appearance, one of which is held together by ionic bonds and the other by van der Waals bonds.   How could you tell them apart?

34.15. Why can metals be deformed with relative ease whereas covalent and ionic solids are quite brittle?

Answers to Supplementary Problems

34.10. The lower the ionization energy of a metal, the more easily one of its electrons can be detached in a chemical reaction and hence the more active it is.   Thus potassium, whose ionization energy is 4.3 eV, is more active chemically than zinc, whose ionization energy is 9.4 eV.

34.11. (a) continuous emission spectrum as described in Chapter 22; (b) emission line spectrum; (c) absorption line spectrum

34.12. The presence of the spectral lines of a particular element in the spectrum of the sun or a star means that this element must be present there.

34.13. A hydrogen sample contains a great many atoms, each of which can undergo a variety of transitions between energy levels.

34.14. The van der Waals solid will be softer and will melt at a much lower temperature.

34.15. Atoms in a metal can be readily rearranged in position because the bonding occurs by means of a sea of freely moving electrons.   In a covalent crystal the bonds are localized between adjacent atoms and must be ruptured to deform the crystal.   In an ionic crystal the bonding process requires a configuration of alternate positive and negative ions whose relative positions cannot be altered without breaking the crystal apart.

Chapter 35

Nuclear Physics

NUCLEAR STRUCTURE

The nucleus of an atom is composed of protons and neutrons whose masses are respectively

$$m_p = 1.673 \times 10^{-27} \text{ kg} = 1.007277 \text{ u}$$

$$m_n = 1.675 \times 10^{-27} \text{ kg} = 1.008665 \text{ u}$$

The proton has a charge of $+e$ and the neutron is uncharged. The *atomic number* of an element is the number of protons in the nucleus of one of its atoms. Protons and neutrons are jointly called *nucleons*.

Although all the atoms of an element have the same number of protons in their nuclei, the number of neutrons may be different. Each variety of nucleus found in a given element is called an *isotope* of the element. Symbols for isotopes follow the pattern

$$^A_Z X$$

where X = chemical symbol of element
Z = atomic number of element = number of protons in nucleus
A = mass number of isotope = number of protons + neutrons in nucleus

BINDING ENERGY

The mass of an atom is always less than the sum of the masses of the neutrons, protons, and electrons of which it is composed. The energy equivalent of the missing mass is called the *binding energy* of the nucleus; the greater its binding energy, the more stable the nucleus. The mass defect Δm of a nucleus with Z protons and N neutrons may be found from its atomic mass m by using the formula

$$\Delta m = (Zm_H + Nm_n) - m$$

where m_H, the mass of the hydrogen atom (which consists of a proton and an electron), is

$$m_H = 1.007825 \text{ u}$$

To find the binding energy in MeV, the usual unit, Δm can be multiplied by the conversion factor 931 MeV/u.

FUNDAMENTAL FORCES

The force between nucleons that holds an atomic nucleus together despite the repulsive electrical forces its protons exert on each other is the result of what is known as the *strong interaction*. This is a fundamental interaction in the same sense as the gravitational and electromagnetic interactions are: none can be explained in terms of any of the others. The strong interaction has only a very short range, unlike the gravitational and electromagnetic interactions, and is only effective within nuclei.

There is another interaction involving nuclei called the *weak interaction* which is responsible for beta decays. Recent evidence indicates that the weak interaction may be electromagnetic in origin and not a fundamental interaction as had hitherto been believed.

NUCLEAR REACTIONS

Nuclei can be transformed into others of a different kind by interaction with each other. Since nuclei are all positively charged, a high-energy collision is necessary between two nuclei if they are to get close enough together to react. Because it has no charge, a neutron can initiate a nuclear reaction even if it is moving slowly. In any nuclear reaction, the total number of neutrons and the total number of protons in the products must be equal to the corresponding total numbers in the reactants.

FISSION AND FUSION

Nuclei of intermediate size have the highest binding energies per nucleon and therefore are more stable than lighter and heavier nuclei. If a heavy nucleus is split into two smaller ones, the greater binding energy of the latter means that energy will be liberated. This process is called *nuclear fission*. Certain very large nuclei, such as $^{235}_{92}U$, undergo fission when they absorb a neutron; since the products of the fission include several neutrons as well as two daughter nuclei, a *chain reaction* can be established in an assembly of a suitable fissionable isotope. If uncontrolled, the result is an atomic bomb; if controlled so that the rate at which fission events occur is constant, the result is a nuclear reactor that can serve as an energy source for generating electricity or for ship propulsion.

In *nuclear fusion*, two light nuclei combine to form a heavier one whose binding energy per nucleon is greater. The difference in binding energies is liberated in the process. To bring about a fusion reaction, the initial nuclei must be moving rapidly when they collide to overcome their electrical repulsion. Nuclear fusion is the source of energy in the sun and stars, where the high temperatures in the interiors mean that nuclei there have sufficiently high velocities and the high pressures mean that nuclear collisions occur frequently. In the operation of a hydrogen bomb, a fission bomb is first detonated to produce the high temperature and pressure necessary for fusion reactions to occur. The problem in constructing a fusion reactor for controlled energy production is to contain a sufficiently hot and dense mixture of suitable isotopes for long enough to yield a net energy output.

RADIOACTIVITY

Certain nuclei are unstable and undergo *radioactive decay* into more stable ones. Four common types of radioactive decay are:

1. *Alpha decay*, in which a helium nucleus that consists of two protons and two neutrons is emitted. Alpha decay occurs in nuclei too large to be stable.

2. *Beta decay*, in which an electron is emitted when one of the neutrons in a nucleus spontaneously turns into a proton. Beta decay occurs in nuclei in which the neutron-proton ratio is too large for stability.

3. *Electron capture*, in which one of the inner electrons in an atom is absorbed by one of the protons in its nucleus to form a neutron. Electron capture occurs in nuclei in which the neutron-proton ratio is too small for stability.

4. *Gamma decay*, in which a gamma ray (an electromagnetic wave of shorter wavelength and hence higher quantum energy than those of X-rays) is emitted by a nucleus with excess energy, as is often the case after one of the other types of decays occurs. Gamma decay does not change the nature of a nucleus.

HALF-LIFE

A nucleus subject to radioactive decay always has a certain definite probability of decay during any time interval. The *half-life* of a radioactive isotope is the time required for half of any initial quantity of it to decay. If an isotope has a half-life of, say, 5 hr and we start with 100 g of it, after 5 hr, 50 g will be left undecayed; after 10 hr, 25 g will be left undecayed; after 15 hr, 12.5 g will be left undecayed; and so on.

Solved Problems

35.1. The largest stable nucleus is that of the bismuth isotope $^{209}_{83}$Bi. Why are larger nuclei unstable?

The range of the strong interaction, which provides the attractive forces that hold nucleons together, is quite short, whereas the electric repulsive forces that act between protons have unlimited range. Hence beyond a certain size the repulsive forces become comparable with the attractive ones and such nuclei are unstable.

35.2. State the number of protons and neutrons in each of the following nuclei:

$$^{6}_{3}\text{Li} \quad ^{12}_{6}\text{C} \quad ^{36}_{16}\text{S} \quad ^{137}_{56}\text{Ba}$$

A nucleus designated $^{A}_{Z}X$ contains Z protons and $A - Z$ neutrons. Accordingly the numbers of protons and neutrons in the given nuclei are as follows:

$^{6}_{3}$Li:	3 protons, 3 neutrons
$^{12}_{6}$C:	6 protons, 6 neutrons
$^{36}_{16}$S:	16 protons, 20 neutrons
$^{137}_{56}$Ba:	56 protons, 81 neutrons

35.3. Ordinary chlorine is a mixture of 75.53% of the $^{35}_{17}$Cl isotope and 24.47% of the $^{37}_{17}$Cl isotope. The atomic masses of these isotopes are respectively 34.969 u and 36.966 u. Find the atomic mass of ordinary chlorine.

The procedure is to multiply the mass of each isotope by the proportion of the whole it represents, and then to add the results together. Thus we obtain

$$0.7553 \times 34.969 \text{ u} + 0.2447 \times 36.966 \text{ u} = 35.458 \text{ u}$$

which is the atomic mass of ordinary chlorine.

35.4. The atomic mass of $^{16}_{8}$O is 15.9949 u. (*a*) What is its binding energy? (*b*) What is its binding energy per nucleon?

(*a*) $^{16}_{8}$O contains 8 protons and 8 neutrons in its nucleus. The mass of 8 H atoms is $8m_\text{H} = 8 \times 1.007825$ u = 8.0626 u and the mass of 8 neutrons is $8m_n = 8 \times 1.008665$ u = 8.0693 u. Hence the mass deficit in $^{16}_{8}$O is

$$\Delta m = (8.0626 + 8.0693) \text{ u} - 15.9949 \text{ u} = 0.1370 \text{ u}$$

and the binding energy is, since 1 u = 931 MeV,

$$\Delta E = 0.1370 \text{ u} \times 931 \text{ MeV/u} = 127.5 \text{ MeV}$$

(*b*) There are 16 nucleons in $^{16}_{8}$O, so the binding energy per nucleon is 127.6 MeV/16 nucleons = 7.97 MeV/nucleon.

35.5. The binding energy of $^{20}_{10}$Ne is 160.6 MeV. Find its atomic mass.

$^{20}_{10}$Ne contains 10 protons and 10 neutrons in its nucleus. The mass of 10 H atoms and 10 neutrons is

$$m_0 = 10.07825 \text{ u} + 10.08665 \text{ u} = 20.1649 \text{ u}$$

The mass equivalent of 160.6 MeV is

$$\Delta m = \frac{160.6 \text{ MeV}}{931 \text{ MeV/u}} = 0.1725 \text{ u}$$

and so the mass of the $_{10}^{20}$Ne atom is

$$m = m_0 - \Delta m = 20.1649 \text{ u} - 0.1725 \text{ u} = 19.9924 \text{ u}$$

35.6. Complete the following nuclear reactions:

$$_{3}^{6}\text{Li} + _{1}^{2}\text{H} \rightarrow _{2}^{4}\text{He} + ?$$

$$_{17}^{35}\text{Cl} + ? \rightarrow _{16}^{32}\text{S} + _{2}^{4}\text{He}$$

$$_{4}^{9}\text{Be} + _{2}^{4}\text{He} \rightarrow _{0}^{1}n + ?$$

In each of these reactions, the number of protons and the number of neutrons must be the same on both sides of the equation. Hence the complete reactions must be as follows

$$_{3}^{6}\text{Li} + _{1}^{2}\text{H} \rightarrow _{2}^{4}\text{He} + _{2}^{4}\text{He}$$

$$_{17}^{35}\text{Cl} + _{1}^{1}\text{H} \rightarrow _{16}^{32}\text{S} + _{2}^{4}\text{He}$$

$$_{4}^{9}\text{Be} + _{2}^{4}\text{He} \rightarrow _{0}^{1}n + _{6}^{12}\text{C}$$

35.7. In a typical fission reaction, a $_{92}^{235}$U nucleus absorbs a neutron and splits into a $_{54}^{140}$Xe nucleus and a $_{38}^{94}$Sr nucleus. How many neutrons are liberated in this process?

In order that the total numbers of protons and neutrons be the same before and after the fission reaction, two neutrons must be liberated. Hence the reaction is

$$_{92}^{235}\text{U} + _{0}^{1}n \rightarrow _{54}^{140}\text{Xe} + _{38}^{94}\text{Sr} + _{0}^{1}n + _{0}^{1}n + \Delta E$$

In this case ΔE is about 200 MeV.

35.8. When $_{92}^{235}$U undergoes fission, about 0.1% of the original mass is released as energy. (a) How much energy is released when 1 kg of $_{92}^{235}$U undergoes fission? (b) How much $_{92}^{235}$U must undergo fission per day in a nuclear reactor that provides energy to a 100-megawatt (10^8-W) electric power plant? Assume perfect efficiency. (c) When coal is burned, about 7800 kcal/kg of heat is liberated. How many kg of coal would be consumed per day by a convential coal-fired 100-MW electric power plant?

(a) $E = mc^2 = 0.001 \text{ kg} \times (3 \times 10^8 \text{ m/s})^2 = 9 \times 10^{13} \text{ J}$

(b)   Energy = power × time, and so here

$$E = Pt = 10^8 \text{ W} \times 3600 \text{ s/hr} \times 24 \text{ hr/day} = 8.64 \times 10^{12} \text{ J/day}$$

Hence the mass of $_{92}^{235}$U required is

$$\frac{8.64 \times 10^{12} \text{ J/day}}{9 \times 10^{13} \text{ J/kg}} = 9.6 \times 10^{-2} \text{ kg/day} = 96 \text{ g/day}$$

(c)   The energy liberated per kg of coal burned is

$$7800 \text{ kcal/kg} \times 4185 \text{ J/kcal} = 3.26 \times 10^7 \text{ J}$$

Hence the mass of coal required is

$$\frac{8.64 \times 10^{12} \text{ J/day}}{3.26 \times 10^7 \text{ J/kg}} = 2.65 \times 10^5 \text{ kg/day}$$

which is 265 metric tons.

35.9. In the sun and most other stars the principal energy-liberating process is the conversion of hydrogen into helium in a series of nuclear fusion reactions in the course of which *positrons* (positively charged electrons) are emitted. (a) Write the equation for the overall process in which four protons form a helium nucleus. (b) How much energy is liberated in each such process?   The masses of $^1_1$H, $^4_2$He, and the electron are respectively 1.007825 u, 4.002603 u, and 0.000549 u.

(a)   Two positrons must be given off in order that charge be conserved.   Hence the overall process is

$$^1_1\text{H} + ^1_1\text{H} + ^1_1\text{H} + ^1_1\text{H} \rightarrow ^4_2\text{He} + e^+ + e^+$$

(b)   Since a helium atom has only two electrons around its nucleus, two electrons as well as two positrons are lost when each helium atom is formed.   The mass change is therefore

$$\Delta m = 4m_\text{H} - (m_\text{He} + 4m_e) = 4 \times 1.007825 \text{ u} - (4.002603 \text{ u} + 4 \times 0.000549 \text{ u})$$
$$= 0.026501 \text{ u}$$

and the energy liberated is 0.026501 u $\times$ 931 MeV/u = 24.7 MeV.

35.10.   What happens to the atomic number and mass number of a nucleus that (a) emits an electron? (b) Undergoes electron capture?   (c) Emits an alpha particle?

(a) Z increases by 1, A is unchanged.   (b) Z decreases by 1, A is unchanged.   (c) Z decreases by 2, A decreases by 4.

35.11.   How many successive alpha decays occur in the decay of thorium isotope $^{228}_{90}$Th into the lead isotope $^{212}_{82}$Pb?

Each alpha decay means a reduction of 2 in atomic number and of 4 in mass number.   Here Z decreases by 8 and A by 16, which means that 4 alpha particles are emitted.

35.12.   Tritium is the hydrogen isotope $^3_1$H whose nucleus contains two neutrons and a proton. Tritium is beta-radioactive and emits an electron.   (a) What does tritium become after beta decay?   (b) The half-life of tritium is 12.5 years.   How much of a 1-g sample will remain undecayed after 25 years?

(a)   In the beta decay of a nucleus, one of its neutrons becomes a proton.   Since the atomic number 2 corresponds to helium, the beta decay of $^3_1$H is given by

$$^3_1\text{H} \rightarrow ^3_2\text{He} + e^-$$

and the new atom is $^3_2$He.

(b)   Twenty-five years is two half-lives of tritium, and so $\frac{1}{2} \times \frac{1}{2} \times 1$ g $= \frac{1}{4}$ g of tritium remains undecayed.

35.13.   The half-life of the sodium isotope $^{24}_{11}$Na against beta decay is 15 hr.   How long does it take for $\frac{7}{8}$ of a sample of this isotope to decay?

After $\frac{7}{8}$ has decayed, $\frac{1}{8}$ is left, and $\frac{1}{8} = \frac{1}{2} \times \frac{1}{2} \times \frac{1}{2}$ which is 3 half-lives.   Hence the answer is 3×15 hr = 45 hr.

35.14.   The carbon isotope $^{14}_6$C (called "radiocarbon") is beta-radioactive with a half-life of 5600 years. Radiocarbon is produced in the earth's atmosphere by the action of cosmic rays on nitrogen atoms, and the carbon dioxide of the atmosphere contains a small proportion of radiocarbon as a result.   All plants and animals therefore contain a certain amount of radiocarbon along with the stable isotope $^{12}_6$C.   When a living thing dies, it stops taking in radiocarbon, and the radiocarbon it already contains decays steadily.   By measuring the ratio between the $^{14}_6$C and $^{12}_6$C contents of the remains of an animal or plant and comparing it with the ratio of these isotopes in living organisms, the time that has passed since the death of the animal or plant can be found.   (a) How old is a piece of wood from an ancient dwelling if its relative radiocarbon content is $\frac{1}{4}$ that of a modern specimen?   (b) If it is $\frac{1}{16}$ that of a modern specimen?

(a)   Since $\frac{1}{4} = \frac{1}{2} \times \frac{1}{2}$, the specimen is two half-lives old, which is 11,200 years old.

(b)   Since $\frac{1}{16} = \frac{1}{2} \times \frac{1}{2} \times \frac{1}{2} \times \frac{1}{2}$, the specimen is four half-lives old, which is 22,400 years old.

35.15.   The rate at which a sample of radioactive substance decays is called its *activity*.   The unit of activity is the *curie*, where 1 curie = 3.7×10^{10} decays/s.   If a luminous watch dial contains 5 microcuries of the radium isotope $^{226}_{88}$Ra, how many decays per second occur in it? (This isotope emits alpha particles which cause flashes of light when they strike a special material the isotope is mixed with.)

A microcurie is 10^{-6} curie, and so the activity of the watch dial is

$$3.7 \times 10^{10} \frac{\text{decays/s}}{\text{curie}} \times 5 \times 10^{-6} \text{ curie} = 1.85 \times 10^5 \text{ decays/s}$$

Supplementary Problems

35.16. (a) Which of the fundamental interactions has the least significance in nuclear physics? (b) Which two apparently are related?

35.17. In experiments involving nuclear fusion, magnetic fields rather than solid containers are used to confine atomic nuclei that are to react. Why?

35.18. What are the similarities and differences between nuclear fission and nuclear fusion?

35.19. What parts of its structure are chiefly responsible for an atom's mass and for its chemical behavior?

35.20. State the numbers of protons and neutrons in each of the following nuclei: $^{15}_{7}$N, $^{35}_{17}$Cl, $^{64}_{30}$Zn, $^{200}_{80}$Hg.

35.21. Ordinary boron is a mixture of 20% of the $^{10}_{5}$B isotope and 80% of the $^{11}_{5}$B isotope. The atomic masses of these isotopes are respectively 10.013 u and 11.009 u. Find the atomic mass of ordinary boron.

35.22. The atomic mass of $^{3}_{2}$He is 3.01603. (a) What is its binding energy? (b) What is its binding energy per nucleon?

35.23. The atomic mass of $^{35}_{17}$Cl is 34.96885 u. (a) What is its binding energy? (b) What is its binding energy per nucleon?

35.24. The binding energy of $^{42}_{20}$Ca is 361.7 MeV. Find its atomic mass.

35.25. Complete the following nuclear reactions:

$$^{14}_{7}N + ^{4}_{2}He \rightarrow ^{1}_{1}H + ?$$

$$^{11}_{5}B + ^{1}_{1}H \rightarrow ^{11}_{6}C + ?$$

$$^{6}_{3}Li + ? \rightarrow ^{7}_{4}Be + ^{1}_{0}n$$

35.26. When $^{235}_{92}$U undergoes fission, about 0.1% of the original mass is released as energy. (a) How much energy is released by an atomic bomb that contains 10 kg of $^{235}_{92}$U? (b) When a ton of TNT is exploded, about 4×10^9 J is released. How many tons of TNT are equivalent in destructive power to the above bomb?

35.27. In some stars three $^{4}_{2}$He nuclei fuse together in sequence to form a $^{12}_{6}$C nucleus ($m = 12.000000$ u). How much energy is liberated each time this happens?

35.28. Radium spontaneously decays into the elements helium and radon. Why is radium itself considered an element and not simply a chemical compound of helium and radon?

35.29. What happens to the atomic number and mass number of a nucleus that emits a gamma-ray photon? What happens to its mass?

35.30. The uranium isotope $^{238}_{92}U$ decays into a stable lead isotope through the successive emission of 8 alpha particles and 6 electrons. What is the symbol of the lead isotope?

35.31. The half-life of $^{238}_{92}U$ against alpha decay is 4.5×10^9 years. How long does it take for 7/8 of a sample of this isotope to decay? For 15/16 to decay?

35.32. The half-life against beta decay of the strontium isotope $^{90}_{38}Sr$ is 28 years. (a) What does $^{90}_{38}Sr$ become after beta decay? (b) What percentage of a sample of $^{90}_{38}Sr$ will remain undecayed after 112 years.

Answers to Supplementary Problems

35.16. (a) the gravitational interaction   (b) the weak and electromagnetic interactions

35.17. In such experiments, the nuclei form a gas at very high temperature that is called a *plasma*. A plasma would be cooled upon contact with a solid container, and atoms of the container would also be dislodged and enter the plasma where they might affect the reaction unfavorably. It is not likely that the container would actually melt, since the total internal energy of the plasma, as distinguished from its temperature, is not very great.

35.18. In fission, a large nucleus splits into smaller ones; in fusion, two small nuclei join to form a larger one. In both processes, the products of the reaction have less mass than the original nucleus or nuclei, with the missing mass being released as energy.

35.19. The number of protons and neutrons in its nucleus determines the mass of an atom, and the number of electrons in the electron cloud surrounding the nucleus governs its chemical behavior.

35.20. $7\,p$, $8\,n$; $17\,p$, $18\,n$; $30\,p$, $34\,n$; $80\,p$, $120\,n$

35.21. 10.81 u

35.22. (a) 7.71 MeV   (b) 2.57 MeV

35.23. (a) 298 MeV   (b) 8.5 MeV

37.24. 41.9586 u

35.25. $^{17}_{8}O$; $^{1}_{0}n$; $^{2}_{1}H$

35.26. (a) 9×10^{14} J   (b) 2.25×10^5 tons

35.27. 7.27 MeV

35.28. Helium and radon cannot be combined to form radium, nor can radium be broken down into helium and radon by chemical means.

35.29. Z and A are unchanged, but the actual mass decreases in proportion to the energy lost.

35.30. $^{206}_{82}Pb$

35.31. (a) 1.35×10^{10} yr   (b) 1.8×10^{10} yr

35.32. (a) $^{90}_{39}Y$ (b) 6.25%

Appendix A

PHYSICAL CONSTANTS AND QUANTITIES

Quantity	Symbol	Value
Absolute zero	0 K, 0 °R	$-273\ °C = -460\ °F$
Acceleration of gravity at earth's surface	g	$9.81\ m/s^2 = 32.2\ ft/s^2$
Avogadro's number	N	6.023×10^{23} atoms/gram-atom or molecules/mole
Boltzmann's constant	k	$1.38 \times 10^{-23}\ J/K$
Coulomb constant	k	$8.99 \times 10^9\ N\text{-}m^2/C^2$
Electron charge	e	$1.60 \times 10^{-19}\ C$
Gravitational constant	G	$6.67 \times 10^{-11}\ N\text{-}m^2/kg^2 = 3.44 \times 10^{-8}\ lb\text{-}ft^2/slug^2$
Molar volume at STP	V_0	22.4 liters/mole
Permeability of free space	μ_0	$4\pi \times 10^{-7}\ T\text{-}m/A$
Permittivity of free space	ε_0	$8.85 \times 10^{-12}\ C^2/N\text{-}m^2$
Stefan-Boltzmann constant	σ	$5.67 \times 10^{-8}\ W/m^2\text{-}K^4$
Universal gas constant	R	$8.31 \times 10^3\ J/mol\text{-}K = 0.0821$ atm-liter/mole-K
Velocity of light in free space	c	$3.00 \times 10^8\ m/s$

Appendix B

CONVERSION FACTORS

Time

1 day = 1.44×10^3 min = 8.64×10^4 s

1 year = 8.76×10^3 hr = 5.26×10^5 min = 3.15×10^7 s

Length

1 meter (m) = 100 cm = 39.4 in. = 3.28 ft

1 centimeter (cm) = 10 millimeters (mm) = 0.394 in.

1 kilometer (km) = 10^3 m = 0.621 mi

1 foot (ft) = 12 in. = 0.305 m = 30.5 cm

1 inch (in.) = 0.0833 ft = 2.54 cm = 0.0254 m

1 mile (mi) = 5280 ft = 1.61 km

Area

1 m^2 = 10^4 cm^2 = 1.55×10^3 in^2 = 10.76 ft^2

1 cm^2 = 10^{-4} m^2 = 0.155 in^2

1 ft^2 = 144 in^2 = 9.29×10^{-2} m^2 = 929 cm^2

Volume

1 m^3 = 10^3 liters = 10^6 cm^3 = 35.3 ft^3 = 6.10×10^4 in^3

1 ft^3 = 1728 in^2 = 2.83×10^{-2} m^3 = 28.3 liters

Velocity

1 m/s = 3.28 ft/s = 2.24 mi/hr = 3.60 km/hr

1 ft/s = 0.305 m/s = 0.682 mi/hr = 1.10 km/hr

(*Note*: It is often convenient to remember that 88 ft/s = 60 mi/hr.)

1 km/hr = 0.278 m/s = 0.913 ft/s = 0.621 mi/hr

1 mi/hr = 1.47 ft/s = 0.447 m/s = 1.61 km/hr

Mass

1 kilogram (kg) = 10^3 grams (g) = 0.0685 slug

(*Note*:   1 kg corresponds to 2.21 lb in the sense that the *weight*
of 1 kg at the earth's surface is 2.21 lb.)

1 slug = 14.6 kg

(*Note*:   1 slug corresponds to 32.2 lb in the sense that the *weight*
of 1 slug at the earth's surface is 32.2 lb.)

1 atomic mass unit (u) = 1.66×10^{-27} kg = 1.49×10^{-10} J = 931 MeV

Force

1 newton (N) = 0.225 lb = 3.60 oz

1 pound (lb) = 16 ounces (oz) = 4.45 N

(*Note*:   1 lb corresponds to 0.454 kg = 454 g in the sense that the *mass*
of something that weighs 1 lb at the earth's surface is 0.454 kg.)

Pressure

1 N/m^2 = 2.09×10^{-2} lb/ft^2 = 1.45×10^{-4} lb/in^2

1 lb/in^2 = 144 lb/ft^2 = 6.90×10^3 N/m^2

1 atm = 1.013×10^5 N/m^2 = 14.7 lb/in^2

Energy

1 joule (J) = 0.738 ft-lb = 2.39×10^{-4} kcal = 6.24×10^{18} eV

1 foot-pound (ft-lb) = 1.36 J = 1.29×10^{-3} Btu = 3.25×10^{-4} kcal

1 kilocalorie (kcal) = 4185 J = 3.97 Btu = 3077 ft-lb

1 Btu = 0.252 kcal = 778 ft-lb

1 electron volt (eV) = 10^{-6} MeV = 10^{-9} GeV = 1.60×10^{-19} J

Power

1 watt (W) = 1 J/s = 0.738 ft-lb/s

1 kilowatt (kW) = 10^3 W = 1.34 hp

1 horsepower (hp) = 550 ft-lb/s = 746 W

1 refrigeration ton = 12,000 Btu/hr

Temperature

$$T_C = \frac{5}{9}\left(T_F - 32°\right)$$

$$T_F = \frac{9}{5}\,T_C + 32°$$

$$T_K = T_C + 273°$$

$$T_R = T_F + 460°$$

Appendix C

NATURAL TRIGONOMETRIC FUNCTIONS

Angle Deg.	Angle Rad.	Sin	Cos	Tan	Angle Deg.	Angle Rad.	Sin	Cos	Tan
0°	0.000	0.000	1.000	0.000					
1°	0.017	0.018	1.000	0.018	46°	0.803	0.719	0.695	1.036
2°	0.035	0.035	0.999	0.035	47°	0.820	0.731	0.682	1.072
3°	0.052	0.052	0.999	0.052	48°	0.838	0.743	0.669	1.111
4°	0.070	0.070	0.998	0.070	49°	0.855	0.755	0.656	1.150
5°	0.087	0.087	0.996	0.088	50°	0.873	0.766	0.643	1.192
6°	0.105	0.105	0.995	0.105	51°	0.890	0.777	0.629	1.235
7°	0.122	0.122	0.993	0.123	52°	0.908	0.788	0.616	1.280
8°	0.140	0.139	0.990	0.141	53°	0.925	0.799	0.602	1.327
9°	0.157	0.156	0.988	0.158	54°	0.942	0.809	0.588	1.376
10°	0.175	0.174	0.985	0.176	55°	0.960	0.819	0.574	1.428
11°	0.192	0.191	0.982	0.194	56°	0.977	0.829	0.559	1.483
12°	0.209	0.208	0.978	0.213	57°	0.995	0.839	0.545	1.540
13°	0.227	0.225	0.974	0.231	58°	1.012	0.848	0.530	1.600
14°	0.244	0.242	0.970	0.249	59°	1.030	0.857	0.515	1.664
15°	0.262	0.259	0.966	0.268	60°	1.047	0.866	0.500	1.732
16°	0.279	0.276	0.961	0.287	61°	1.065	0.875	0.485	1.804
17°	0.297	0.292	0.956	0.306	62°	1.082	0.883	0.470	1.881
18°	0.314	0.309	0.951	0.325	63°	1.100	0.891	0.454	1.963
19°	0.332	0.326	0.946	0.344	64°	1.117	0.899	0.438	2.050
20°	0.349	0.342	0.940	0.364	65°	1.134	0.906	0.423	2.145
21°	0.367	0.358	0.934	0.384	66°	1.152	0.914	0.407	2.246
22°	0.384	0.375	0.927	0.404	67°	1.169	0.921	0.391	2.356
23°	0.401	0.391	0.921	0.425	68°	1.187	0.927	0.375	2.475
24°	0.419	0.407	0.914	0.445	69°	1.204	0.934	0.358	2.605
25°	0.436	0.423	0.906	0.466	70°	1.222	0.940	0.342	2.747
26°	0.454	0.438	0.899	0.488	71°	1.239	0.946	0.326	2.904
27°	0.471	0.454	0.891	0.510	72°	1.257	0.951	0.309	3.078
28°	0.489	0.740	0.883	0.532	73°	1.274	0.956	0.292	3.271
29°	0.506	0.485	0.875	0.554	74°	1.292	0.961	0.276	3.487
30°	0.524	0.500	0.866	0.577	75°	1.309	0.966	0.259	3.732
31°	0.541	0.515	0.857	0.601	76°	1.326	0.970	0.242	4.011
32°	0.559	0.530	0.848	0.625	77°	1.344	0.974	0.225	4.331
33°	0.576	0.545	0.839	0.649	78°	1.361	0.978	0.208	4.705
34°	0.593	0.559	0.829	0.675	79°	1.379	0.982	0.191	5.145
35°	0.611	0.574	0.819	0.700	80°	1.396	0.985	0.174	5.671
36°	0.628	0.588	0.809	0.727	81°	1.414	0.988	0.156	6.314
37°	0.646	0.602	0.799	0.754	82°	1.431	0.990	0.139	7.115
38°	0.663	0.616	0.788	0.781	83°	1.449	0.993	0.122	8.144
39°	0.681	0.629	0.777	0.810	84°	1.466	0.995	0.105	9.514
40°	0.698	0.643	0.766	0.839	85°	1.484	0.996	0.087	11.43
41°	0.716	0.658	0.755	0.869	86°	1.501	0.998	0.070	14.30
42°	0.733	0.669	0.743	0.900	87°	1.518	0.999	0.052	19.08
43°	0.751	0.682	0.731	0.933	88°	1.536	0.999	0.035	28.64
44°	0.768	0.695	0.719	0.966	89°	1.553	1.000	0.018	57.29
45°	0.785	0.707	0.707	1.000	90°	1.571	1.000	0.000	∞

INDEX

Catalog

If you are interested in a list of SCHAUM'S
OUTLINE SERIES send your name
and address, requesting your free catalog, to:

SCHAUM'S OUTLINE SERIES, Dept. C
McGRAW-HILL BOOK COMPANY
1221 Avenue of Americas
New York, N.Y. 10020